Fundamentals of Geosynthetic Engineering

 BALKEMA - Proceedings and Monographs in Engineering, Water and Earth Sciences

Fundamentals of Geosynthetic Engineering

Sanjay Kumar Shukla
Department of Civil Engineering,
Institute of Technology, Banaras Hindu University,
Varanasi, India

Jian-Hua Yin
Department of Civil and Structural Engineering,
The Hong Kong Polytechnic University,
Hung Hom, Kowloon, Hong Kong

Taylor & Francis
Taylor & Francis Group
LONDON / LEIDEN / NEW YORK / PHILADELPHIA / SINGAPORE

© 2006 Taylor & Francis Group, London, UK

Typeset in Times New Roman by
Newgen Imaging Systems (P) Ltd, Chennai, India
Printed and bound in Great Britain by
TJ International, Padstow, Cornwall

All rights reserved. No part of this publication or the
information contained herein may be reproduced, stored in
a retrieval system, or transmitted in any form or by any means,
electronic, mechanical, by photocopying, recording or otherwise,
without written prior permission from the publishers.

Although all care is taken to ensure integrity and the quality
of this publication and the information herein, no
responsibility is assumed by the publishers nor the author
for any damage to the property or persons as a result of
operation or use of this publication and/or the information
contained herein.

Published by: Taylor & Francis/Balkema
P.O. Box 447, 2300 AK Leiden, The Netherlands
e-mail: Pub.NL@tandf.co.uk
www.balkema.nl, www.tandf.co.uk, www.crcpress.com

British Library Cataloguing in Publication Data
A catalogue record for this book is available from the British Library

Library of Congress Cataloging in Publication Data
Fundamentals of geosynthetic engineering: Sanjay Kumar Shukla, Jian-Hua Yin.
p. cm.
Includes bibliographical references and index.
1. Geosynthetics. 2. Civil engineering. I. Shukla, Sanjay Kumar.
II. Yin, Jian-Hua.

TA455.G44F86 2006
624.1′5136–dc22   2005036367

ISBN10 0–415–39444–9   ISBN13 978–0–415–39444–4

To our parents

Contents

About the authors xi
Preface xiii
Acknowledgements xv

1 General description 1

   1.1 Introduction 1
   1.2 Geosynthetics 1
   1.3 Basic characteristics 6
   1.4 Raw materials 10
   1.5 Manufacturing processes 17
   1.6 Geosynthetic engineering 25
   Self-evaluation questions 25

2 Functions and selection 29

   2.1 Introduction 29
   2.2 Functions 29
   2.3 Selection 39
   Self-evaluation questions 43

3 Properties and their evaluation 47

   3.1 Introduction 47
   3.2 Physical properties 47
   3.3 Mechanical properties 50
   3.4 Hydraulic properties 69
   3.5 Endurance and degradation properties 80
   3.6 Test and allowable properties 92
   3.7 Description of geosynthetics 96
   Self-evaluation questions 97

4 Application areas — 105

- 4.1 Introduction — 105
- 4.2 Retaining walls — 105
- 4.3 Embankments — 112 ✓
- 4.4 Shallow foundations — 115
- 4.5 Roads — 117
  - 4.5.1 Unpaved roads — 117
  - 4.5.2 Paved roads — 119
- 4.6 Railway tracks — 127
- 4.7 Filters and drains — 129
- 4.8 Slopes — 135
  - 4.8.1 Erosion control — 135
  - 4.8.2 Stabilization — 143
- 4.9 Containment facilities — 148
  - 4.9.1 Landfills — 148
  - 4.9.2 Ponds, reservoirs, and canals — 152
  - 4.9.3 Earth dams — 153
- 4.10 Tunnels — 157
- 4.11 Installation survivability requirements — 157
- Self-evaluation questions — 164

5 Analysis and design concepts — 171

- 5.1 Introduction — 171
- 5.2 Design methodologies — 171
- 5.3 Retaining walls — 175
- 5.4 Embankments — 182 ✓
- 5.5 Shallow foundations — 189
- 5.6 Roads — 196
  - 5.6.1 Unpaved roads — 196
  - 5.6.2 Paved roads — 207
- 5.7 Railway tracks — 211
- 5.8 Filters and drains — 215
- 5.9 Slopes — 226
  - 5.9.1 Erosion control — 226
  - 5.9.2 Stabilization — 234
- 5.10 Containment facilities — 239
  - 5.10.1 Landfills — 239
  - 5.10.2 Ponds, reservoirs, and canals — 249
  - 5.10.3 Earth dams — 251
- 5.11 Tunnels — 254
- Self-evaluation questions — 255

6 Application guidelines 263

- 6.1 Introduction 263
- 6.2 General guidelines 263
  - 6.2.1 Care and consideration 263
  - 6.2.2 Geosynthetic selection 264
  - 6.2.3 Identification and inspection 265
  - 6.2.4 Sampling and test methods 265
  - 6.2.5 Protection before installation 267
  - 6.2.6 Site preparation 268
  - 6.2.7 Geosynthetic installation 268
  - 6.2.8 Joints/seams 269
  - 6.2.9 Cutting of geosynthetics 274
  - 6.2.10 Protection during construction and service life 275
  - 6.2.11 Damage assessment and correction 277
  - 6.2.12 Anchorage 278
  - 6.2.13 Prestressing 279
  - 6.2.14 Maintenance 279
  - 6.2.15 Certification 279
  - 6.2.16 Handling the refuse of geosynthetics 280
- 6.3 Specific guidelines 280
  - 6.3.1 Retaining walls 280
  - 6.3.2 Embankments 282
  - 6.3.3 Shallow foundations 286
  - 6.3.4 Unpaved roads 286
  - 6.3.5 Paved roads 288
  - 6.3.6 Railway tracks 290
  - 6.3.7 Filters and drains 291
  - 6.3.8 Slopes – erosion control 294
  - 6.3.9 Slopes – stabilization 297
  - 6.3.10 Containment facilities 305
  - 6.3.11 Tunnels 308
- Self-evaluation questions 309

7 Quality and field performance monitoring 315

- 7.1 Introduction 315
- 7.2 Concepts of quality and its evaluation 315
- 7.3 Field performance monitoring 320
- Self-evaluation questions 325

8 Economic evaluation — 329

- 8.1 Introduction — 329
- 8.2 Concepts of cost analysis — 329
- 8.3 Experiences of cost analyses — 330
- Self-evaluation questions — 336

9 Case studies — 339

- 9.1 Introduction — 339
- 9.2 Selected case studies — 339
- 9.3 Concluding remarks — 368
- Self-evaluation questions — 368

Appendix A Answers to multiple choice type questions and selected numerical problems — 373

Appendix B Standards and codes of practice — 377

- B.1 Introduction — 377
- B.2 General information — 377
- B.3 Standards on test methods — 378
- B.4 Codes of practice — 382

Appendix C Some websites related to geosynthetics — 385

References — 387
Index — 403

Authors

Sanjay Kumar Shukla obtained his PhD degree from the Indian Institute of Technology, Kanpur, India in 1995. He has worked as a faculty member at several Indian universities and institutes and has been a visiting research scholar at Hong Kong Polytechnic University. Dr Shukla is currently Reader in Civil Engineering at the Department of Civil Engineering of the Institute of Technology, Banaras Hindu University, Varanasi, India. His expertise is in the area of geosynthetics, ground improvement, soil–structure interaction and foundation modelling.

Jian-Hua Yin obtained a PhD degree from the University of Manitoba, Canada in 1990. He then worked for a major geotechnical consulting firm in Dartmouth, Nova Scotia, and later at C-CORE in St John's, Newfoundland, Canada before joining an international consulting firm in Hong Kong in 1994. He is currently a professor at the Department of Civil and Structural Engineering, Hong Kong Polytechnic University. Dr Yin specializes in constitutive modelling, soil and geosynthetics testing, analytical and numerical analysis, ground improvement, and geotechnical instrumentation.

Preface

The development of polymeric materials in the form of geosynthetics has brought major changes in the civil engineering profession. Geosynthetics are available in a wide range of compositions appropriate to different applications and environments. Over the past three to four decades, civil engineers have shown an increasing interest in geosynthetics and in understanding their correct uses. Simultaneously, significant advances have been made regarding the use of geosynthetics in civil engineering applications. These developments have occurred because of ongoing dialogue among engineers and researchers in both the civil engineering field and the geosynthetic industry.

Every four years since 1982, the engineering community has held an international conference on geosynthetics. There are presently two official journals, namely *Geotextiles and Geomembranes*, and *Geosynthetics International*, from the International Geosynthetics Society. A few books have also been written to meet the demands of students, researchers, and practising civil engineers. However, we feel that there should be a textbook on geosynthetics that deals with the basic concepts of the subject, especially for meeting the requirements of students as well as of practising civil engineers who have not been exposed to geosynthetics during their university education. The book also covers major aspects related to field applications including application guidelines and description of case studies to generate full confidence in the engineering use of geosynthetics. We have made every effort in this direction and do hope that this book will be of value to both civil engineering students and practising civil engineers. In fact, the aim of this book is to assist all those who want to learn about the fundamentals of geosynthetic engineering.

The subject is divided into nine chapters and presented in a sequence intended to appeal to students and other learners reading the book as an engineering subject. Chapter 1 deals with the general description of geosynthetics including their basic characteristics and manufacturing processes. Any application may require one or more functions from the geosynthetic that will be installed. Such functions are described in Chapter 2. This chapter also discusses aspects of the geosynthetic selection. Chapter 3 covers the properties of geosynthetics along with the basic principles of their measurement. Chapter 4 presents physical descriptions of major applications of geosynthetics in civil engineering and identifies the geosynthetic functions involved. The analysis and design concepts for selected applications are described in Chapter 5. The general application guidelines for various applications are provided in Chapter 6 along with some application-specific guidelines. Chapter 7 addresses important aspects related to quality control and field performance monitoring. Chapter 8 includes some aspects related to cost analysis and provides some available economic experiences. Selected case studies are presented in Chapter 9. Answers to multiple choice questions and selected numerical problems included as self evaluation questions are given in Appendix A. A list of the latest common test standards and codes of practice are included in Appendix B. Some important websites related to geosynthetics are given in Appendix C. A list of all references and subject index are included at the end of the book.

It should be noted that there is good continuity in the presentation of the topics and their concepts described. Line drawings, sketches, graphs, photographs and tables have been included, as required for an engineering subject. Therefore, readers will find this book lively and interactive, and they will learn the basic concepts of most of the topics. However, we welcome suggestions from readers of this book for improving its contents in a future edition.

Sanjay Kumar Shukla
Jian-Hua Yin

Acknowledgements

The authors wish to acknowledge and thank Terram Limited, Gwent, UK; Netlon Limited, Blackburn, UK; Naue Fasertechnik GmbH & Co., Lubbecke, Germany; Huesker Synthetic GmbH & Co. Gescher, Germany; GSE Lining Technology Inc., USA; and Netlon India, Vadodara, India for providing useful information and materials. The authors would also like to thank other manufacturers of geosynthetics for their contributions in the development of geosynthetic engineering, which have been used in the present book.

The authors would like to thank the following individuals for inspiring us through their published works for authoring this book for its use by all those who want to learn the fundamentals of geosynthetic engineering: Dr J.P. Giroud, Professor T.S. Ingold, Profesor R.M. Koerner, Professor R.J. Bathurst, Professor F. Schlosser, Professor R.D. Holtz, Professor R.K. Rowe, Professor P.L. Bourdeau, Professor D.E. Daniel, Professor B.B. Broms, Dr B.R. Christopher, Professor R.A. Jewell, Professor N.W.M. John, Professor M.R. Madhav, Professor S.W. Perkins, Professor K.W. Pilarczyk, Professor G.P. Raymond, Dr G.N. Richardson, Professor S.A. Tan, Professor B.M. Das, Professor M.L. Lopes, Professor E.M. Palmeira, Professor C. Duquennoi, Professor F. Tatsuoka, and Mr H. Zanzinger. The authors are also grateful to all other research workers for their contributions in the field of geosynthetic engineering, which have been used in various forms in the present book for the benefit of readers and users of this book.

The authors wish to acknowledge and thank the American Association of State Highway and Transportation Officials, USA; ASTM International, USA; British Standards Institution, UK; and Standards Australia International Ltd, Australia for their permission to reprint some tables and figures from their standards and codes of practice in the present book.

The authors extend special thanks to Mr John Clement and the staff of A.A. Balkema Publishers – Taylor & Francis, The Netherlands for their cooperation at all the stages of production of this book.

The first author would like to thank his wife, Sharmila; daughter, Sakshi; and son, Sarthak for their encouragement and full support whilst working on this book. He also wishes to acknowledge The Hong Kong Polytechnic University, Hong Kong in providing financial support for his visits to Hong Kong for the book-related works. The second author owes the deepest gratitude to his mother Li Feng-Yi for her love and support in his education and study.

Sanjay Kumar Shukla
Jian-Hua Yin

Chapter 1

General description

1.1 Introduction

In the past four decades, considerable development has taken place in the area of geosynthetics and their applications (see Table 1.1). Geosynthetics are now an accepted civil engineering construction material and have unique characteristics like all other construction materials such as steel, concrete, timber, etc. This chapter provides a general description of geosynthetics including their basic characteristics and manufacturing processes.

1.2 Geosynthetics

The term 'Geosynthetics' has two parts: the prefix 'geo', referring to an end use associated with improving the performance of civil engineering works involving earth/ground/soil and the suffix 'synthetics', referring to the fact that the materials are almost exclusively from man-made products. The materials used in the manufacture of geosynthetics are primarily synthetic polymers generally derived from crude petroleum oils; although rubber, fiberglass, and other materials are also sometimes used for manufacturing geosynthetics. Geosynthetics is, in fact, a generic name representing a broad range of planer products manufactured from polymeric materials; the most common ones are *geotextiles, geogrids, geonets, geomembranes* and *geocomposites*, which are used in contact with soil, rock and/or any other civil engineering-related material as an integral part of a man-made project, structure or system (Figs 1.1–1.5).

Products, based on natural fibres (jute, coir, cotton, wool, etc.), are also being used in contact with soil, rock and/or other civil engineering-related material, especially in temporary civil engineering applications. Such products, that may be called *geonaturals*, have a short life span when used with earth materials due to their biodegradable characteristics, and therefore, they have not many field applications as geosynthetics have (Shukla, 2003a). Though geonaturals are significantly different from geosynthetics in material characteristics, they can be considered a complementary companion of geosynthetics, rather than a replacement, mainly because of some common field application areas. In fact, geonaturals are also polymeric materials, since they contain a large proportion of naturally occurring polymers such as lignin and cellulose.

Geotextile: It is a planar, permeable, polymeric textile product in the form of a flexible sheet (Fig. 1.1). Currently available geotextiles are classified into the following categories based on the manufacturing process:

- *Woven geotextile*: A geotextile produced by interlacing, usually at right angles, two or more sets of yarns (made of one or several fibres) or other elements using a conventional weaving process with a weaving loom.

2   General description

- *Nonwoven geotextile*: A geotextile produced from directionally or randomly oriented fibres into a loose web by bonding with partial melting, needle-punching, or chemical binding agents (glue, rubber, latex, cellulose derivative, etc.).
- *Knitted geotextile*: A geotextile produced by interlooping one or more yarns (or other elements) together with a knitting machine, instead of a weaving loom.
- *Stitched geotextile*: A geotextile in which fibres or yarns or both are interlocked/bonded by stitching or sewing.

Geogrid: It is a planar, polymeric product consisting of a mesh or net-like regular open network of intersecting tensile-resistant elements, called ribs, integrally connected at the

Table 1.1 Historical developments in the area of geosynthetics and their applications

Decades	Developments
Early decades	The first use of fabrics in reinforcing roads was attempted by the South Carolina Highway Department in 1926 (Beckham and Mills, 1935). Polymers which form the bulk of geosynthetics did not come into commercial production until thirty years later starting with polyvinyl chloride (PVC) in 1933, low density polyethylene (LDPE) and polyamide (PA) (a.k.a. nylon) in 1939, expanded polystyrene (EPS) in 1950, polyester (PET) in 1953, and high density polyethylene (HDPE) and polypropylene (PP) in 1955 (Hall, 1981). The US Bureau of Reclamation has been using geomembranes in water conveyance canals since the 1950s (Staff, 1984).
Late 1950s	A range of fabrics was manufactured for use as separation and filter layers between granular fills and weak subsoils. Woven fabrics (nowadays called *geotextiles*) played critical filtration functions in coastal projects in The Netherlands and in the USA.
1960s	Rhone-Poulenc Textiles in France began working with nonwoven needle-punched geotextiles for quite different applications. Geotextiles found a role as beds for highway and railway track support systems. Chlorosulfonated polyethylene (CSPE) was developed around 1965.
1970s	The first geotextile used in a dam, in 1970, was a needle-punched nonwoven geotextile used as a filter for the aggregate downstream drain in the Valcross Dam (17 m high), France (Giroud, 1992). Geotextiles were incorporated as reinforcement in retaining walls, steep slopes, etc. The beginning of the ongoing process of standards development started with the formation of the ASTM D-13-18 joint committee on geosynthetics and the formation of industry task forces. The first samples of Tensar grid were made in the Blackburn laboratories of Netlon Ltd, UK, in July 1978. The first conference on geosynthetics was held in Paris in 1977. The geofoam was originally applied as a lightweight fill in Norway in 1972.
1980s	The beginning of the use of geosynthetics occurred in the construction of safe containment of environmentally hazardous wastes. Soil confinement systems based on cellular geotextile nets were first developed and evaluated in France during 1980. Netlon developed a similar concept, but on a larger scale, with the introduction of the Tensar Geocell Mattress in 1982. The first known environmental application of geonet was in 1984 for leak detection in a double-lined hazardous liquid-waste impoundment in Hopewell, Virginia. Koerner and Welsh wrote the first book on geosynthetics in 1980. The International Geosynthetics Society was established in 1983. The first volume of international journal entitled *Geotextiles and Geomembranes* was published in 1984.
1990s	Many standards on geosynthetics were published by the American Society of Testing Materials (ASTM), USA; the International Organization for Standardization (ISO), Switzerland; the British Standards Institution (BSI), UK; the Bureau of Indian Standards (BIS), India, etc. The second international journal entitled *Geosynthetics International* was first published in 1995.

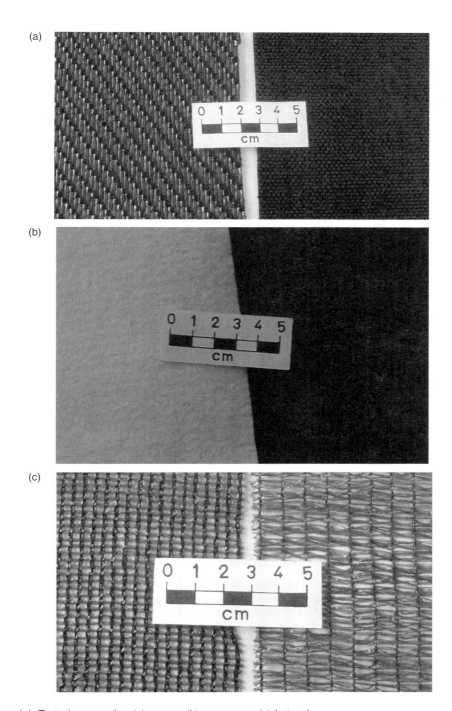

Figure 1.1 Typical geotextiles: (a) woven; (b) nonwoven; (c) knitted.

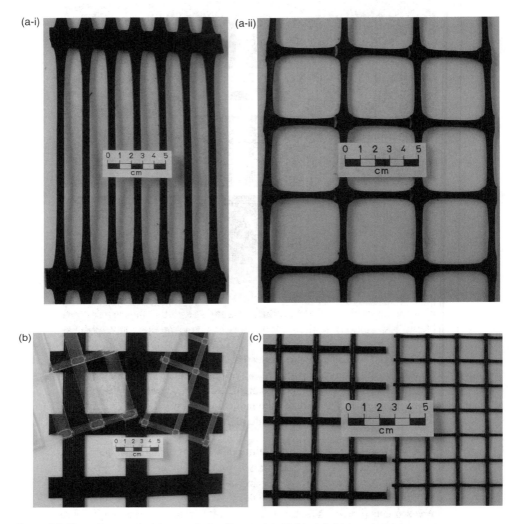

Figure 1.2 Typical geogrids: (a) extruded – (i) uniaxial; (ii) biaxial; (b) bonded; (c) woven.

junctions (Fig. 1.2). The ribs can be linked by extrusion, bonding or interlacing: the resulting geogrids are respectively called *extruded geogrid*, *bonded geogrid* and *woven geogrid*. Extruded geogrids are classified into the following two categories based on the direction of stretching during their manufacture:

- *Uniaxial geogrid*: A geogrid produced by the longitudinal stretching of a regularly punched polymer sheet, and therefore it possesses a much higher tensile strength in the longitudinal direction than the tensile strength in the transverse direction.
- *Biaxial geogrid*: A geogrid produced by stretching in both the longitudinal and the transverse directions of a regularly punched polymer sheet, and therefore it possesses equal tensile strength in both the longitudinal and the transverse directions.

Figure 1.6 Continued.

Table 1.2 Abbreviations and graphical symbols of geosynthetic products as recommended by the International Geosynthetics Society

Abbreviations	Graphical symbols	Geosynthetic products
GTX	— — — — — — — — — —	Geotextile
GMB	———————	Geomembrane
GBA	┼┼┼┼┼┼┼┼┼┼┼┼	Geobar
GBL	∩∪∩∪∩∪∩∪∩∪	Geoblanket
GCD	⋈⋈⋈⋈⋈⋈⋈⋈	Geocomposite drain with geotextile on both sides
GCE	IIIIIIIIIIIIIIIIIIIIIIII	Geocell
GCL	⌐⌐⌐⌐⌐⌐⌐⌐⌐⌐⌐⌐	Geocomposite clay liner
GEC	#################	Surficial geosynthetic erosion control
GEK	ZZZZZZZZZZZZZZZZ	Electrokinetic geosynthetic
GGR	—•—•—•—•—•—	Geogrid
GMA	∼∼∼∼∼∼	Geomat
GMT	▓▓▓▓▓▓▓▓▓▓	Geomattress
GNT	××××××××××××××	Geonet
GSP	⊓⊔⊓⊔⊓⊔⊓⊔	Geospacer
GST	—•—•—•—•—•—	Geostrip

Table 1.3 Some product names and types of geosynthetics

Product names and types	Product classification
Tensar 160RE	Uniaxial geogrid
Tensar SS40	Biaxial geogrid
Secutex 301 GRK5	Needle-punched geotextile
Naue Fasertechnik PEHD 406 GL/GL, 1.5 mm	Geomembrane both sides smooth
Terram W/20-4	Woven polypropylene geotextile
Terram PW4	Composite reinforced drainage separator
Netlon CE 131	Geonet
Terram ParaLink 600S	Welded geogrid
Terram Grid 4/2-W	Coated polyester woven geogrid
Secumat ES 601 G4	Single layer erosion control layer with woven fabric

- high flexibility
- minimum volume
- lightness
- ease of storing and transportation
- simplicity of installation
- speeding the construction process
- making economical and environment-friendly solution
- providing good aesthetic look to structures.

The importance of geosynthetics can also be observed in their ability to partially or completely replace natural resources such as gravel, sand, bentonite clay, etc. In fact, geosynthetics can be used for achieving better durability, aesthetics and environment of the civil engineering projects.

1.4 Raw materials

Almost exclusively, the raw materials from which geosynthetics are produced are polymeric. Polymers are materials of very high molecular weight and are found to have multifarious applications in the present society. The polymers used to manufacture geosynthetics are generally thermoplastics, which may be amorphous or semi-crystalline. Such materials melt on heating and solidify on cooling. The heating and cooling cycles can be applied several times without affecting the properties.

Any polymer, whether amorphous or semi-crystalline, consists of long chain molecules containing many identical chemical units bound together by covalent bonds. Each unit may be composed of one or more small molecular compounds called *monomers*, which are most commonly hydrocarbon molecules. The process of joining monomers, end to end, to form long polymer chains is called *polymerization*. In Figure 1.10, it is observed that during polymerization, the double carbon bond of the ethylene monomer forms a covalent bond with the carbon atoms of neighbouring monomers. The end result is a long polyethylene (PE) chain molecule in which one carbon atom is bonded to the next. If the chains are packed in a regular form and are highly ordered, the resulting configuration will have a crystalline structure, otherwise amorphous structure. No polymers used for manufacturing geosynthetics are

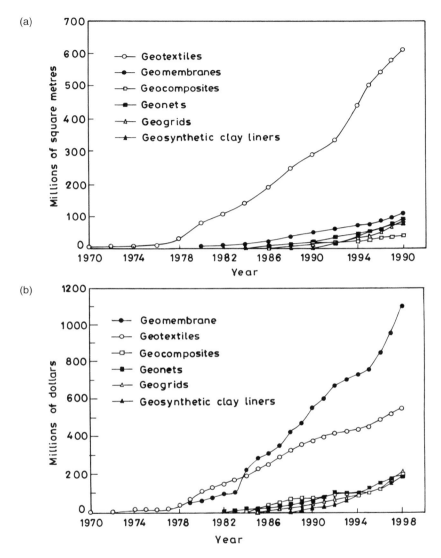

Figuer 1.7 Growth of geosynthetics in North America based on (a) quantity; (b) sales (after Koerner, 2000).

completely crystalline, although HDPE can attain 90% or so crystallinity, but some are completely amorphous. Manufacture of polymers is generally carried out by chemical or petrochemical companies who produce polymers in the form of solid pellets, flakes or granules.

The number of monomers in a polymer chain determines the length of the polymeric chain and the resulting molecular weight. Molecular weight can affect physical and mechanical properties, heat resistance and durability (resistance to chemical and biological attack)

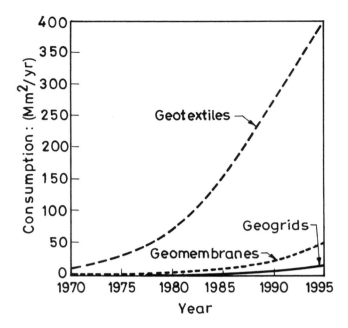

Figure 1.8 Estimated consumption of geosynthetics in Western Europe (after Lawson and Kempton, 1995).

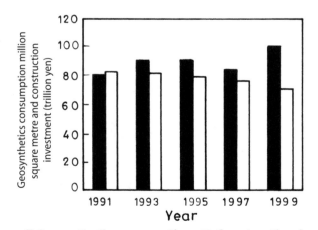

Figure 1.9 Geosynthetic consumption and construction investment in Japan (after Akagi, 2002).

properties of geosynthetics. The physical and mechanical properties of the polymers are also influenced by the bonds within and between chains, the chain branching and the degree of crystallinity. An increase in the degree of crystallinity leads directly to an increase in rigidity, tensile strength, hardness, and softening point and to a decrease in chemical permeability.

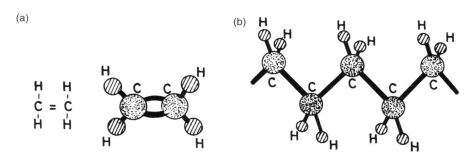

Figure 1.10 The process of polymerization: (a) an ethylene monomer; (b) a polyethylene molecule.

Table 1.4 Polymers commonly used for the manufacture of geosynthetics

Types of polymer	Abbreviations
Polypropylene	PP
Polyester (polyethylene terephthalate)	PET
Polyethylene	
Low density polyethylene	LDPE
Very low density polyethylene	VLDPE
Linear low density polyethylene	LLDPE
Medium density polyethylene	MDPE
High density polyethylene	HDPE
Chlorinated polyethylene	CPE
Chlorosulfonated polyethylene	CSPE
Polyvinyl chloride	PVC
Polyamide	PA
Polystyrene	PS

Notes
The basic materials consist mainly of the elements carbon, hydrogen, and sometimes nitrogen and chlorine; they are produced from coal and petroleum oil.

If the polymer is stretched in the melt, or in solid form above its final operating temperature, the molecular chains become aligned in the direction of stretch. This alignment, or *molecular orientation*, can be permanent if, still under stress, the material is cooled to its operating temperature. The orientation of molecules of the polymers by mechanical drawing results in higher tensile properties and improved durability of the fibres. The properties of polymers can also be altered by including the introduction of side branches, or grafts, to the main molecular chain.

The polymers used for manufacturing geosynthetics are listed in Table 1.4 along with their commonly used abbreviations. The more commonly used types are polypropylene (PP), high density polyethylene (HDPE) and polyester (polyethylene terephthalate (PET)).

Most of the geotextiles are manufactured from PP or PET (see Table 1.5). Polypropylene is a semi-crystalline thermoplastic with a melting point of 165°C and a density in the range of 0.90–0.91 g/cm³. Polyester is also a thermoplastic with a melting point of 260°C and a density in the range of 1.22–1.38 g/cm³.

Table 1.5 Major geosynthetics and the most commonly used polymers for their manufacture

Geosynthetics	Polymers used for manufacturing
Geotextiles	PP, PET, PE, PA
Geogrids	PET, PP, HDPE
Geonets	MDPE, HDPE
Geomembranes	HDPE, LLDPE, VLDPE, PVC, CPE, CSPE, PP
Geofoams	EPS
Geopipes	HDPE, PVC, PP

The primary reason for PP usage in geotextile manufacturing is its low cost. For non-critical structures, PP provides an excellent, cost-effective raw material. It exhibits a second advantage in that it has excellent chemical and pH range resistance because of its semi-crystalline structure. Additives and stabilizers (such as carbon black) must be added to give PP ultraviolet (UV) light resistance during processing. As the critical nature of the structure increases, or the long-term anticipated loads go up, PP tends to lose its effectiveness. This is because of relatively poor creep deformation characteristics under long-term sustained load.

Polyester is increasingly being used to manufacture reinforcing geosynthetics such as geogrids because of high strength and resistance to creep. Chemical resistance of polyester is generally excellent, with the exception of very high pH environments. It is inherently stable to UV light.

Polyethylene is one of the simplest organic polymers used extensively in the manufacture of geomembranes. It is used in its low density and less crystalline form (low density polyethylene (LDPE)), which is known for its excellent pliability, ease of processing and good physical properties. It is also used as HDPE, which is more rigid and chemically more resistant.

Polyvinyl chloride (PVC) is the most significant commercial member of the family of vinyl-based resins. With plasticizers and other additives it takes up a great variety of forms. Unless PVC has suitable stabilizers, it tends to become brittle and darkens when exposed to UV light over time and can undergo heat-induced degradation.

Polyamides (PA), better known as nylons, are melt processable thermoplastics that contain an amide group as a recurring part of the chain. Polyamide offers a combination of properties including high strength at elevated temperatures, ductility, wear and abrasion resistance, low frictional properties, low permeability by gases and hydrocarbons and good chemical resistance. Its limitations include a tendency to absorb moisture, with resulting changes in physical and mechanical properties, and limited resistance to acids and weathering.

The raw material for the manufacture of geofoams is polystyrene (PS), which is known as packaging material and insulating material by the common people. Its genesis is from ethylene and is available in two forms: expanded polystyrene (EPS) and extruded polystyrene (XPS).

There are several environmental factors that affect the durability of polymers. Ultraviolet component of solar radiation, heat and oxygen, and humidity are the factors above ground that may lead to degradation. Below ground the main factors affecting the durability of

Table 1.6 A comparison of the resistance of polymers, commonly used in the production of geosynthetics (adapted from John, 1987; Shukla, 2002a)

Influencing factors	Resistance of polymers			
	PP	PET	PE	PA
Ultraviolet light (unstabilized)	Medium	High	Low	Medium
Ultraviolet light (stabilized)	High	High	High	Medium
Alkalis	High	Low	High	High
Acids	High	Low	High	Low
Salts	High	High	High	High
Detergents	High	High	High	High
Heat, dry (up to 100°C)	Medium	High	Low	Medium
Steam (up to 100°C)	Low	Low	Low	Medium
Hydrolysis (reaction with water)	High	High	High	High
Micro-organisms	High	High	High	Medium
Creep	Low	High	Low	Medium

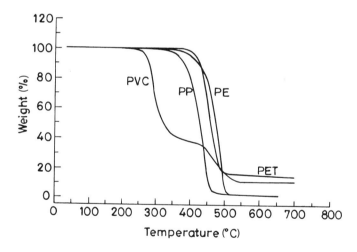

Figure 1.11 Effect of temperature on some geosynthetic polymers (after Thomas and Verschoor, 1988).

polymers are soil particle size and angularity, acidity/alkalinity, heavy metal ions, presence of oxygen, moisture content, organic content and temperature. The resistance of commonly used polymers to some environmental factors is compared in Table 1.6. It must be emphasized that the involved reactions are usually slow and can be retarded even more by the use of suitable additives. When polymers are subjected to higher temperature, they lose their weight. What remains above 500°C is probably carbon black and ash (Fig. 1.11). Note that the ash content of a polymeric compound is the remains of inorganic ingredients used as fillers or cross-linking agents and ash from the base polymer.

The formulation of a polymeric material is a complex task. No geosynthetic material is 100% of the polymer resin associated with its name, because pure polymers are not suitable for production of geosynthetics. The primary resins are always formulated with additives, fillers, and/or other agents as UV light absorbers, antioxidants, thermal stabilizers, etc. to produce a plastic with the required properties. For example, PE, PET and PA have 97% resin, 2–3% carbon black (or pigment), and 0.5–1.0% other additives.

If the long molecular chains of the polymer are cross-linked to one another, the resulting material is called *thermoset*, which, once cooled, remains solid upon the subsequent application of heat. Though a thermoset does not melt on reheating, it degrades. The thermoset materials (such as Ethylene vinyl acetate (EVA), butyl, etc.), alone or in combination with thermoplastic materials, are sometimes used to manufacture geomembranes.

Although most of the geosynthetics are made from synthetic polymers, a few specialist geosynthetics, especially geotextiles, may also incorporate steel wire or natural biodegradable fibres such as jute, coir, paper, cotton, wool, silk, etc. Biodegradable geotextiles are usually limited to erosion control applications where natural vegetation will replace the geotextile's role as it degrades. Jute nets are marketed under various trade names, including *geojute, soil saver, and anti-wash*. They are usually in the form of a woven net with a typical mesh open size of about 10 by 15 mm, a typical thickness of about 5 mm and an open area of about 65%. Vegetation can easily grow through openings and use the fabric matrix as support. The jute, which is about 80% natural cellulose, should completely degrade in about two years. An additional advantage of these biodegradable products is that the decomposed jute improves the quality of the soil for vegetation growth. Some non-polymeric materials like sodium or calcium bentonite are also used to make a few geosynthetic products.

It is sometimes important to know the polymer compound present in the geosynthetic being used. A quantitative assessment requires the use of a range of identification test techniques, as listed in Table 1.7 along with a brief description. More detailed descriptions can be found in the works of Halse *et al.* (1991), Landreth (1990) and Rigo and Cuzzuffi (1991).

Table 1.7 Chemical identification tests (based on Halse et al., 1991)

Method	Information obtained
Thermogravimetric analysis (TGA)	Polymer, additives, and ash contents; carbon black content; decomposition temperatures
Differential scanning calorimetry (DSC)	Melting point, degree of crystallinity, oxidation time, glass transition
Thermomechanical analysis (TMA)	Coefficient of linear thermal expansion, softening point, glass transition
Infrared spectroscopy (IR)	Additives, fillers, plasticizers, and rate of oxidation reaction
Chromatography: gas chromatography (GC) and high pressure liquid chromatography (HPLC)	Additives and plasticizers
Density determination (ρ)	Density and degree of crystallinity
Melt index (MI)	Melt index and flow rate ratio
Gel permeation chromatography (GPC)	Molecular weight distribution

1.5 Manufacturing processes

Geotextiles are manufactured in many different ways, partly using traditional textile procedures and partly using procedures not commonly recognized as textile procedures. The manufacturing process of a geotextile basically includes two steps (Giroud and Carroll, 1983): the first step consists in making linear elements such as *fibres or yarns* from the polymer pellets, under the agency of heat and pressure and the second step consists in combining these linear elements to make a planar structure usually called a *fabric*.

The basic elements of a geotextile are its fibres. A *fibre* is a unit of matter characterized by flexibility, fineness, and high ratio of length to thickness. There are four main types of synthetic fibres: the *filaments* (produced by extruding melted polymer through dies or spinnerets and subsequently drawing it longitudinally), *staple fibres* (obtained by cutting filaments to a short length, typically 2–10 cm), *slit films* (flat tape-like fibres, typically 1–3 mm wide, produced by slitting with blades an extruded plastic film and subsequently drawing it) and *strands* (a bundle of tape-like fibres that can be partially attached to each other). During the drawing process, the molecules become oriented in the same direction resulting in increase of modulus of the fibres. A *yarn* consists of a number of fibres from the particular polymeric compound selected. Several types of yarn are used to construct woven geotextiles: *monofilament yarn* (made from a single filament), *multifilament yarn* (made from fine filaments aligned together), *spun yarn* (made from staple fibres interlaced or twisted together), *slit film yarn* (made from a single slit film fibre) and *fibrillated yarn* (made from strands). It should be noted that synthetic fibres are very efficient load carrying elements, with tensile strengths equivalent to prestressing steel in some cases (e.g. in case of polyaramid fibres). Fibre technology in itself is a well-advanced science with an enormous database. It is in the fibre where control over physical and mechanical properties first takes place in a well-prescribed and fully automated manner.

As the name implies, woven geotextiles are obtained by conventional weaving processes. Although modern weaving looms are extremely versatile and sophisticated items, they operate on the basic principles embodied in a mechanical loom illustrated in Figure 1.12. The weaving process gives these geotextiles their characteristic appearance of two sets of parallel yarns interlaced at right angles to each other as shown in Figure 1.13. The terms '*warp and weft*' are used to distinguish between the two different directions of yarn. The longitudinal yarn, running along the length of the weaving machine or loom and hence running lengthwise in a woven geotextile roll, is called the *warp*. The transverse yarn, running across the width of the loom and hence running widthwise in a woven geotextile roll, is called the *weft*. Since the warp direction coincides with the direction in which the geotextile is manufactured on the mechanical loom, this is also called the *machine direction (MD)* (a.k.a. *production direction or roll length direction*), whereas at right angles to the machine direction in the plane of the geotextile is the *cross machine direction (CMD)*, which is basically the weft direction.

In Figure 1.13, the type of weave described is a *plain weave*, of which there are many variations such as *twill*, *satin* and *serge*; however, plain weave is the one most commonly used in geotextiles.

Nonwoven geotextiles are obtained by processes other than weaving. The processing involves continuous laying of the fibres on to a moving conveyor belt to form a loose web slightly wider than the finished product. This passes along the conveyor to be bonded by mechanical bonding (obtained by punching thousands of small barbed needles through the loose web), thermal bonding (obtained by partial melting of the fibres) or chemical

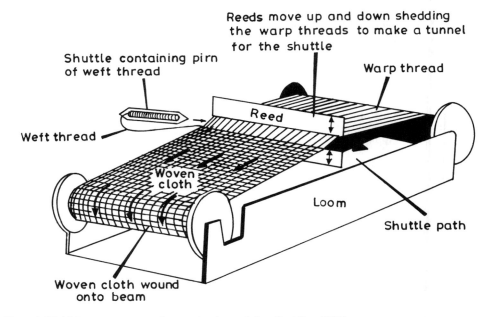

Figure 1.12 Main components of a weaving loom (after Rankilor, 1981).

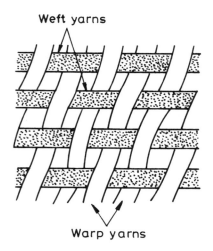

Figure 1.13 A typical woven geotextile having a plain weave.

bonding (obtained by fixing the fibres with a cementing medium such as glue, latex, cellulose derivative, or synthetic resin) resulting in the following three different types:

1   mechanically bonded nonwoven geotextile (or needle-punched nonwoven geotextiles)
2   thermally bonded nonwoven geotextile
3   chemically bonded nonwoven geotextile, respectively.

Figure 1.14 shows the diagrammatic representation of the production of needle-punched geotextiles. These geotextiles are usually relatively thick, with a typical thickness in the region of 0.5–5 mm.

Knitted geotextiles are manufactured using a knitting process, which involves interlocking a series of loops of one or more yarns together to form a planar structure. There is a wide range of different types of knits used, one of which is illustrated in Figure 1.15. These geotextiles are very extensible and therefore used in very limited quantities.

Stitch-bonded geotextiles are produced from multi-filaments by a stitching process. Even strong, heavyweight geotextiles can be produced rapidly. Geotextiles are sometimes manufactured in a tubular or cylindrical fashion without longitudinal seam. Such geotextiles are called *tubular geotextiles*.

A geotextile can be saturated with bitumen, resulting in a *bitumen impregnated geotextile*. Impregnation aims at modifying the geotextile, to protect it against external forces and, in some cases, to make it fluid impermeable.

All the geogrids share a common geometry comprising two sets of orthogonal load carrying elements which enclose substantially rectangular or square patterns. Due to the

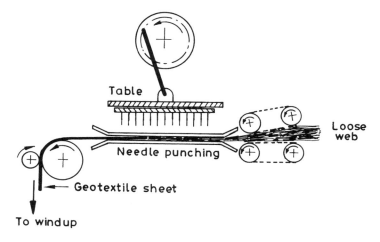

Figure 1.14 Manufacturing process of the needle-punched nonwoven geotextile.

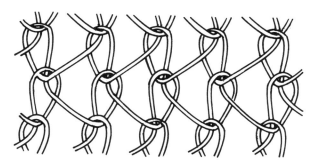

Figure 1.15 A typical knitted geotextile.

requirement of high tensile properties and acceptable creep properties, all geogrids are produced from molecularly oriented plastic. The main difference between different grid structures lies in how the longitudinal and transverse elements are joined together.

Extruded geogrids are manufactured from polymer sheets in two or three stages of processing: the first stage involves feeding a sheet of polymer, several millimetres thick, into a punching machine, which punches out holes on a regular grid pattern. Following this, the punched sheet is heated and stretched, or drawn, in the machine direction. This distends the holes to form an elongated grid opening known as *aperture*. In addition to changing the initial geometry of the holes, the drawing process orients the randomly oriented long-chain polymer molecules in the direction of drawing. The degree of orientation will vary along the length of the grid; however, the overall effect is an enhancement of tensile strength and tensile stiffness. The process may be halted at this stage, in which case the end product is a *uniaxially oriented geogrid*. Alternatively, the uniaxially oriented grid may proceed to a third stage of processing to be warm drawn in the transverse direction, in which case a *biaxially oriented geogrid* is obtained (Fig. 1.16). Although the temperatures used in the drawing process are above ambient, this is effectively a cold drawing process, as the temperatures are significantly below the melting point of the polymer. It should be noted that the ribs of geogrids are often quite stiff compared to the fibres of geotextiles.

Woven geogrids are manufactured by weaving or knitting processes from polyester multi-filaments. Where the warp and weft filaments cross they are interlaced at multiple levels to form a competent junction. The skeletal structure is generally coated with acrylic or PVC

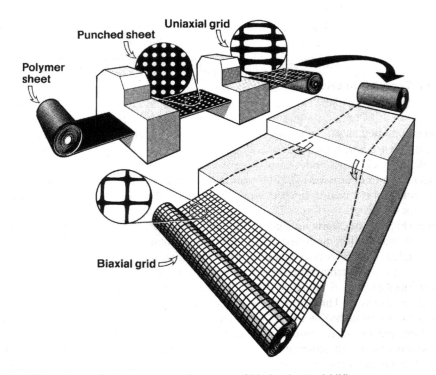

Figure 1.16 'Tensar' manufacturing process (courtesy of Netlon Limited, UK).

or bitumen to provide added protection against environmental attack and construction-induced damages.

Bonded geogrids are manufactured by bonding the mutually perpendicular PP or PET strips together at their crossover points using either laser or ultrasonic welding. There are several bonded geogrids, which are extremely versatile, because they can be used in isolated strip form and as multiple strips for ground reinforcement.

Geonets are manufactured typically by an extrusion process in which a minimum of two sets of strands (bundles of tape-like fibres that can be attached to each other) are overlaid to yield a three-dimensional structure. A counterrotating die, with a simplified section as shown in Figure 1.17, is fed with hot plastic by a screw extruder. The die consists of an inner mandrel mounted concentrically inside a heavy tubular sleeve. When both the inner and outer sections of the die are rotated then the two sets of spirals are produced simultaneously; however, at the instant that inner and outer slots align with one another there is only one, double thickness, set of strands extruded. It is at this instant that the crossover points of the two spirals are formed as extruded junctions. Consequently the extrudate takes the form of a tubular geonet. This continuously extruded tube is fed coaxially over a tapering mandrel which stretches the tube to the required diameter. This stretching process results in inducing a degree of molecular orientation and it also controls the final size and geometry of the finished geonet. To convert the tubular geonet to flat sheet, the tube is cut and laid flat. If the tube is slit along its longitudinal axis, the resulting geonet appears to have a diamond shaped aperture. Alternatively, the tube may be cut on the bias, for example parallel to one of the strand arrays, in which case the apertures appear to be almost square.

Unlike geogrids, the intersecting ribs of geonets are generally not perpendicular to one another. In fact they intersect at typically 60°–80° to form a diamond shaped aperture. It can be seen that one parallel array of elements sits on top of the underlying array so creating a structure with some depth. The geonets are typically 5–10 mm in thickness.

Most of the geomembranes are made in a plant using one of the following manufacturing processes: (i) extrusion, (ii) spread coating or (iii) calendering. The extrusion process is a method whereby a molten polymer is extruded into a non-reinforced sheet using an extruder. Immediately after extrusion, when the sheet is still warm, it can be laminated with a geotextile; the geomembrane thus produced is *reinforced*. The spread coating process usually consists in coating a geotextile (woven, nonwoven, knitted) by spreading a polymer or asphalt compound on it. The geomembranes thus produced are therefore also reinforced. *Non-reinforced geomembranes* can be made by spreading a polymer on a sheet of paper, which is removed and discarded at the end of the manufacturing process. Calendering is the most frequently

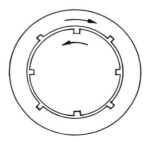

Figure 1.17 Rotating die.

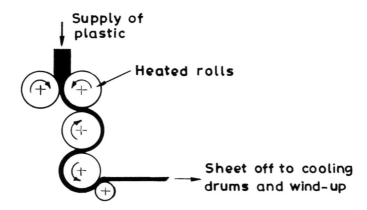

Figure 1.18 Calendering process of manufacturing geomembranes (after Ingold, 1994).

used manufacturing process in which a heated polymeric compound is passed through a series of heated rollers of the calender, rotating under mechanical or hydraulic pressure (Fig. 1.18).

By utilizing auxiliary extruders, both HDPE and linear low density polyehylene (LLDPE) geomembranes are sometimes coextruded, using flat dies or blown film methods. Coextruded sheet is manufactured such that a HDPE–LLDPE–HDPE geomembrane results. The HDPE on the upper and lower surfaces is approximately 10–20% of the total sheet thickness. The objective is to retain the excellent chemical resistance of HDPE on the surfaces of geomembranes, with the flexibility of LLDPE in the core. Typical thickness of geomembranes ranges from 0.25 to 7.5 mm (10–300 mils, 1 mil = 0.001 in. ≈ 0.025 mm).

Textured surfaces (surfaces with projections or indentations) can be made on one or both sides of a geomembrane by blown film coextrusion or impingement by hot PE particles or any other suitable method. A geomembrane with textured surfaces on one or both surfaces is called *textured geomembrane*. The textured surface greatly improves stability, particularly on sloping grounds, by increasing interface friction between the geomembrane and the soils or the geosynthetics. Textured surfaces are generally produced with about 6-in (150-mm) nontextured border on both sides of the sheet. The smooth border provides a better surface for welding than a textured surface. The smooth edges also permit quick verification of the thickness and strength before installation.

Geocomposites can be manufactured from two or more of the geosynthetic types described in Sec. 1.2. A geocomposite can therefore combine the properties of the constituent members in order to meet the needs of a specific application. Some examples of geocomposites are band/strip/wick drains and geosynthetic clay liners (GCLs).

Band drains, also called fin drains, usually consist of a plastic fluted or nubbed water conducting core (drainage core) wrapped in geotextie sleeve (Fig. 1.19). They are designed for easy installation in either a slot or trench dug in the soil along the edge of the highway pavement or railway track or any other civil engineering structures that require drainage measures.

Geotextiles can be attached to geomembranes to form geocomposites. Geotextiles are commonly used in conjunction with geomembranes for puncture protection, drainage and improved tensile strength.

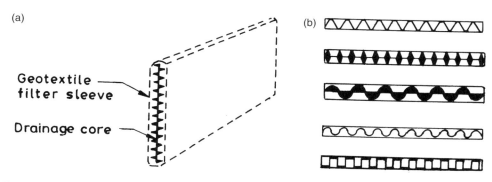

Figure 1.19 Strip drain: (a) components; (b) various shapes of drainage cores.

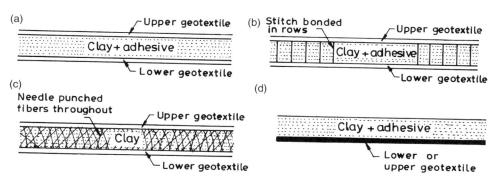

Figure 1.20 Types of geosynthetic clay liners (after Koerner and Daniel, 1997).

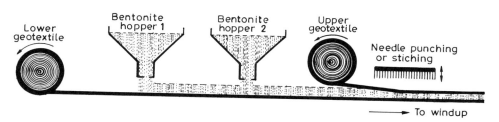

Figure 1.21 Major steps of the manufacturing process for needle-punched and stitch-bonded geosynthetic clay liners.

A geosynthetic clay liner is a manufactured hydraulic barrier used as alternative material to substitute a conventional compacted soil layer for the low-permeability soil component of various environmental and hydraulic projects including landfill and remediation projects. It consists of a thin layer of sodium or calcium bentonite (mass per unit area ≈ 5 kg/m^2), which is either sandwiched between two sheets of woven or nonwoven geotextiles (Fig. 1.20) or mixed with an adhesive and attached to a geomembrane. The sodium bentonite has a lower hydraulic conductivity. Figure 1.21 shows the major steps of the manufacturing process for a needle-punched or stitch bonded GCL. It should be noted that the GCLs are also known by several other names such as *clay blankets*, *bentonite blankets* and *bentonite mats*.

The combination of the geotextile (filtering action), geomembranes (waterproofing properties) and geonets (drainage and load distribution) offers a complete system of filter-drainage-protection, which is very compact and easy to install.

The geosynthetics manufactured in the factory environment have specific properties whose uniformity is far superior to soils, which are notoriously poor in homogeneity, as well as in isotropy. Based on many years experience of manufacturing and the development of quality assurance procedures, geosynthetics are made in such a way that good durability properties are obtained.

Most geosynthetics are supplied in rolls. Although there is no standard width for geosynthetics, most geotextiles are provided in widths of around 5 m while geogrids are generally narrower and geomembranes may be wider. Geotextiles are supplied typically with an area of 500 m² per roll for a product of average mass per unit area. In fact, depending on the mass per unit area, thickness, and flexibility of the product, roll lengths vary between a few tens of metres up to several hundreds of metres with the majority of roll lengths falling in the range 100–200 m. The product of mass per unit area, roll width and length gives the mass of the roll. Rolls with a mass not exceeding 100 kg can usually be handled manually. If allowed to become wet, the weight of a roll, particularly of geotextiles, can increase dramatically. Geonets are commercially available in rolls up to 4.5 m wide. The HDPE and very low density polyethylene (VLDPE) geomembranes are supplied in roll form with widths of approximately 4.6–10.5 m and lengths of 200–300 m. Geosynthetic clay liners are manufactured in panels that measure 4–5 m in width and 30–60 m in length and are placed on rolls for shipment to the jobsite.

ILLUSTRATIVE EXAMPLE 1.1

Consider a roll of geotextile with the following parameters:
Mass, $M = 100$ kg
Width, $B = 5$ m
Mass per unit area, $m = 200$ g/m².
Determine the length of the roll.

SOLUTION

Mass per unit area of the geotextile can be expressed as

$$m = \frac{M}{L \times B} \tag{1.1}$$

where L is the roll length.
From Equation (1.1),

$$L = \frac{M}{m \times B} = \frac{100 \text{ kg}}{(0.2 \text{ kg/m}^2) \times (5 \text{ m})} = 100 \text{ m} \qquad \textbf{Answer}$$

1.6 Geosynthetic engineering

Geosynthetics technology is a composite science involving the skills of polymer technologists, chemists, production engineers and application engineers. Applications of geosynthetics fall mainly within the discipline of civil engineering and the design of these applications, due to the use of geosynthetics mostly with soils and rocks, is closely associated with geotechnical engineering. For a given application, knowledge of the geotechnical engineering only serves to define and enumerate the functions and properties of a geosynthetic (Ingold, 1994).

The engineering involved with geosynthetics and their applications may be called *geosynthetic engineering*. In the present-day civil engineering practice, this engineering has become so vast that its study is required as a separate subject in civil engineering. This new and emerging subject can be defined as follows:

> Geosynthetic engineering deals with the engineering applications of scientific principles and methods to the acquisition, interpretation, and use of knowledge of geosynthetic products for the solutions of geotechnical, transportation, environmental, hydraulic and other civil engineering problems.

Self-evaluation questions

(Select the most appropriate answers to the multiple-choice type questions from 1 to 16)

1. A planar, polymeric product consisting of a mesh or net-like regular open network of intersecting tensile-resistant elements, integrally connected at the junctions, is called

    (a) Geotextile.
    (b) Geogrid.
    (c) Geonet.
    (d) Geocell.

2. Which one of the following basic characteristics is not found in geosynthetics?

    (a) Non-corrosiveness.
    (b) Lightness.
    (c) Long-term durability under soil cover.
    (d) High rigidity.

3. The materials used in the manufacture of geosynthetics are primarily synthetic polymers generally derived from

    (a) Rubber.
    (b) Fiberglass.
    (c) Crude petroleum oils.
    (d) Jute.

4. Molecular weight of a polymer can affect

    (a) Only physical property of geosynthetics.
    (b) Only mechanical property of geosynthetics.
    (c) Heat resistance and durability of geosynthetics.
    (d) All of the above.

5. The most widely used polymers for manufacturing geosynthetics are

   (a) Polypropylene and polyamide.
   (b) Polyester and polyethylene.
   (c) Polypropylene and polyester.
   (d) Polypropylene and polyethylene.

6. The resistance to creep is high for

   (a) Polypropylene (PP).
   (b) Polyester (PET).
   (c) Polyethylene (PE).
   (d) Polyamide (PA).

7. Carbon black content in geosynthetics can be determined by

   (a) Thermogravimetric analysis (TGA).
   (b) Differential scanning calorimetry (DSC).
   (c) Thermomechanical analysis (TMA).
   (d) None of the above.

8. The synthetic fibres produced by extruding melted polymer through dies or spinnerets and subsequently drawn longitudinally are called

   (a) Filaments.
   (b) Staple fibres.
   (c) Slit films.
   (d) Strands.

9. The term 'weft' refers to

   (a) The longitudinal yarn of the geotextile.
   (b) The transverse yarn of the geotextile.
   (c) Both (a) and (b).
   (d) None of the above.

10. Most of the geotextiles are commercially available in rolls of width of around

    (a) 1 m.
    (b) 5 m.
    (c) 10 m.
    (d) None of the above.

11. Geotextiles are commonly used in conjunction with geomembranes for

    (a) Puncture protection.
    (b) Drainage.
    (c) Improved tensile strength.
    (d) All of the above.

12. Which one of the following geosynthetics is a geocomposite?

    (a) Geogrid.
    (b) Geonet.

Chapter 2

Functions and selection

2.1 Introduction

For any given application of a geosynthetic, there can be one or more functions that the geosynthetic will be expected to serve during its performance life. The selection of a geosynthetic for any field application is highly governed by the function(s) to be performed by the geosynthetic in that specific application. This chapter describes all such functions that can be performed by commercially available geosynthetics and several aspects related to selection of geosynthetics.

2.2 Functions

Geosynthetics have numerous application areas in civil engineering. They always perform one or more of the following basic functions when used in contact with soil, rock and/or any other civil engineering-related material:

- Reinforcement
- Separation
- Filtration
- Drainage
- Fluid barrier
- Protection.

Reinforcement A geosynthetic performs the reinforcement function by improving the mechanical properties of a soil mass as a result of its inclusion. When soil and geosynthetic reinforcement are combined, a composite material, 'reinforced soil', possessing high compressive and tensile strength (and similar, in principle, to the reinforced concrete) is produced. In fact, any geosynthetic applied as reinforcement has the main task of resisting applied stresses or preventing inadmissible deformations in geotechnical structures. In this process, the geosynthetic acts as a tensioned member coupled to the soil/fill material by friction, adhesion, interlocking or confinement and thus maintains the stability of the soil mass (Fig. 2.1).

Different concepts have been advanced to define the basic mechanism of reinforced soils. The effect of inclusion of relatively inextensible reinforcements (such as metals, fibre-reinforced plastics, etc. having a high modulus of deformation) in the soil can be explained using either an induced stresses concept (Schlosser and Vidal, 1969) or an induced deformations concept (Basset and Last, 1978). According to the induced stresses concept, the tensile strength of the

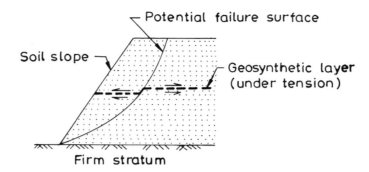

Figure 2.1 Basic mechanism involved in the reinforcement function.

reinforcements and friction at soil–reinforcement interfaces give an apparent cohesion to the reinforced soil system. The induced deformations concept considers that the tensile reinforcements involve anisotropic restraint of the soil deformations. The behaviour of the soil reinforced with extensible reinforcements, such as geosynthetics, does not fall within these concepts. The difference, between the influences of inextensible and extensible reinforcements, is significant in terms of the load-settlement behaviour of the reinforced soil system (Fig. 2.2). The soil reinforced with extensible reinforcement (termed *ply-soil* by McGown and Andrawes (1977)) has greater extensibility and smaller losses of post peak strength compared to soil alone or soil reinforced with inextensible reinforcement (termed *reinforced earth* by Vidal (1969)). However, some similarity between ply-soil and reinforced earth exists in that they inhibit the development of internal tensile strains in the soil and develop tensile stresses.

Fluet (1988) subdivided the reinforcement function into the following two categories:

1 A tensile member, which supports a planar load, as shown in Figure 2.3(a).
2 A tensioned member, which supports not only a planar load but also a normal load, as shown in Figure 2.3(b).

Jewell (1996) and Koerner (2005) consider not two but three mechanisms for soil reinforcement, because when the geosynthetic works as a tensile member it might be due to two different mechanisms: shear and anchorage. Therefore, the three reinforcing mechanisms, concerned simply with the types of load that are supported by the geosynthetic, are

1 *Shear*, also called *sliding*: The geosynthetic supports a planar load due to slide of the soil over it.
2 *Anchorage*, also called *pullout*: The geosynthetic supports a planar load due to its pullout from the soil.
3 *Membrane*: The geosynthetic supports both a planar and a normal load when placed on a deformable soil.

Shukla (2002b, 2004) describes reinforcing mechanisms that take into account the reinforcement action of the geosynthetic, that is, how the geosynthetic reinforcement takes

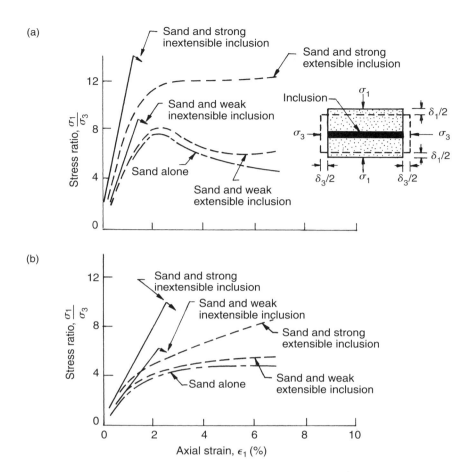

Figure 2.2 Postulated behaviour of a unit cell in plane strain conditions with and without inclusions: (a) dense sand with inclusions; (b) loose sand with inclusions (after McGown *et al.*, 1978).

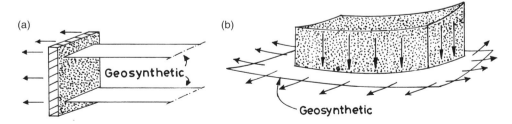

Figure 2.3 Reinforcement function: (a) tensile member; (b) tensioned member (Fluet, 1988).

the stresses from the soil and which type of stresses are taken. This concept can be observed broadly in terms of the following roles of geosynthetics:

1. A geosynthetic layer reduces the outward horizontal stresses (shear stresses) transmitted from the overlying soil/fill to the top of the underlying foundation soil. This action of geosynthetics is known as *shear stress reduction effect*. This effect results in a general-shear, rather than a local-shear failure (Fig. 2.4(a)), thereby causing an increase in the load-bearing capacity of the foundation soil (Bourdeau et al., 1982; Guido et al., 1985; Love et al., 1987; Espinoza, 1994; Espinoza and Bray, 1995; Adams and Collin, 1997). Through the shear interaction mechanism the geosynthetic can therefore improve the performance of the system with very little or no rutting. In fact, the reduction in shear stress and the change in the failure mode is the primary benefit of the geosynthetic layer at small deformations.

2. A geosynthetic layer redistributes the applied surface load by providing restraint of the granular fill if embedded in it, or by providing restraint of the granular fill and the soft foundation soil, if placed at their interface, resulting in reduction of applied normal stress on the underlying foundation soil (Fig. 2.4(b)). This is referred to as *slab effect* or *confinement effect* of geosynthetics (Bourdeau, et al., 1982; Giroud et al., 1984; Madhav and Poorooshasb, 1989; Hausmann, 1990; Sellmeijer, 1990). The friction mobilized between the soil and the geosynthetic layer plays an important role in confining the soil.

3. The deformed geosynthetic, sustaining normal and shear stresses, has a membrane force with a vertical component that resists applied loads, that is, the deformed geosynthetic provides a vertical support to the overlying soil mass subjected to loading. This action of geosynthetics is popularly known as its *membrane effect* (Fig. 2.4(c)) (Giroud and Noiray, 1981; Bourdeau et al., 1982; Sellmeijer et al., 1982; Love et al., 1987; Madhav and Poorooshasb, 1988; Bourdeau, 1989; Sellmeijer, 1990; Shukla and Chandra, 1994a, 1995). Depending upon the type of stresses – normal stress and shear stress – sustained by the geosynthetic during their action, the membrane support may be classified as 'normal stress membrane support' and 'interfacial shear stress membrane support' respectively (Espinoza and Bray, 1995). Edges of the geosynthetic layer are required to be anchored in order to develop the membrane support contribution resulting from normal stresses, whereas the membrane support contribution resulting from mobilized interfacial membrane shear stresses does not require any anchorage. The membrane effect of geosynthetics causes an increase in the load-bearing capacity of the foundation soil below the loaded area with a downward loading on its surface to either side of the loaded area, thus reducing its heave potential. It is to be noted that both the geotextile and the geogrid can be effective in membrane action in case of high-deformation systems.

4. The use of geogrids has another benefit owing to the interlocking of the soil through the apertures (openings between the longitudinal and transverse ribs, generally greater than 6.35 mm (1/4 inch)) of the grid known as *interlocking effect* (Guido et al., 1986) (Fig. 2.4(d)). The transfer of stress from the soil to the geogrid reinforcement is made through bearing (passive resistance) at the soil to the grid cross-bar interface. It is important to underline that owing to the small surface area and large apertures of geogrids, the interaction is due mainly to interlocking rather than to friction. However, an exception occurs when the soil particles are small. In this situation the interlocking effect is negligible because no passive strength is developed against the geogrid (Pinto, 2004).

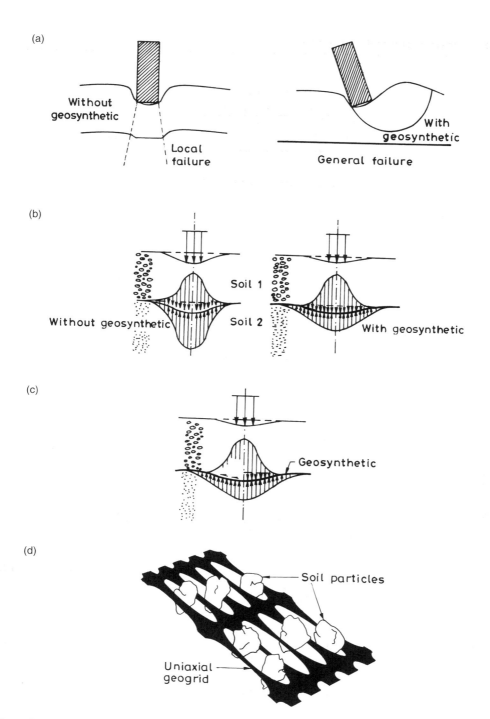

Figure 2.4 Roles of a geosynthetic reinforcement: (a) causing change of failure mode (shear stress reduction effect); (b) redistribution of the applied surface load (confinement effect); (c) providing vertical support (membrane effect) (after Bourdeau et al., 1982 and Espinoza, 1994); (d) providing passive resistance through interlocking of the soil particles (interlocking effect).

Separation If the geosynthetic has to prevent intermixing of adjacent dissimilar soils and/or fill materials during construction and over a projected service lifetime of the application under consideration, it is said to perform a separation function. Figure 2.5 shows that the geosynthetic layer prevents the intermixing of soft soil and granular fill, thereby keeping the structural integrity and functioning of both materials intact. This function can be observed if a geotextile layer is provided at the soil subgrade level in pavements or railway tracks to prevent pumping of soil fines into the granular subbase/base course and/or to prevent intrusion of granular particles into soil subgrade.

In many geosynthetic applications, especially in roads, rail tracks, shallow foundations, and embankments, a geosynthetic layer is placed at the interface of soft foundation soil and the overlying granular layer (Fig. 2.6). In such a situation, it becomes a difficult task to identify the major function of reinforcement and separation. Nishida and Nishigata (1994) have suggested that the separation can be a dominant function over the reinforcement function

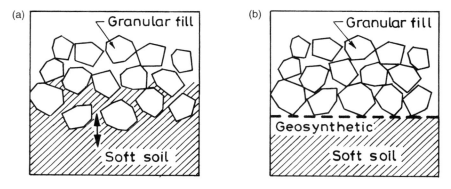

Figure 2.5 Basic mechanism involved in the separation function: (a) granular fill–soft soil system without the geosynthetic separator; (b) granular fill–soft soil system with the geosynthetic separator.

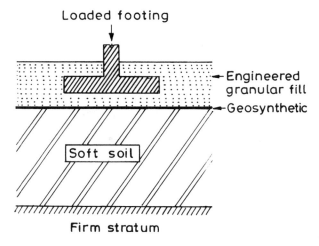

Figure 2.6 A loaded geosynthetic-reinforced granular fill–soft soil system.

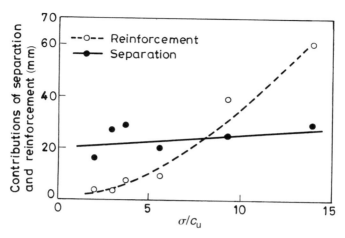

Figure 2.7 Relationship between the separation and the reinforcement functions (after Nishida and Nishigata, 1994).

when the ratio of the applied stress (σ) on the subgrade soil to the shear strength (c_u) of the subgrade soil has a low value (less than 8), and it is basically independent of the settlement of the reinforced soil system (Fig. 2.7).

It is important to note that separation depends on the grain size of the soils involved. Most low-strength foundation soils are composed of small particles, whereas the placed layers (for roads, railways, foundations and embankments) are of coarser materials. In these situations separation is always needed, quite independent of the ratio of the applied stress to the strength of the subgrade soil, as Figure 2.7 clearly shows. In general, reinforcement will increase in importance as that ratio increases. Fortunately, separation and reinforcement are compatible functions. Furthermore, they work together, interacting: reinforcement reduces deformation and therefore reduces mixing of the particles (performing indirectly and to some extent the separation function); on the other hand, separation prevents mixing and consequently prevents progressive loss of the strength of the subsequent layers. The ideal material to be used for roads, railways, foundations and embankments (i.e. when a coarse-grained soil is placed on top of a fine-grained soil with low strength) would be a continuous material such as a high-strength geotextile or a composite of a stiffer geogrid combined with a geotextile. In this way, the necessary separation and reinforcement functions can be performed simultaneously.

The selection of primary function from reinforcement and separation can also be done on the basis of available empirical knowledge, if the California Bearing Ratio (CBR) of the subgrade soil is known. If the subgrade soil is soft, that is, the CBR of the subgrade soil is low, say its unsoaked value is less than 3 (or soaked value is less than 1), then the reinforcement can generally be taken as the primary function because of adequate tensile strength mobilization in the geosynthetic through large deformation, that is, deep ruts (say, greater than 75 mm) in the subgrade soil. Geosynthetics, used with subgrade soils with an unsoaked CBR higher than 8 (or soaked CBR higher than 3), will have generally negligible amount of reinforcement role, and in such cases the primary function will uniquely be separation. For soils with intermediate unsoaked CBR values between 3 and 8 (or soaked

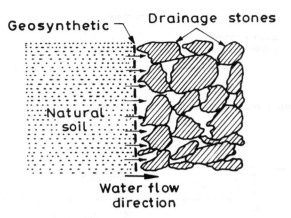

Figure 2.8 Basic mechanism involved in the filtration function.

CBR values between 1 and 3), the selection of the primary function is totally based on the site-specific situations.

Filtration A geosynthetic may function as a filter that allows for adequate fluid flow with limited migration of soil particles across its plane over a projected service lifetime of the application under consideration. Figure 2.8 shows that a geosynthetic allows passage of water from a soil mass while preventing the uncontrolled migration of soil particles.

When a geosynthetic filter is placed adjacent to a base soil (the soil to be filtered), a discontinuity arises between the original soil structure and the structure of the geosynthetic. This discontinuity allows some soil particles, particularly particles closest to the geosynthetic filter and having diameters smaller than the *filter opening size* (see Chapter 3 for more explanation), to migrate through the geosynthetic under the influence of seepage flows. For a geosynthetic to act as a filter, it is essential that a condition of equilibrium is established at the soil/geosynthetic interface as soon as possible after installation to prevent soil particles from being piped indefinitely through the geosynthetic. At equilibrium, three zones may generally be identified: the undisturbed soil, a 'soil filter' layer which consists of progressively smaller particles as the distance from the geosynthetic increases and a bridging layer which is a porous, open structure (Fig. 2.9). Once the stratification process is complete, it is actually the soil filter layer, which actively filters the soil.

It is important to understand that the filtration function also provides separation benefits. However, a distinction may be drawn between filtration function and separation function with respect to the quantity of fluid involved and to the degree to which it influences the geosynthetic selection. In fact, if the water seepage across the geosynthetic is not a critical situation, then the separation becomes the major function. It is also a practice to use the separation function in conjunction with reinforcement or filtration; accordingly separation is not specified alone in several applications.

Drainage If a geosynthetic allows for adequate fluid flow with limited migration of soil particles within its plane from surrounding soil mass to various outlets over a projected service lifetime of the application under consideration, it is said to perform the drainage (a.k.a. fluid transmission) function.

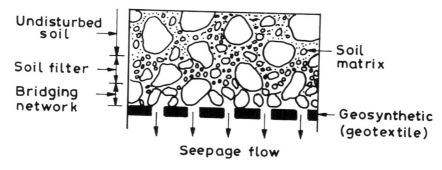

Figure 2.9 An idealized interface conditions at equilibrium between the soil and the geosynthetic filter.

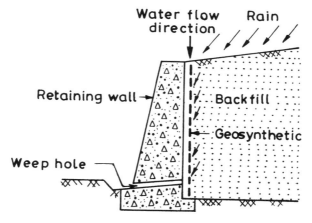

Figure 2.10 Basic mechanism involved in the drainage function.

Figure 2.10 shows that the geosynthetic layer adjacent to the retaining wall collects water from the backfill and transports it to the weep holes constructed in the retaining wall.

It should be noted that while performing the filtration and drainage functions, a geosynthetic dissipates the excess pore water pressure by allowing flow of water in plane and across its plane.

Fluid barrier A geosynthetic performs the fluid barrier function, if it acts like an almost impermeable membrane to prevent the migration of liquids or gases over a projected service lifetime of the application under consideration.

Figure 2.11 shows that a geosynthetic layer, installed at the base of a pond, prevents the infiltration of liquid waste into the natural soil.

Protection A geosynthetic, placed between two materials, performs the protection function when it alleviates or distributes stresses and strains transmitted to the material to be protected against any damage (Fig. 2.12). In some applications, a geosynthetic layer is needed as a localized stress reduction layer to prevent or reduce local damage to a geotechnical system.

It should be noted that the basic functions of geosynthetics described above can be quantitatively described by standard tests or design techniques or both. Geosynthetics can also perform some other functions that are, in fact, qualitative descriptions, mostly dependent on

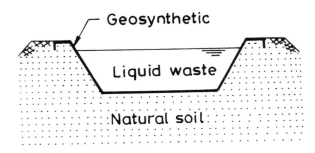

Figure 2.11 Basic mechanism involved in the fluid barrier function.

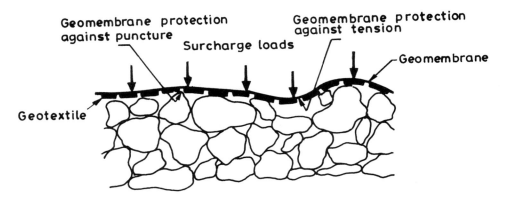

Figure 2.12 Basic mechanism involved in the protection function.

basic functions, and are not yet supported by standard tests or generally accepted design techniques. Such functions of geosynthetics, basically describing their performance characteristics, are the following:

- *Absorption* A geosynthetic provides absorption if it is used to assimilate or incorporate a fluid. This may be considered for two specific environmental aspects: water absorption in erosion control applications and the recovery of floating oil from surface waters following ecological disasters.
- *Containment* A geosynthetic provides containment when it is used to encapsulate or contain a civil engineering related material such as soil, rock or fresh concrete to a specific geometry and prevent its loss.
- *Cushioning* A geosynthetic provides cushioning when it is used to control and eventually to damp dynamic mechanical actions. This function has to be emphasized particularly for the geosynthetic applications in canal revetments, shore protections, pavement overlay protection from reflective cracking and seismic base isolation of earth structures.
- *Insulation* A geosynthetic provides insulation when it is used to reduce the passage of electricity, heat or sound.
- *Screening* A geosynthetic provides screening when it is placed across the path of a flowing fluid carrying fine particles in suspension to retain some or all particles while allowing the fluid to pass through. After some period of time, particles accumulate against

Table 2.1 Functions of geosynthetics and their symbols

Functions	Symbols
Reinforcement	R
Separation	S
Filtration	F
Drainage (a.k.a. fluid transmission)	D (or FT)
Fluid barrier	FB
Protection	P
Absorption	A
Containment	C
Cushioning	Cus
Insulation	I
Screening	Scr
Surface stabilization	SS
Vegetative reinforcement	VR

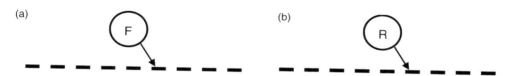

Figure 2.13 Symbolic representations: (a) a filtration geotextile; (b) a reinforcement geotextile.

the geosynthetic, and hence it is required that the geosynthetic be able to withstand pressures generated by the accumulated particles and the increasing fluid pressure.
- *Surface stabilization* (*surficial erosion control*) A geosynthetic provides surface stabilization when it is placed on a soil surface to restrict movement and prevent dispersion of surface soil particles subjected to erosion actions of rain and wind, often while allowing or promoting growth of vegetation.
- *Vegetative reinforcement* A geosynthetic provides vegetative reinforcement when it extends the erosion control limits and performance of vegetation.

The relative importance of each function is governed by the site conditions, especially soil type and groundwater drainage, and the construction application. In many cases, two or more basic functions of the geosynthetic are required in a particular application.

All the functions of geosynthetics described in this section are listed along with their symbols/abbreviations in Table 2.1. The suggested symbols may help in making the drawing or diagram of a geosynthetic application with clarity. For example, if a geotextile is required to be represented for reinforcement function or filtration function, then this can be done as shown in Figure 2.13.

2.3 Selection

Geosynthetics are available with a variety of geometric and polymer compositions to meet a wide range of functions and applications. Depending on the type of application, geosynthetics may have specific requirements.

Table 2.2 Selection of geosynthetics based on their functions

Functions to be performed by the geosynthetics		Geosynthetics that can be used
Separation	Primary	GTX, GCP, GFM
	Secondary	GTX, GGR, GNT, GMB, GCP, GFM
Reinforcement	Primary	GTX, GGR, GCP
	Secondary	GTX, GCP
Filtration	Primary	GTX, GCP
	Secondary	GTX, GCP
Drainage	Primary	GTX, GNT, GCP, GPP
	Secondary	GTX, GCP, GFM
Fluid barrier	Primary	GMB, GCP
	Secondary	GCP
Protection	Primary	GTX, GCP
	Secondary	GTX, GCP

Notes
GTX = Geotextile, GGR = Geogrid, GNT = Geonet, GMB = Geomembrane, Geofoam = GFM, Geopipe = GPP, GCP = Geocomposite

When installed, a geosynthetic may perform more than one of the listed functions (see Table 2.1) simultaneously, but generally one of them will result in the lower factor of safety; thus it becomes a primary function. The use of a geosynthetic in a specific application needs classification of its functions as primary or secondary. Table 2.2 shows such a classification, which is useful while selecting the appropriate type of geosynthetic for solving the problem at hand. Each function uses one or more properties of the geosynthetic (see Chapter 3), such as tensile strength or water permeability, referred to as *functional properties*. The function concept is generally used in the design with the formulation of a factor of safety, FS, in the traditional manner as:

$$FS = \frac{\text{Allowable (or test) functional property}}{\text{Required (or design) functional property}} \quad (2.1)$$

The allowable functional property is available property, measured by the *performance test* or the *index test* (explained in Chapter 3), possibly factored down to account for uncertainties in its determination or in other site-specific conditions during the design life of the soil–geosynthetic system; whereas the required value of functional property is established by the designer or specifier using accepted methods of analysis and design or empirical guidelines for the actual field conditions. The entire process, generally called 'design-by-function' is widespread in its use. The actual magnitude of the factor of safety depends upon the implication of failure, which is always site-specific. If the factor of safety is sufficiently larger than one, the geosynthetic is acceptable for utilization because it ensures the stability and serviceability of the structure. However, as might be anticipated with new technology, universally accepted values of a minimum factor of safety have not yet been established, and conservation in this regard is still warranted.

It is observed that only geotextiles and geocomposites perform most of the functions, and hence they are used in many applications. Geotextiles are porous across their manufactured planes and also within their planes. Thick, nonwoven needle-punched geotextiles have considerable void volume in their structure, and thus they can transmit fluid within the structure to a very high degree. The degree of porosity, which may vary widely, is used to determine the selection of specific geotextiles. Geotextiles can also be used as a fluid barrier on impregnation with materials like bitumen. The geotextiles vary with the type of polymer used, the type of fibre and the fabric style.

Geogrids are used to primarily function as reinforcement, and separation may be an occasional function, especially when soils having very large particle sizes are involved. The performance of the geogrid as reinforcement relies on its rigidity or high tensile modulus and on its open geometry, which accounts for its high capacity for interlocking with soil particles.

It has been observed that for geotextiles to function properly as reinforcement, friction must develop between the soil and the reinforcement to prevent sliding, whereas for geogrids, it is the interlocking of the soil through the apertures of the geogrid that achieves an efficient interlocking effect. In this respect, geotextiles are frictional resistance dependent reinforcement, whereas geogrids are passive resistance dependent reinforcement. The laboratory studies have shown that geogrids are a superior form of reinforcement owing to the interlocking of the soil with the grid membrane.

Geonets, unlike geotextiles, are relatively stiff, net-like materials with large open spaces between structural ribs. They are used exclusively as fluid conducting cores in prefabricated drainage geocomposites. Geonets currently play a major role in landfill leachate collection and leak detection systems in association with geomembranes or geotextiles. For a fair drainage function, geonets should not be laid in contact with soils or waste materials but used as drainage cores with geotextile, geomembrane or other material on their upper and lower surfaces, thus avoiding the soil particles from obstructing the drainage net channels. Geonets as a drainage material generally fall intermediate in their flow capability between thick needle-punched nonwoven geotextiles and numerous drainage geocomposites.

Geomembranes are mostly used as a fluid barrier or liner. Sometimes, a geomembrane is also known as a flexible membrane liner (FML), especially in landfill applications. The permeability of typical geomembranes range from 0.5×10^{-12} to 0.5×10^{-15} m/s. Thus, the geomembranes are from 10^3 to 10^6 times lower in permeability than compacted clay. In this context, we speak of geomembranes as being essentially impermeable. The recommended minimum thickness for all geomembranes is 30 mil (0.75 mm), with the exception of HDPE, (High-density polyethylene) which should be at least 60 mil (1.5 mm) to allow for extrusion seaming (Qian et al., 2002). The most widely used geomembrane in the waste management industry is HDPE, because this offers excellent performance for landfill liners and covers, lagoon liners, wastewater treatment facilities, canal linings, floating covers, tank linings and so on. If greater flexibility than HDPE is required, then linear low density polyethylene (LLDPE) is used because it has lower molecular weight resin that allows LLDPE to conform to non-uniform surfaces, making it suitable for landfill caps, pond liners, lagoon liners, potable water containment, tunnels and tank linings.

Geofoams are available in slab or block form. Since these geosynthetic products are very light-weight material with unit weight ranging from 0.11 to 0.48 kN/m^3, they can be selected as highway fill over compressible subgrade soils and frost-sensitive soils and as backfill material for retaining walls to reduce lateral earth pressure, thereby functioning basically as a separator. They can also function as thermal insulator beneath buildings and as drainage channels beneath building slabs.

Geopipes are available in a wide range of diameters and wall dimensions for carrying liquid and gas. For applications like subdrainage systems and leachate collection systems, the geopipes should have perforations through the wall section to allow for the inflow of water and gas. Standard dimension ratio, defined as the ratio of outside pipe diameter to minimum pipe wall thickness, varies from a minimum of about 10 to a maximum of 40. This ratio can be related to external strength and internal pressure capability. Compared to steel pipes, they are cheap, light and easy to install and join together along with better durability.

It may happen that the geotextile, geogrid, geonet, geomembrane, geofoam or geopipe chosen to meet the requirements of a particular function does not match any other function, which has to be served simultaneously in an application. In such a situation, geocomposites can be used. In fact, geocomposites can be manufactured to perform a combination of the functions described above. For example, a geomembrane–geonet–geomembrane composite can be made where the interior net acts as a drain to the leak detection system. Similarly a geotextile–geonet composite improves the separation, filtration and drainage. A geocomposite consisting of a geotextile cover and drainage core (called band drain or wick drain or vertical strip drain or prefabricated vertical drain (PVD)) provides drainage for accelerating consolidation of soil when installed vertically into the consolidating soil.

Geocomposites are generally, but certainly not always, completely polymeric. Other options include using fiberglass or steel for tensile reinforcement, sand in compression or as a filler, dried clay for subsequent expansion as a liner or bitumen as a waterproofing agent. Geomembrane–clay composites are used as the liners, where the geomembrane decreases the leakage rate while the clay layer increases the breakthrough time. In addition, the clay layer reduces the leakage rate from any holes that might develop in the geomembrane, while the geomembrane will prevent cracks in clay layer due to changes in moisture content.

The selection of a geosynthetic for a particular application is governed by several other factors such as specification, durability, availability, cost and construction. The durability and other properties including the cost of geosynthetics are dependent on the type of polymers used as raw materials for their manufacture. To be able to accurately specify a geosynthetic, which will provide the required properties, it is essential to have at least a basic understanding of how polymers and production processes affect the properties of the finished geosynthetic products, as described in Chapter 1. Table 1.5 (see Chapter 1) lists major types of geosynthetics and the most commonly used polymers for their manufacture. Table 2.3 provides the basic properties of some of these polymers, helping in the selection of geosynthetics.

For example, geotextiles can perform several basic functions – separation, reinforcement, filtration, drainage and protection (see Table 2.1). They are manufactured using polypropylene, polyester, polyethylene or polyamide (see Table 1.5, Chapter 1). Geotextiles as a reinforcement requires a strong, relatively stiff and a preferably water-permeable material.

Table 2.3 Typical properties of polymers used for the manufacture of geosynthetics

Polymer	Specific gravity	Melting temperature (°C)	Tensile strength at 20°C (MN/m^2)	Modulus of elasticity (MN/m^2)	Strain at break (%)
PP	0.90–0.91	165	400–600	2000–5000	10–40
PET	1.22–1.38	260	800–1200	12,000–18,000	8–15
PE	0.91–0.96	130	80–600	200–6000	10–80
PVC	1.3–1.5	160	20–50	10–100	50–150
PA	1.05–1.15	220–250	700–900	3000–4000	15–30

Table 2.3 indicates that the polyester has a very high tensile strength at relatively low strain. Thus a woven geotextile of polyester is a logical choice for reinforcement applications. For separation/filtration applications, the geotextile has to be flexible, water-permeable and soil-tight. A nonwoven geotextile or a lightweight woven geotextile of polyethylene can be a logical choice for separation and filtration applications. It may be noted that the environmental factors and the site conditions also greatly govern the selection of geosynthetics for a particular application (Shukla, 2003b).

Sometimes, during the selection process, one finds that several geosynthetics satisfy minimum requirements for the particular application. In such a situation, the geosynthetic should be selected on the basis of cost–benefit ratio, including the value of field experience and product documentation.

The properties of geosynthetics can change unfavourably in several ways such as ageing, mechanical damage (particularly by installation stresses), creep, hydrolysis (reaction with water), chemical and biological attack, ultraviolet light exposure, etc., which will be discussed in Chapter 3. These factors have to be taken into account when geosynthetics are selected. In permanent installations, there must be proper care for maintaining the long-term satisfactory performance of geosynthetics, i.e. *durability*.

Considerating the risks and consequences of failure, especially for critical projects, great care is required in the selection of the appropriate geosynthetic. One should not try to economage by eliminating soil–geosynthetic performance testing when such testing is required for the selection.

For some applications, geosynthetics are selected on the basis of empirical guidelines. In these cases proper care should be taken to clearly define the required properties of the geosynthetic, in physical as well as in statistical terms.

Self-evaluation questions

(Select the most appropriate answers to the multiple-choice questions from 1 to 15)

1. If a geosynthetic allows for adequate fluid flow with limited migration of soil particles across its plane over a projected service lifetime of the application under consideration, this function of geosynthetic is called

    (a) Separation.
    (b) Filtration.
    (c) Drainage.
    (d) Protection.

2. A geosynthetic as a reinforcement

    (a) Resists applied stresses.
    (b) Prevents inadmissible deformations in the geotechnical structures.
    (c) Maintains the stability of the soil mass.
    (d) All of the above.

3. The deformed geosynthetic provides a vertical support to the overlying soil mass subjected to a loading. This action of the geosynthetic is known as

    (a) Shear stress reduction effect.
    (b) Membrane effect.

(c) Interlocking effect.
(d) None of the above.

4. For a geotextile, separation can be a dominant function over the reinforcement function when the ratio of the applied stress (σ) on the subgrade soil to the shear strength (c_u) of the subgrade soil has generally a value

   (a) Less than 8.
   (b) Equal to 8.
   (c) More than 8.
   (d) None of the above.

5. Filtration function also provides

   (a) Reinforcement benefits.
   (b) Separation benefits.
   (c) Fluid barrier benefits.
   (d) None of the above.

6. Which one of the following is a basic function of geosynthetics?

   (a) Absorption.
   (b) Insulation.
   (c) Screening.
   (d) Protection.

7. In geosynthetic engineering, most of the functions are served by

   (a) Geotextiles and geogrids.
   (b) Geogrids and geonets.
   (c) Geotextiles and geocomposites.
   (d) None of the above.

8. Which one of the following geosynthetics can serve the *protection* function?

   (a) Geotextiles.
   (b) Geogrids.
   (c) Geomembrane.
   (d) Geonets.

9. Which geosynthetic can serve the fluid barrier as a primary function?

   (a) Geotextile and geocomposite.
   (b) Geotextile and geogrid.
   (c) Geotextile and geonet.
   (d) None of the above.

10. The following geosynthetics are used as a drainage medium:
    (A) Thick needle-punched nonwoven geotextiles.
    (B) Geonets.
    (C) Drainage geocomposites.
    The correct decreasing order of flow capability is generally

(a) (A), (B), (C).
(b) (B), (A), (C).
(c) (C), (A), (B).
(d) (C), (B), (A).

11. Which one of the following geocomposites is used as a drain to leak detection system of a landfill?

    (a) Geotextile–geonet.
    (b) Geomembrane–geotextile.
    (c) Geomembrane–geonet–geomembrane.
    (d) None of the above.

12. The specific gravity is less than 1 for

    (a) Polypropylene and polyester.
    (b) Polypropylene and polyethylene.
    (c) Polyethylene and polyester.
    (d) Polyethylene and polyamide.

13. The melting temperature for polyester is

    (a) 130°C.
    (b) 160°C.
    (c) 165°C.
    (d) 260°C.

14. Which one of the following polymers has the highest modulus of elasticity?

    (a) Polypropylene.
    (b) Polyethylene.
    (c) Polyester.
    (d) Polyvinyl chloride.

15. Which one of the following statements is wrong?

    (a) For some applications, geosynthetics are selected on the basis of empirical guidelines.
    (b) The environmental factors and the site conditions greatly govern the selection of geosynthetics for a particular application.
    (c) Polymers and production processes are required to be taken into consideration while making the selection of geosynthetics for a particular field application.
    (d) None of the above.

16. List the basic functions that the geosynthetics perform.
17. Explain the basic mechanisms involved in the separation and filtration functions with the help of neat sketches.
18. Explain the mechanism involved in drainage function compared with the mechanism involved in filtration function?
19. What are the performance characteristics of geosynthetics, other than their basic functions? How do they differ from the basic functions?

20. What are the characteristics of a soil reinforced with an extensible reinforcement? Are the characteristics similar for the soil reinforced with an inextensible reinforcement? Can you list the differences, if any?
21. What are the different mechanisms for soil reinforcement? Explain briefly.
22. Describe the various roles of geosynthetic reinforcement.
23. Is there any difference between tensile member and tensioned member? Justify your answer.
24. How will you decide the primary function of reinforcement and separation in any field application?
25. Describe an idealized interface condition at equilibrium between the soil and geosynthetic filter.
26. What is the difference between *protection* and *cushioning* functions of geosynthetics?
27. What do you mean by functional properties? Explain with some examples.
28. Define the factor of safety required for the acceptance of a geosynthetic for a specific application.
29. Which manufactured style of a geotextile is best suited for its application as a drainage medium?
30. Describe the basic similarities and differences between geotextiles and geogrids.
31. How does a geonet differ from a geogrid in terms of functions?
32. If a geotextile is placed adjacent to a geonet, what function(s) does the geotextile provide? How does the combination of geotextile and geonet accommodate flow?
33. What are the advantages of geomembrane–clay composite liner?
34. List the major factors to be considered in the selection of a geosynthetic for field applications.
35. What is the role of cost–benefit ratio in the selection of geosynthetics?
36. Give the symbolic representation for a reinforcement geotextile.

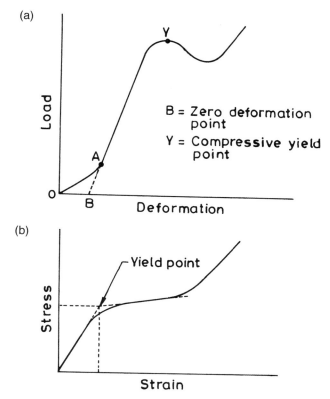

Figure 3.2 Compression behaviour of geosynthetics: (a) typical load–deformation curve; (b) typical stress–strain curve.

of load and plotted as shown in Figure 3.2(a). Being an artifact caused by the alignment or seating of the specimen, the toe region OA may not represent a compressive property of the material. The yield point and strain should be calculated considering the zero deformation point as shown in Figure 3.2 (a). Many geosynthetics exhibit compressive deformation, but all may not exhibit a well-defined compressive yield point; however, the significant change in the slope of the stress–strain curve can be used to determine yield point for comparative purposes (Fig. 3.2(b)). Variable inclined plates or set angled blocks, as described by ASTM D 6364-99, may be used to evaluate the deformation of the geosynthetic(s) under loading at various angles. The compressive loading test is generally used for quality control to evaluate uniformity and consistency within a *lot* (a unit of production) or between lots where sample geometry factors such as thickness or materials may change.

Tensile properties The determination of tensile properties, mainly *tensile strength* and *tensile modulus*, of geosynthetics is important when they need to resist tensile stresses transferred from the soil in reinforcement type applications, for example design of reinforced embankments over soft subgrades, reinforced soil retaining walls and reinforcement of slopes. The *tensile strength* is the maximum resistance to deformation developed for a geosynthetic when it is subjected to tension by an external force. Due to specific geometry

52 Properties and their evaluation

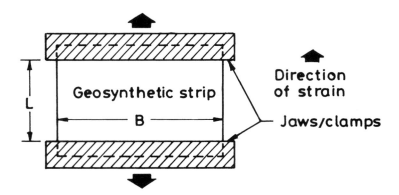

Figure 3.3 Wide-width strip tensile test.
Note
B = 200 mm; L = 100 mm.

and irregular cross-sectional area that cannot be easily defined, the tensile strength of geosynthetics cannot be expressed conveniently in terms of stress. It is, therefore, defined as the peak (or maximum) load that can be applied per unit length along the edge of the geosynthetic in its plane. Tensile properties of a geosynthetic are studied using a tensile strength test in which the geosynthetic specimen is loaded and the corresponding force–elongation curve is obtained.

Tensile strength is usually determined by the wide-width strip tensile test on a 200-mm wide geosynthetic strip with a gauge length of 100 mm (Fig. 3.3). The entire width of a 200-mm wide geosynthetic specimen is gripped in the jaws of a tensile strength testing machine and it is stretched in one direction at a prescribed constant rate of extension until the specimen ruptures (breaks). During the extension process, both load and deformations are measured. The width of the specimen is kept greater than its length, as some geosynthetics have a tendency to contract ('neck down') under load in the gauge length area. The greater width reduces the contraction effect of such geosynthetics, and by approximating plane strain conditions, it more closely simulates the deformation experienced by a geosynthetic when embedded in soil under field conditions. The test provides parameters such as peak strength, elongation and tensile modulus. The tensile properties depend on the geosynthetic polymer and manufacturing process leading to the structure of the finished product. The measured strength and the rupture strain are a function of many test variables, including sample geometry, gripping method, strain rate, temperature, initial preload, conditioning and the amount of any normal confinement applied to the geosynthetic.

Figure 3.4 shows the influence of the geotextile specimen width on the tensile strength. To minimize the effects, the test specimen should have a width-to-gauge length ratio (a.k.a. *aspect ratio*) of at least two, and the test should be carried out at a standard temperature. The actual temperature has a great influence on the strength properties of many polymers (Fig. 3.5). The tensile strength of geosynthetics is closely related to mass per unit area (Fig. 3.6). A heavyweight geotextile, with a higher mass per unit area, will usually be stronger than a lightweight geotextile. For a given geosynthetic, the tensile strength is also

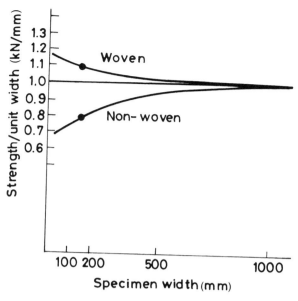

Figure 3.4 Influence of geotextile specimen width on its tensile strength (after Myles and Carswell, 1986).

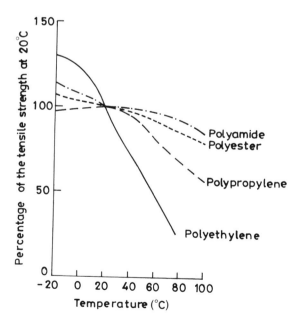

Figure 3.5 Influence of temperature on the tensile strength of some polymers (after Van Santvoort, 1995).

54 Properties and their evaluation

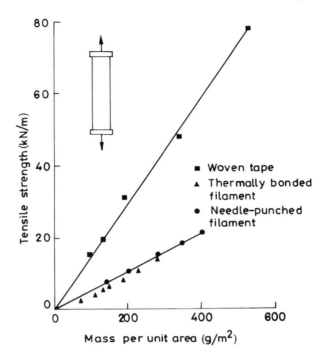

Figure 3.6 Variation of tensile strength with mass per unit area for PP geotextiles (after Ingold and Miller, 1988).

a function of the rate of strain at which the specimen is tested. At a low strain rate, the measured strength tends to be lower and occurs at a higher failure strain. Conversely, at a high strain rate, the measured strength tends to be higher and occurs at a lower failure strain.

Other forms of tensile strength tests such as grab test, biaxial test, plain strain test and multi-axial test are shown schematically in Figure 3.7. The grab tensile test is used to determine the strength of the geosynthetic in a specific width, together with the additional strength contributed by adjacent geosynthetic or other material. This test is basically a uniaxial tensile test in which only the central portion of the geosynthetic specimen is gripped in the jaws (Fig. 3.7(a)). The test normally uses 25.4-mm (1 in.) wide jaws to grip a 101.6-mm (4 in.) wide geosynthetic specimen. A continually increasing load is applied longitudinally to the specimen and the test is carried to rupture. It is not clear how the force is distributed across the width of the specimen. This test simulates the field situation as shown in Figure 3.8. It is difficult to relate grab tensile strength to wide-width strip tensile strength in a simple manner without direct correlation tests. Therefore, the grab tensile test is useful as a quality control or acceptance test for geotextiles. Typical range of grab tensile strength of geotextiles is 300–3000 N.

The plain strain tensile test is a uniaxial tensile test in which the entire width of the specimen is gripped in the jaws with the specimen being restrained from necking during the tensile load application (Fig. 3.7(c)). This test can be carried out to assess the strength of the geotextile when buried under excessive soil.

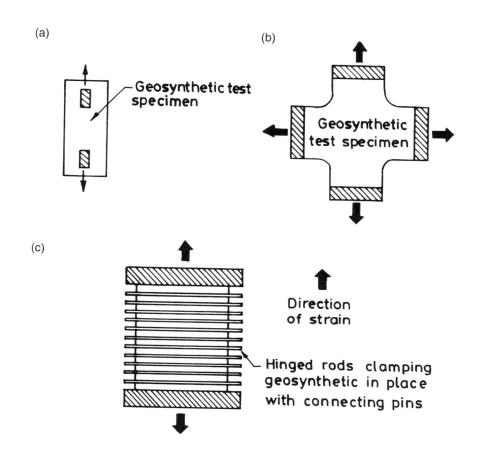
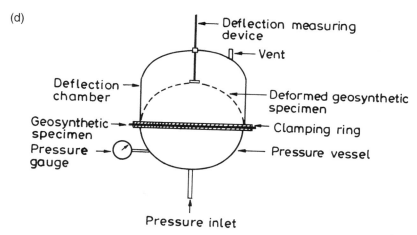

Figure 3.7 Typical arrangements of tensile strength tests: (a) grab tensile strength test; (b) biaxial tensile strength test; (c) plain strain tensile strength test; (d) multi-axial tensile strength test.

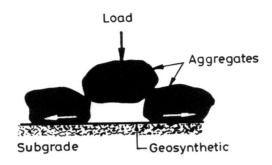

Figure 3.8 Field situation that can be simulated by grab tensile strength test.

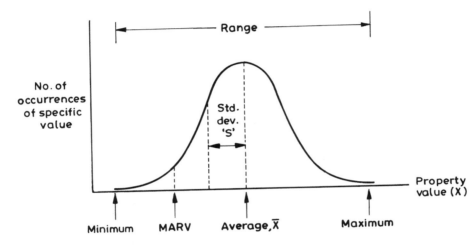

Figure 3.9 Statistical distribution of geotextile properties.

Installed geosynthetics are subjected to forces from more than one direction including forces perpendicular to the surfaces of the geosynthetic causing out-of-plane deformation. The multi-axial tensile test can be carried out to measure the out-of-plane response of geosynthetics to a force that is applied perpendicular to the initial plane of the geosynthetic specimen. In this test, the geosynthetic specimen is clamped to the edges of a large diameter, generally 0.6 m, pressure vessel (Fig. 3.7(d)). Pressure is applied to the specimen to cause out-of-plane deformation and failure. This deformation with pressure information is then analysed to evaluate the geosynthetic. When the geosynthetic deforms to a simplified geometric shape (arc of a sphere or ellipsoid), the data obtained from the test can be converted to biaxial tensile stress–strain values. In geosynthetic applications where local subsidence is expected, the multi-axial tensile test can be considered a performance test.

During manufacturing process, the variability in geosynthetic properties may occur as happens with other civil engineering construction materials. Based on quality control tests, a manufacturer of geosynthetics can represent properties statistically in normal distribution curve as shown in Figure 3.9. Project specifications tend to include several qualifiers such

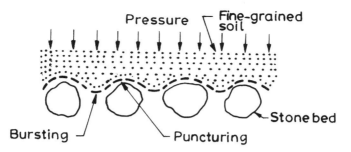

Figure 3.12 Field situations showing puncturing and bursting of the geosynthetic (after Giroud, 1984).

closely as practicable simulation of selected field conditions which can be used in the design. Performance testing, if possible, should also be carried out full scale at site. Since, geosynthetics vary randomly in thickness and weight in a given sample roll due to normal manufacturing techniques, tests must be conducted on representative samples collected as per the guidelines of available standards, which ensure that all areas of the sample roll and a full variation of the product are represented within each sample group. For applications in more severe environments such as soil treated with lime or cement, landfills or industrial waste, or highly acidic volcanic soils, for applications with indeterminate design lives, for applications of high temperature, or for unusual site-specific conditions, performance tests with site-specific parameters may be required.

Survivability properties   There are some mechanical properties of geosynthetics, which are related to geosynthetic survivability (constructability) and separation function. Tests to determine such properties are generally treated as integrity/index tests. These properties are as follows:

- *Tearing strength*: The ability of a geosynthetic to withstand stresses causing to continue or propagate a tear in it, often generated during their installation.
- *Static puncture strength*: The ability of a geosynthetic to withstand localized stresses generated by penetrating or puncturing objects such as aggregates or roots, under quasi-static conditions (Fig. 3.12).
- *Impact strength* (*dynamic puncture strength*): The ability of a geosynthetic to withstand stresses generated by the sudden impact and penetration of falling objects such as coarse aggregates, tools, and other construction items during installation process.
- *Bursting strength*: The ability of a geosynthetic to withstand a pressure applied normal to its plane while constrained in all directions in that plane (Fig. 3.12).
- *Fatigue strength*: The ability of a geosynthetic to withstand repetitive loading before undergoing failure.

The *tearing strength test* aims to measure the propensity of a geosynthetic to tearing force once a tear has been initiated. The tearing strength of geotextiles under in-plane loading is determined by *trapezoid tearing strength test*. In this test, a trapezoidal outline is marked centrally on a rectangular test specimen (Fig. 3.13). Note that an initial 15-mm cut is made to start the tearing process. The specimen is gripped along the two non-parallel sides of the

62 Properties and their evaluation

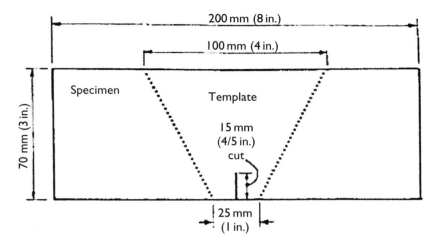

Figure 3.13 Trapezoidal template for trapezoid tearing strength test (Reprinted, with permission, from ASTM D4533-91 (1996), *Standard Test Method for Trapezoid Tearing Strength of Geotextiles*, copyright ASTM International, 100 Barr Harbor Drive, West Conshohocken, PA 19428).

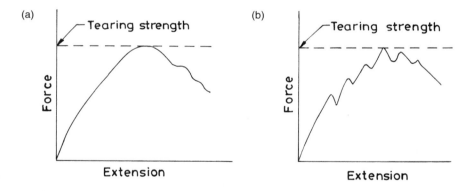

Figure 3.14 Typical tearing force–extension curves for individual test specimens: (a) geotextile exhibiting single maximum; (b) geotextile exhibiting several maxima (Reprinted, with permission, from AS 3706.3 (2000), *Determination of Tearing strength of Geotextiles – Trapezoidal Method*, copyright Standards Australia International Ltd, Sydney, NSW 2001).

trapezoid in the jaws of a tensile testing machine. A continuously increasing force is applied in such a way that the tear propagates across the width of the specimen. The load actually stresses the individual fibres gripped in the clamps rather than stressing the geosynthetic structure. The value of tearing strength of the specimen is obtained from the force–extension curve and is taken as the maximum force thus recorded (Fig. 3.14). The failure pattern in tear is different in nonwoven geotextiles from that in woven geotextiles. A failure of woven geotextiles occurs essentially through the sequential rupture of yarns in tension, whereas the failure of a nonwoven geotextile is significantly affected by the inter-fibre friction forces. A typical range of trapezoid tearing strength of geotextiles is 90–1300 N.

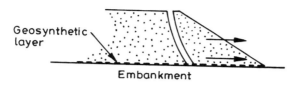

Figure 3.20 A reinforcing geosynthetic application with sliding failure mode.

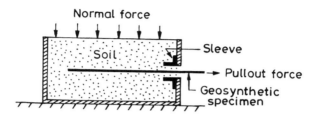

Figure 3.21 Details of pullout test.

It should be noted that the direct shear test is not suited for the development of exact stress–strain relationships for the test specimen due to the non-uniform distribution of shearing forces and displacement. Total resistance may be a combination of sliding, rolling, interlocking of soil particles and geosynthetic surfaces, and shear strain within the geosynthetic specimen. Shearing resistance may be different on the two faces of a geosynthetic and may vary with direction of shearing relative to orientation of the geosynthetic. The direct shear test data can be used in the design of geosynthetic applications in which sliding may occur between the soil and the geosynthetic (Fig. 3.20). Note that the direct shear test can also be conducted to study the geosynthetic–geosynthetic interface frictional behaviour by placing the lower geosynthetic specimen flat over a rigid medium in the lower half of the direct shear box and the upper geosynthetic specimen over the previously placed lower specimen.

In the *pullout test*, a geosynthetic specimen, embedded between two layers of soil in a rigid box, is subjected to a horizontal force, keeping the normal stress applied to the upper layer of soil constant and uniform. Figure 3.21 depicts the general test arrangement of the pullout test. The force required to pull the geosynthetic out of the soil is recorded. Pullout resistance is calculated by dividing the maximum load by the test specimen width.

The ultimate pullout resistance, P, of the geosynthetic reinforcement is given by

$$P = 2 \times L_e \times W \times \sigma'_n \times C_i \times F \tag{3.5}$$

where, L_e is the embedment length of the test specimen; W is the width of the test specimen; σ'_n is the effective normal stress at the soil–test specimen interfaces; C_i is the coefficient of interaction (a scale effect correction factor) depending on the geosynthetic type, soil type and normal load applied; and F is the pullout resistance (or friction bearing interaction) factor. For preliminary design or in the absence of specific geosynthetic test data, F may be conservatively taken as $F = (2/3) \tan\phi$ for geotextiles and $F = 0.8 \tan\phi$ for geogrids. Equation (3.5) is known as *pullout capacity formula*.

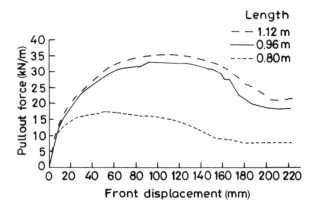

Figure 3.22 Influence of the specimen embedment length on the pullout behaviour of a geogrid (after Lopes and Ladeira, 1996).

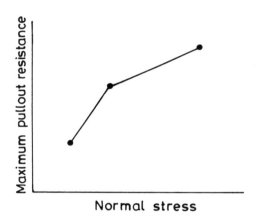

Figure 3.23 Typical pullout resistance versus normal stress plot.

The pullout resistance versus normal stress plot is a function of soil gradation, plasticity, as-placed dry unit weight, moisture content, embedment length and surface characteristics of the geosynthetic, displacement rate, normal stress and other test parameters. Therefore, the results should be expressed in terms of the actual test conditions. Figure 3.22 shows the effect of specimen embedment length on the pullout behaviour of a geogrid. A typical plot of maximum pullout resistance versus normal stress is shown in Figure 3.23. The pullout test data can be used in the design of geosynthetic applications in which pullout may occur between the soil and the geosynthetic as shown in Figure 3.24.

A designer of geosynthetic-reinforced soil structures must consider the potential failure mode, and then the appropriate test procedure should be used to evaluate the soil–geosynthetic interaction properties. In the case of an unpaved road with a geosynthetic layer at the subgrade level, the recommended test should be a combination of direct shear and pullout

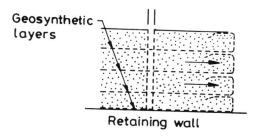

Figure 3.24 A reinforcing geosynthetic application with pullout failure mode.

tests conducted simultaneously (Giroud, 1980). It is common to assume a soil–geotextile friction angle between 2/3 and 1 of the angle of shearing resistance of soil.

3.4 Hydraulic properties

The hydraulic properties of geosynthetics influence their ability to function as filters and drains. Hydraulic testing of geosynthetics is completely based on new and original concepts, methods, devices, interpretation and databases, unlike the physical and mechanical testing, as discussed in previous sections of this chapter. The reason behind this is that the traditional textile tests rarely have hydraulic applications. *Porosity*, *permittivity* and *transmissivity* are the most important hydraulic properties of geosynthetics, mainly of geotextiles, geonets and many drainage geocomposites, which are commonly used in filtration and drainage applications.

Geosynthetic pore (or opening) characteristics   The voids (or holes) in a geosynthetic are called pores or openings. The measurement of sizes of pores and the study of their distribution is known as *porometry*.

Geosynthetic porosity is related to the ability of the geosynthetic to allow fluid to flow through it and is defined as the ratio of the void volume (volume of void spaces) to the total volume of the geosynthetic, usually expressed as a percentage. It may be indirectly calculated for geotextiles using the relationship derived below:

$$\eta = \frac{V_v}{V} = \frac{V - V_s}{V} = 1 - \frac{V_s}{V} = 1 - \frac{\frac{mA}{\rho_s}}{A\Delta x} = 1 - \frac{m}{\rho_s \Delta x} \quad (3.6)$$

where: η is the porosity; V_v is the void volume; V_s is the volume of solid polymer; $V (= V_v + V_s)$ is the total volume; A is the surface area of geotextile; m is the mass per unit area; ρ_s is the density of solid polymer; and Δx is the thickness of geotextile.

ILLUSTRATIVE EXAMPLE   3.2

Calculate the porosity of the geotextile with the following properties:
  Thickness, $\Delta x = 2.7$ mm
  Mass per unit area, $m = 300$ g/m²
  Density of polymer solid, $\rho_s = 900$ kg/m³.

70 Properties and their evaluation

SOLUTION
Using Equation (3.6), the porosity, η, of the geotextile is calculated as follows:

$$\eta = 1 - \frac{m}{\rho_s \Delta x} = 1 - \frac{300 \text{ g/m}^2}{(900 \times 1000 \text{ g/m}^3) \times (0.0027 \text{ m})} = 0.876 \text{ or } 87.6\% \qquad \textbf{Answer}$$

Percent open area (POA) of a geosynthetic is the ratio of the total area of its openings to the total area, expressed as a percentage. This characteristic is considered to be a design parameter only for woven geotextiles, which have area of openings as the void spaces between adjacent filaments and yarns. It is to be noted that a higher POA generally indicates a greater number of openings per unit area in the geotextile. For filter applications of a geotextiles, its POA should be higher to avoid any *clogging* phenomenon (see Sec. 4.7 for explanation) to occur throughout the design life of the particular application.

The pores in a geotextile are not of one size but are of a range of sizes. The pore size distribution can be represented in much the same way as the particle size distribution for a soil. In fact, a geotextile is similar to a soil in that it has voids (pores) and particles (filaments and fibres). However, because of the shape and arrangement of filaments and the compressibility of the structure with geotextiles, the geometric relationships between filaments and voids are more complex than in soils. Therefore, in geotextiles the pore size is measured directly, rather than using particle size as an estimate of pore size, as is done with soils. In the determination of the particle size distribution of soil, the soil, which initially has particles of unknown sizes, is passed through a series of sieves of different known sizes to determine the percentages of soil particles of the various sizes present. In determining the pore size distribution of a geotextile the process is reversed. The geotextile is used as a sieve, of unknown sizes, and the particles of different known sizes are passed through the geotextile as a sieve. From the measured weights of particles of various known sizes which either pass through the geotextile or are retained on the geotextile, the pore size distribution of the geotextile can be obtained. Due to the importance of pore size distribution in the design of geotextiles for use as filters and separators, various test methods have been developed for measuring the size of openings in the geotextiles. Bhatia *et al.* (1994) made a comparison of six methods as presented in Table 3.1.

Table 3.1 Comparison of methods for determining pore size distribution of geotextiles

Test method	Test mechanism	Test material	Sample size (cm²)	Time for 1 test
Dry sieving	Sieving–dry	Glass beads fraction	434	2 h
Hydrodynamic sieving	Alternating water flow	Glass beads mixture	257	24 h
Wet sieving	Sieving–wet	Glass beads mixture	434	2 h
Bubble point	Comparison of air flow, dry vs. saturated	Pore wick	22.9	20 min
Mercury intrusion	Intrusion of a liquid in a pore	Mercury	1.77	35 min
Image analysis	Direct measurement of pore spaces in cross-section of the geotextile	None	1.5	2–3 days

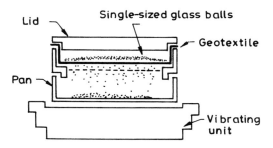

Figure 3.25 Diagram showing details of dry sieving method (courtesy of Terram Ltd, UK).

In the *dry sieving test method*, known-sized spherical solid glass beads (or calibrated quartz sand particles) are sieved in dry condition through a screen made of the geotextile specimen, being tested, in a sieve frame (see Fig. 3.25) for a constant period of time, generally 10 min. Sieving is done by allowing beads of successively coarser size until they are 5% or less in weight, to pass through the geotextile. A mechanical sieve shaker, which imparts lateral and vertical motion to the sieve, causing the particles thereon to bounce and turn so as to present different orientations to the sieving surface, should be used for carrying out the sieving operations. It may be noted that for the measurement of fine pores, difficulties are encountered in the dry sieving of sand particles through geotextiles, particularly through thick nonwovens, due to the particles being trapped in the geotextile. On the other hand with the use of glass beads, electrostatic forces can affect the sieving, but no practical alternative dry methods of determining pore sizes for these types of geotextiles are available at the present time.

The *hydrodynamic test method* is based on hydrodynamic filtration, where a glass bead mixture is sieved with a basket with geotextile bottom by alternating water flow that occurs as a result of the immersion and emersion of the basket several times in water. In the *wet sieving method*, a glass bead mixture is sieved through a screen made of geotextile while a continuous water spray is applied. The *bubble point method* is based on a process in which (i) a dry porous material passes air through all of its pores when any amount of air pressure is applied to one side of the material; and (ii) a saturated porous material will only allow a fluid to pass when the pressure applied exceeds the capillary attraction of the fluid in the largest pore. The *mercury intrusion method* is based on the relationship between the pressure required to force a non-wetting fluid (mercury) into the pores of a geotextile and the radius of the pores intruded. *Image analysis* is a technique used for the direct measurement of pore spaces within a cross-sectional plane of the geotextile with the help of a microscope. The pore openings, which are obtained experimentally, are dependent on the technique used for their determination. It is believed that, despite some limitations, both wet and hydrodynamic sieving methods are better techniques than dry sieving. Note that the pore sizes measured by all these methods are not actual dimensions of the openings through the geotextile.

In the case of most of the geogrids, the open areas of the grids are greater than 50% of the total area. In this respect, a geogrid may be looked upon as a highly permeable polymeric structure.

Figure 3.26 shows pore size distribution curves for typical woven and nonwoven geotextiles. The pore size (or opening size), at which 95% of the pores in the geotextile are finer, is

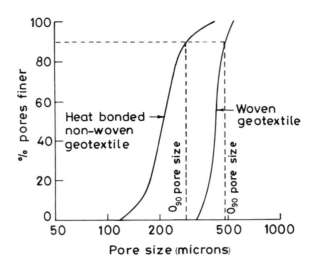

Figure 3.26 Pore size distributions of typical geotextiles (after Ingold and Miller, 1988).

originally termed the *equivalent opening size* (*EOS*) designated as O_{95}. In the USA, this pore size is determined by dry sieving method and is termed *apparent opening size* (*AOS*), whereas in Europe and Canada this is determined by wet and hydrodynamic sieving methods and is termed *filtration opening size* (*FOS*). If a geotextile has an O_{95} value of 300 μm, then 95% of geotextile pores are 300 μm or smaller. In other words, 95% of particles with a diameter of 300 μm are retained on the geotextile during sieving. This notation is similar to that used for soil particle size distributions where, for instance, D_{10} is the sieve size through which 10%, by weight, of the soil passes. AOS or FOS is, in fact, considered as the property that indicates the approximately largest particle that would effectively pass through the geo-textile and thus reflects the approximately largest opening dimension available in the geotextile for soil to pass through. The opening size is also quoted for other percentages retained, such as O_{50} or O_{90}, to determine the pore size distribution of a geotextile. It should be noted that the meaning of opening size values and their determination in the laboratory are still not uniform throughout the engineering profession and hence filter criteria developed in different countries may not be directly comparable.

In Figure 3.26, it is noted that the pores in a woven geotextile tend to be fairly uniform in size and regularly distributed. In general, nonwoven geotextiles exhibit smaller O_{90} pore sizes than wovens; however, there is a degree of overlap in the commonly employed O_{90} sizes, which vary from approximately 50 μm to 350 μm for the nonwovens and from 150 μm to 600 μm for the wovens (Ingold and Miller, 1988). For filtration application, a geotextile high in POA should be selected, with a controlled opening size to suit the soil being filtered. Most nonwoven geotextiles and some woven geotextiles will suit this application.

Permeability characteristics The ability of a geosynthetic to transmit a fluid is called *permeability*. The permeability (or hydraulic conductivity) of a geosynthetic to fluid flow may be expressed by *Darcy's coefficient*, by *permittivity* (as defined here) or by a *volume flow rate*. The Darcy's coefficient is the volume rate of flow of fluid under laminar flow

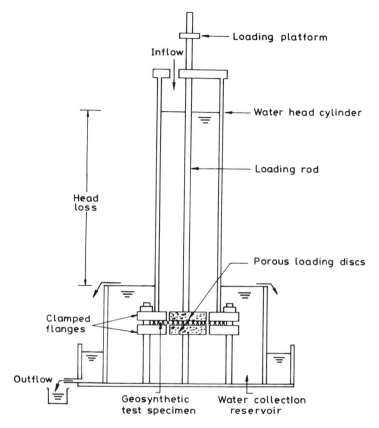

Figure 3.27 A typical test arrangement of constant head cross-plane water flow apparatus.

conditions through a unit cross-sectional area of a geosynthetic under a unit hydraulic gradient and standard temperature conditions (generally $22 \pm 3°C$). The advantage in expressing geosynthetic permeability in terms of Darcy's coefficient is that it is easy to relate geosynthetic permeability directly with soil permeability. A major disadvantage is that Darcy's law assumes laminar flow, whereas geosynthetics, especially, geotextiles, are often characterized as exhibiting semi-turbulent or turbulent flows.

The simplest method of describing the permeability characteristics of geosynthetics is in terms of volume flow rate at a specific constant water head (generally 10 cm) (Fig. 3.27). The advantage of this method is that it is the simplest test to carry out, it does not rely on Darcy's law for its authenticity, and it can easily be used to compare different geosynthetics used for drainage and filtration applications. In this method due to high hydraulic gradient, turbulent flow can occur in many geotextiles. Thus, the measured permeability value cannot be compared with the actual permeability value measured for laminar flow conditions.

The measurement of in-plane water permeability is important if the geosynthetic, such as a geotextile or a band/fin drain, is being used to carry water within itself and parallel to its plane; that is, its water transporting capacity is of prime importance. The in-plane water

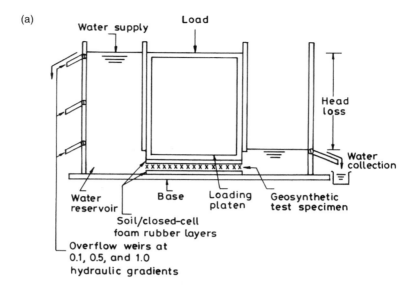

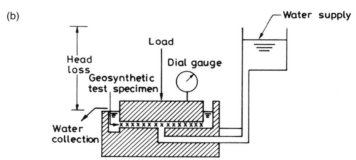

Figure 3.28 Typical test arrangements of constant head in-plane water flow apparatus: (a) full width flow; (b) radial flow.

permeability, normally described in terms of *transmissivity* (as defined in this section later on), is determined by measuring the volume of water that passes along the test specimen in a known time and under specified normal stress and hydraulic gradient. The test used to measure the in-plane drainage characteristics of geosynthetics is essentially the same as that used to measure water permeability normal to the plane of the geosynthetic (Fig. 3.27), except that the hydraulic gradient is applied along the length of the geosynthetic (Fig. 3.28) rather than across the thickness of the geosynthetic. The test can be conducted to model particular field conditions, for example, by employing specific contact surfaces and varying compressive stresses and typical hydraulic gradients.

Permittivity of a geosynthetic (generally geotextile) is simply the coefficient of permeability for water flow normal to its plane (Fig. 3.29 (a)) divided by its thickness. This property is the preferred measure of water flow capacity across the geosynthetic plane and

the hydraulic gradient; and $A_p = B\Delta x$ is the area of cross-section of geosynthetic for in-plane flow, in m². Transmissivity may thus be defined as the volumetric flow rate of water per unit width of the geosynthetic per unit hydraulic gradient, under laminar conditions of flow within the plane of the geosynthetic (Fig. 3.30(b)). To exhibit a large transmissivity, a geotextile must be thick and/or have a large permeability in its plane.

Equations (3.7) and (3.8) indicate that once permittivity (ψ) and transmissivity (θ) are successfully determined, the flow rates, Q_n and Q_p, do not depend on the thickness of the strip of geosynthetic, Δx, which is highly dependent on the applied pressures and is therefore difficult to measure accurately in the case of some types of geotextiles. Thus, it is preferable to determine and report the permittivity and transmissivity values of geotextiles rather than their coefficients of in-plane and cross-plane permeability respectively.

ILLUSTRATIVE EXAMPLE 3.3
In a laboratory constant head in-plane permeability test on a 300-mm length (flow direction) by 200-mm width geotextile specimen, the following parameters were measured:

Nominal thickness, $\Delta x = 2.0$ mm
Flow rate of water in the plane of the geotextile, $Q_p = 52$ cm³/min
Head loss in the plane of the geotextile, $\Delta h = 200$ mm.

Calculate the transmissivity (θ) and the in-plane coefficient of permeability (k_p) of the geotextile.

SOLUTION
From Equation (3.8),

$$\theta = \frac{Q_p}{iB} = \frac{Q_p}{\frac{\Delta h}{L} \times B} = \frac{\left(\frac{52 \times 10^{-6}}{60}\right) \text{m}^3/\text{s}}{\frac{0.2\,\text{m}}{0.3\,\text{m}} \times 0.2\,\text{m}} = 6.5 \times 10^{-5}\,\text{m}^2/\text{s} \qquad \text{Answer}$$

Now,

$$k_p = \frac{\theta}{\Delta x} = \frac{6.5 \times 10^{-5}\,\text{m}^2/\text{s}}{0.002\,\text{m}} = 3.25 \times 10^{-2}\,\text{m/s} \qquad \text{Answer}$$

If the transmissivity of a geotextile is determined by the radial transmissivity method (see Fig. 3.28(b) for the schematic diagram of the test) using a circular specimen, Darcy's law in terms of transmissivity can be expressed as follows:

$$Q_p = k_p \frac{dh}{dr}(2\pi r \Delta x), \qquad (3.9a)$$

where r is any radius between the outer radius, r_o, and the inner radius, r_i, of the geotextile specimen, and dh is the head loss across the radial distance dr.

On rearranging the parameters and integrating within proper limits, Equation (3.9a) reduces to

$$\theta \int_{h_i}^{h_0} dh = \frac{Q_p}{2\pi} \int_{r_i}^{r_0} \frac{dr}{r}$$

$$\Rightarrow \theta(h_0 - h_i) = \frac{Q_p \ln(r_0/r_i)}{2\pi}$$

$$\Rightarrow \theta = \frac{Q_p \ln(r_0/r_i)}{2\pi(h_0 - h_i)} = \frac{Q_p \ln(r_0/r_i)}{2\pi \Delta h}, \tag{3.9b}$$

where h_i and h_0 are the hydraulic heads at the inner and outer edges of the geotextile specimen respectively, and $\Delta h \ (= h_0 - h_i)$ is the head loss across the radial distance $\Delta r \ (= r_0 - r_i)$. Equation (3.9b) can be directly used to calculate the transmissivity by the radial method.

Note that the determination of permittivity and transmissivity of geotextiles is based on Darcy's law of water flow. This means that permittivity and transmissivity are the only constants for a particular geotextile of given thickness and confining pressure if laminar flow conditions exist, which is likely in a typical soil environment where geotextiles are used. It appears that for most geotextiles, Darcy's law holds if the approach velocity, that is the velocity of the water approaching the geotextile, is kept at or below 0.035 m/s (AS 3706.9-2001). The permittivity and transmissivity, if determined for the region of transient or turbulent flow conditions, are called permittivity and transmissivity under nonlinear flow conditions.

Typical values of permeability are 10^{-5}–1 m/s for geotextiles and 10^{-13} m/s or less for geomembranes. The permeability of geotextiles is of the same order of magnitude as the permeability of highly permeable soils, such as sand and gravel. Woven geotextiles and thermally bonded nonwoven geotextiles have almost no transmissivity and cannot be used as drains. The permeability of geomembranes is much smaller than the permeability of clay, which is the least permeable soil. Needle-punched geotextiles have permeability values of the order of 10^{-4} or 10^{-3} m/s and geonets have permeability values of the order of 10^{-2} or 10^{-1} m/s. A maximum saturated hydraulic conductivity ranging from 5×10^{-11} to 1×10^{-12} m/s is typical of geosynthetic clay liners (GCLs) over the range of confining pressures typically encountered in practice.

It is further stressed that Darcy's law is valid only for laminar flow. This means that permeability, permittivity and transmissivity are constants, that is, independent of the gradient only if the water flow is not turbulent. These properties are governed by several other factors such as fibre type, size and orientation; porosity or void ratio; confining pressure; repeated loading; contamination; and ageing. When dry, some fabrics exhibit resistance to wetting. In such cases initial permeability is low but rises until the fabric reaches saturation. Permeability may also be reduced through air bubbles trapped in the geosynthetic. This is the reason why testing standards usually require careful saturation of the geosynthetic specimens before they are subjected to water flow. In addition, permeability measurements will be more consistent with the use of deaired water rather than tap water. Woven geotextiles are much less affected by stress level, but their permeability is dramatically controlled by the structure of the fabric. The common, and generally less expensive, tape-on-tape fabrics have a low open area ratio and, in consequence, exhibit water permeabilities typically in the range 10–30 l/s/m² for a 10-cm head. In contrast, the woven monofilament-on-monofilament geotextiles have much larger open area ratios,

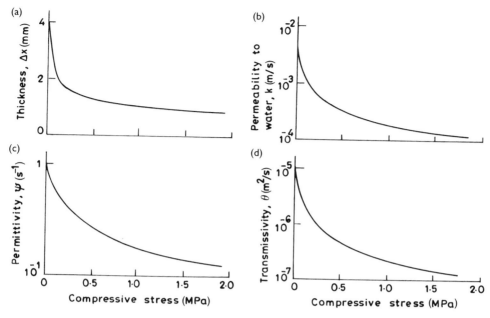

Figure 3.31 Influence of compressive stress on (a) thickness; (b) permeability; (c) permittivity; (d) transmissivity of a needle-punched nonwoven geotextile (after Giroud, 1980).

giving water permeabilities in the range 100–1000 l/s/m² for a 10 cm head (Ingold and Miller, 1988). Tests performed at the University of Grenoble (France) have shown that the thickness and the permeability of needle-punched nonwoven geotextiles are significantly affected by confining pressure as shown in Figure 3.31. In this figure the values of k_n and k_p were close for the considered geotextile; therefore only an average value, k, is presented. It may be noted that the flow in the plane of the geotextile is more affected by the confining pressure than a normal flow. It should be noted that permittivity, transmissivity, and apparent opening size of certain geosynthetics (such as geotextiles) may also change if they are subjected to tension or creep deformation. Thus, for example, the ability of geotextiles to drain water, retain soil particles, and resist clogging may get altered by such applied loads.

Geomembranes are nonporous homogeneous materials that are permeable in varying degrees to gases, vapours, and liquids on a molecular scale in a three-step process: (1) absorption of the permeant, (2) diffusion of the dissolved species, and (3) desorption (evaporation) of the permeant, controlled mainly by the chemical potential gradient (or concentration gradient) that decreases continuously in the direction of the permeation. They are mostly used as a fluid barrier or liner. Note that there are other liner materials that are porous, such as soils and concretes, in which the driving force for permeation is hydraulic gradient. Sometimes, a geomembrane is also known as a flexible membrane liner (FML), especially in landfill applications. The wide range of uses of geomembranes under different service conditions to many different permeating species requires determination of permeability by test methods that relate to and simulate as closely as possible the actual environmental conditions in which the geomembrane will be in service. Various test methods for the

80 Properties and their evaluation

measurement of permeation and transmission through geomembranes of individual constituents in complex mixtures such as waste liquids are recommended by ASTM D 5886-95 (reapproved 2001).

It is found that inorganic salts do not permeate geomembranes but some organic species do. The rate of diffusion of an organic within a geomembrane is governed by several factors, including solubility of the permeant in the geomembrane, microstructure of the polymer, size and shape of the diffusing molecules, temperature at which the diffusion is taking place, the thickness of the geomembrane and the chemical potential across the geomembrane. A steady state of the flow of constituents will be established when, at every point within the geomembrane, flow can be defined by *Fick's first law of diffusion*:

$$Q_i = -D_i \times \frac{dc_i}{dx}, \qquad (3.10)$$

where Q_i is the mass flow of constituent 'i', in (g/cm²/s); D_i is the diffusivity of constituent 'i', in cm²/s; c_i is the concentration of constituent 'i' within the mass of the geomembrane, in g/cm³; and x is the thickness of the geomembrane, in cm.

3.5 Endurance and degradation properties

The endurance and degradation properties (e.g. creep behaviour, abrasion resistance, long-term flow capability, durability – construction survivability and longevity, etc.) of geosynthetics are related to their behaviour during service conditions, including time.

Creep Creep is the time-dependent increase in accumulative strain or elongation in a geosynthetic resulting from an applied constant load. Depending on the type of polymer and ambient temperature, creep may be significant at stress levels as low as 20% of the ultimate tensile strength. In the test for determining the creep behaviour of a geosynthetic, the specimen of wide-width variety (say, 200 mm wide) is subjected to a sustained load using weights, or mechanical, hydraulic or pneumatic systems, in one step while maintaining constant ambient conditions of temperature and humidity. The longitudinal extensions/strains are recorded continuously or are measured at specified time intervals. Unless otherwise specified, the duration of testing is generally not less than 10,000 h, or to failure if this occurs in a shorter time. A test duration of 100 h is useful for monitoring of products, but for a full analysis of creep properties, durations of up to 10,000 h will be necessary. The percent strain versus log of time is plotted for each stress increment to calculate the creep rate, defined as the slope of the creep–time curve at a given time. Figure 3.32 compares strain versus time behaviour of various yarns of different polymers. As shown, both the total strain and the rate of strain differ markedly.

More recent developments allow for accelerated determination of creep characteristics via stepped isothermal methods (SIM). Curves are developed such that a prediction of the total likely creep effect over significant time intervals can be extrapolated (may be 100 years design life).

Creep is an important factor in the design and performance of some geosynthetic-reinforced structures, such as retaining walls, steep-sided slopes, embankments over weak foundations, etc. In all these applications, geosynthetic reinforcements may be required to endure exposure to high tensile stresses for long periods of time – typically 75-plus years.

Properties and their evaluation 81

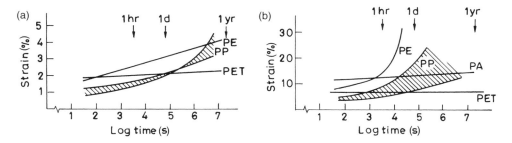

Figure 3.32 Results of creep tests on various yarns of different polymers: (a) creep at 20% load; (b) creep at 60% load (after den Hoedt, 1986).

Table 3.2 Factors of safety (after den Hoedt, 1986)

Polymers	Factor of safety
Polypropylene	4.0
Polyester	2.0
Polyamide (nylon)	2.5
Polyethylene	4.0

Creep should also be carefully considered as a relevant design criterion in some drainage applications and some containment applications when the geosynthetic is under load and is expected to perform in a specified manner for a defined time period. At higher loads, creep leads ultimately to *stress rupture*, also known as *creep rupture* or *static fatigue*. The higher the applied load, the shorter the time to rupture. Thus the design load will itself limit the lifetime of the geosynthetic. The understanding of the geosynthetic creep thus helps the design engineer in the selection of the allowable load to be used in designs. In design, it is generally accepted that creep data should not be extrapolated beyond one order of magnitude. Two approaches to evaluate the allowable load are given below:

(a) *Allowable load based on limiting creep strains*: This requires the analysis of creep strains versus time plots for various stress levels. Details of this procedure were described by Jewell (1986), and Bonaparte and Berg (1987).
(b) *Allowable loads using factors of safety*: It is required to reduce the geosynthetic strength by a factor of safety corresponding to the specific polymer type to obtain the allowable load. Values of factor of safety are given in Table 3.2. Although not as technically accurate as the previous method, this approach is sometimes the only one available to the designer.

It must be noted that creep is more pronounced in PE and PP than it is in polyester (Polyethylene terephthalate (PET)).

Since polymers are viscoelastic materials, strain rate and temperature are important while testing geosynthetics (Andrawes et al., 1986). When a low strain is applied in a wide-width tensile test, the geosynthetic sample takes longer to come to failure and, therefore, the creep strain is greater. High rates of strain (which can be as much as 100% per minute) tend to

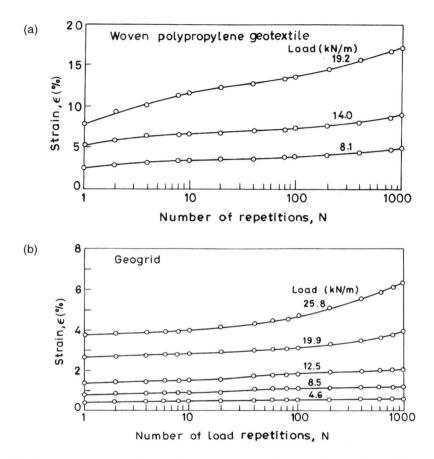

Figure 3.33 Total strain versus log N plot: (a) woven geotextile; (b) geogrid (after Kabir and Ahmed, 1994).

produce lower failure strains and sometimes yield higher strengths than the strengths caused by low rates of strain. Creep rate of geosynthetics depends on temperature. Higher creep rates are associated with higher temperatures resulting in larger strains of geosynthetics to rupture. The rate of creep is also related to the level of load to which the polymer is subjected (Greenwood and Myles, 1986; Mikki et al., 1990). Chang et al. (1996) reported that under the same confining pressure on geotextiles, the amount of creep increases as the creep load rises; and where the creep load is the same, the increases in confining pressure decrease the amount of creep, which may even be reduced to nil. The creep is minimized by the pre-stretching operation of the 'Tensar' process.

It has been found that the tensile and creep properties of some nonwoven geotextiles can be improved by confinement in soil (McGown et al., 1982). This has a greater effect on the tensile properties of the mechanically bonded geotextile and those of the heat-bonded geotextile. The creep of a geosynthetic is likely to be reduced in soil because of load transfer to the soil through a significant increase in frictional resistance between the soil and the geosynthetic. However, there is little effect from confining pressure on the performance of woven geotextiles. Note that the confined or in-soil testing may model the field behaviour of the geosynthetic more accurately. The results from the creep tests under unconfined environment are conservative with regard to the behaviour of the material in service.

Soil pressure can cause compressive creep in geocomposite drains that have an open internal structure to allow flow in the plane of the product. Compressive creep can lead to a reduction in thickness, restriction of the flow or ultimately to collapse of the geocomposite drain or like structure.

In some applications, increase in soil strength is accompanied by the reduction in geosynthetic stress with time. An example of this type of application is foundation support for a permanent embankment over soft deposits. The phenomenon of the decrease in stress, at constant strain, with time is called *stress relaxation*, which is closely related to creep.

Dynamic creep and repeated loading behaviour of geosynthetics are of paramount importance in a number of applications. These include reinforcement in paved and unpaved roads, reinforced retaining structures and slopes under large repetitive live loads, such as traffic and wave action. Behaviour of the geogrid under repeated loading is generally different from those of the geotextiles (Fig. 3.33). Due care must be paid to such applications.

In the recent past, constitutive models have been developed to describe direction and time-dependent, nonlinear, inelastic stress–strain behaviour. More details on this type of model can be found in the works of Perkins (2000).

Abrasion Abrasion of a geosynthetic is defined as the wearing away of any part of it by rubbing against a stationary platform by an abradant with specified surface characteristics. The ability of a geosynthetic to resist wear due to friction or rubbing is called *abrasion resistance*. The abrasion tester used for determining abrasion resistance consists of two parallel smooth plates, one of which makes a reciprocating motion along a horizontal axis. Both the plates are equipped with clamps at each end to hold the test specimen and the abrasive medium (generally emery cloth) without any slippage. Under controlled conditions of pressure and abrasive action, the abradant, generally attached to the lower plate, is moved against the geosynthetic test specimen attached to the upper stationary plate. Resistance to abrasion is expressed as the percentage loss of tensile strength or weight of the test specimen as a result of abrasion. In testing the abrasion resistance of geosynthetics, it is important to simulate the actual type of abrasion, which a geosynthetic would meet in the field. Van Dine *et al.* (1982) and Gray (1982) suggested the test procedures for evaluation of resistance to abrasion caused by different processes such as wear and impact.

Geosynthetics used under pavements, railway tracks or in coastal erosion protection may be subject to dynamic loading, which will lead to mechanical damage of the product in a manner similar to mechanical damage on installation. While geosynthetics are susceptible to mechanical fatigue, the principal cause of degradation is abrasion and frictional rubbing.

Long-term flow characteristics Long-term flow capability of geosynthetics (generally geotextiles) with respect to the hydraulic load coming from the upstream soil is of significant practical interest. The compatibility between the pore size openings of a geotextile and retained soil particles in filtration and/or drainage applications can be assessed by the *gradient ratio test*. This test is basically used to evaluate the clogging resistance of geotextiles with cohesionless soils (having a hydraulic conductivity/permeability greater than 5×10^{-4} m/s) under unidirectional flow conditions. It is best suited for evaluating the movement of finer solid particles in coarse grained or gap-graded soils where internal stability from differential hydraulic gradients may be a problem. Figure 3.34 shows the

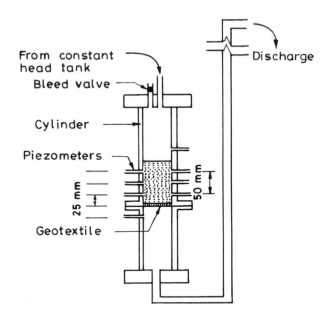

Figure 3.34 Gradient ratio permeameter developed by US Army Corps of Engineers (after Haliburton and Wood, 1982).

constant-head-type permeameter developed by the US Army Corps of Engineers. This permeameter allows the measurement of the head loss along a soil–geotextile system while passing water through the system at different time intervals. After the test is run for some hours (or days), the piezometer readings stabilize and the so-called gradient ratio (GR) is determined. It is defined as the ratio of the hydraulic gradient through the lower 25 mm of the soil plus geotextile thickness to the hydraulic gradient through the adjacent 50 mm of soil alone.

A gradient ratio of one or slightly less is preferred. A value less than one is an indication that some soil particles have moved through the system and a more open filter bridge has developed in the soil adjacent to the geotextile. A continued decrease in the GR indicates piping and may require quantitative evaluation to determine filter effectiveness. Although the GR values of higher than one mean that some system clogging and flow restriction has occurred, if system equilibrium is present, the resulting flow may well satisfy design requirements. Note that the allowable GR values and related flow rates for various soil–geotextile systems will be dependent on the specific site application. One should establish these allowable values on a case-by-case basis. For cohesionless soils, ASTM D5101-01 provides the standard test method for evaluation of permeability and clogging behaviour of the soil–geotextile system by the GR. The long-term flow rate behaviour of geotextile filters can also be assessed by an accelerated filtration test method.

The filtration behaviour of soil–geotextile systems with cohesive soils having a hydraulic conductivity/permeability less than or equal to 5×10^{-4} m/s can be studied by the Hydraulic Conductivity Ratio Test as per ASTM D5567-94 (Reapproved 2001). This test can be used to determine the hydraulic conductivity ratio (HCR), which is defined as below:

$$\text{HCR} = \frac{k_{sg}}{k_{sgo}}, \tag{3.11}$$

where k_{sg} is the hydraulic conductivity of the soil–geotextile system at any time during the test, and k_{sgo} is the initial hydraulic conductivity of the soil–geotextile system measured at the beginning of the test. The hydraulic conductivity ratio test is used only when water is the permeant liquid. Since the hydraulic conductivity varies with void ratio, which in turn varies with effective stress, the test is carried out as a means of determining hydraulic conductivity at a controlled level of effective stress to simulate field conditions. HCR value indicates the performance of geotextile as a filter when used with a particular cohesive soil.

A reduction in HCR with time is representative of significant retention of soil particles. This condition may be desirable in certain drainage applications or it may be undesirable in other applications. The undesirable development of low-permeability conditions within the geotextile filter resulting in the filter's inability to perform the intended drainage function is called *clogging*. The drainage designer must provide the quantitative definition of clogging on a case-by-case basis. A stable value of HCR indicates that excessive transport of soil particles up against, or through, the geotextile filter does not occur. Note that in the case of continued transport of soil particles through the geotextile filter (a phenomenon known as *piping*), a stable HCR value can also be obtained. Thus, it becomes necessary to provide a quantitative definition of stabilized filter conditions and the level of acceptable piping in a specific field application.

For soils containing more than 5% non-plastic fines, Richardson and Christopher (1997) suggested a simple field jar test to empirically assess the clogging potential of a geotextile filter. To perform this test, a small amount of soil is placed in a jar (approximately 1/4 full). The jar should preferably have a removable centre lid (e.g. a Mason jar). The jar is filled with water, and the lid is replaced and secured. It is then shaken to form a soil–water slurry. The jar opening is covered with a specimen of the candidate geotextile and secured, the jar is allowed to stand for about one minute to allow coarser particles to settle. The liquid is then poured from the jar through the geotextile, tilting the jar such that trapped air does not impede water flow. If the fines pass through the geotextile, it should not clog. If very little fine soil passes and a significant buildup of fines is observed on the surface of the geotextile, a clogging potential may exist. While certainly not a standardized test, this has been found to be very useful. It is essentially a fine fraction filtration test and permits a qualitative evaluation of the ability of fines to pass through a geotextile.

Some liquids, such as landfill leachates, may create biological activity on geosynthetic filters thereby reducing their flow capability. In such applications, the general practice is to determine the potential for, and relative degree of, biological growth which can accumulate on geosynthetic filters by measuring flow rates over an extended period of time (e.g. up to 1000 h) under aerobic or anaerobic conditions on the basis of either a constant head test procedure (Fig. 3.35(a)) or a falling head test procedure (Fig. 3.35(b)). It has been observed that once biological clogging initiates, constant head test often passes inadequate quantities of liquid for being measured accurately. It thus becomes necessary to use a falling head test, which works on the basis of the measurement of the time of movement of a relatively small

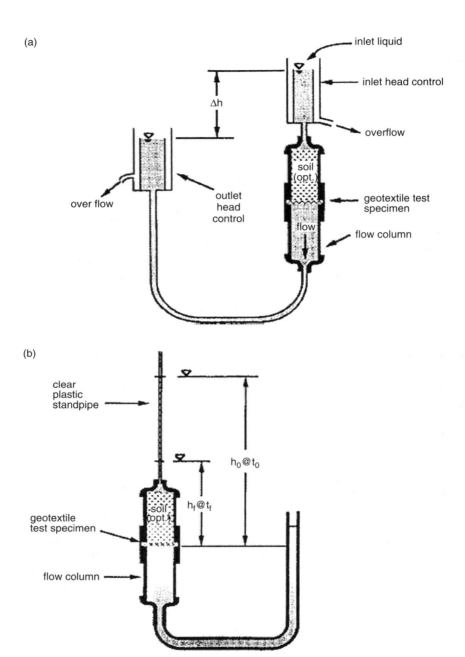

Figure 3.35 Test arrangements for evaluation of biological clogging of geotextile filters: (a) constant head test; (b) variable (falling) head test (Reprinted, with permission, from ASTM D 1987-95 (Reapproved 2002), *Standard Test Method for Biological Clogging of Geotextile or Soil/ Geotextile Filter*, copyright ASTM International, 100 Barr Harbor Drive, West Conshohocken, PA 19428).

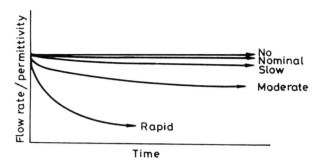

Figure 3.36 Various degrees of biological clogging.

quantity of liquid between two selected points on a transparent standpipe. The various degrees of biological clogging are shown in Figure 3.36. The understanding of the biological clogging helps in making an effective design of filtration and drainage systems and remediation schemes in civil engineering applications such as landfill projects.

The long-term water flow capacity of geotextiles is also assessed in conjunction with the long-term compressive creep behaviour. In fact, the compressibility of the geotextile over time substantially influences the permittivity and transmissivity of geotextiles in service conditions.

Durability The *durability* of a geosynthetic may be regarded as its ability to maintain requisite properties against environmental or other influences over the selected design life. It can be thought of as relating to changes over time of both the polymer microstructure and the geosynthetic macrostructure. The former involves molecular polymer changes and the later assesses geosynthetic bulk property changes. The durability of a geosynthetic is dependent to a great extent upon the composition of the polymers from which it is made. To quantify the properties of polymers, knowledge of their structures at the chemical, molecular and supermolecular level is necessary, which was described by Cassidy *et al.* (1992). The durability of geosynthetics can be assessed by visual examination or microscopic examination with a specified magnification factor to give a qualitative prediction of differences between the exposed and unexposed specimens, for example, discolourations, damage to the individual fibres (due to chemical or microbiological attack, surface degradation, or environmental stress cracking), etc. It is traditionally assessed on the basis of mechanical property test results and not on the microstructural changes that cause the changes in the mechanical properties. It may be assessed in terms of percentage retained tensile strength, R_T and/or percentage retained strain, R_ε, defined as below:

$$R_T = \frac{T_e}{T_u} \times 100\%, \tag{3.12}$$

where T_e is the mean tensile strength of the exposed geosynthetic specimen and T_u is the mean tensile strength of the unexposed geosynthetic specimen.

$$R_\varepsilon = \frac{\varepsilon_e}{\varepsilon_u} \times 100\%, \tag{3.13}$$

where ε_e is the mean strain at maximum load of the exposed geosynthetic specimen and ε_u is the mean strain at maximum load of the unexposed geosynthetic specimen. The durability can also be assessed by determining changes in the mass per unit area of the geosynthetic.

The object of the durability assessment is to provide the design engineer with the necessary information generally in terms of property changes or partial safety factors so that the expected design life can be achieved with confidence. The durability study consists of the following (HB 154-2002):

1. listing significant environmental factors
2. defining the possible degradation phenomena with regard to the selected geosynthetics and the environment
3. estimating the available property as a function of time
4. supplying the designer with suitable reduction factors or available properties at the end of the design life of the soil–geosynthetic system.

The effects of a given application environment on the durability of a geosynthetic must be determined through appropriate testing. Selection of appropriate tests for durability assessment requires consideration of design parameters and determination of the primary function(s) and/or performance characteristics of the geosynthetic in the specific field application and the associated degradation processes caused by the application environment. Note that the physical structure of the geosynthetic, the type of the polymer used, the manufacturing process, the application environment, the conditions of storage and installation and the different loads supported by the geosynthetic are all parameters that govern the durability of the geosynthetic.

From the engineering point of view, the durability of geosynthetics is studied as *construction survivability* and *longevity*. *Construction survivability* addresses the geosynthetics survival during installation. Geosynthetics may suffer mechanical damage (e.g. abrasion, cuts or holes) during installation due to placement and compaction of the overlying fill. In some cases, the installation stresses might be more severe than the actual design stresses for which the geosynthetic is intended. The susceptibility of some geosynthetics to mechanical damage during installation can increase under frost conditions. The severity of the damage increases with the coarseness and angularity of the fill in contact with the geosynthetic and with the applied compactive effort, and it generally decreases with the increasing thickness of the geosynthetic. This damage may reduce the mechanical strength of the geosynthetic, and when holes are present it will affect the hydraulic properties as well.

The occurrence and consequences of mechanical damage caused during installation can be assessed by carrying out a site test or by simulating the effects of damage through a trial. The effect of mechanical damage should be expressed as the ratio of the mechanical properties of the damaged material to that of the undamaged material, as explained earlier. The ratio may be used as a partial safety factor in the design of reinforcement applications. The partial safety factor is used to reduce the characteristic strength of the geosynthetic selected for the application. It is to be noted that the installation day is the most difficult day in the life of a geosynthetic. In general, the stronger the geosynthetic, the greater its resistance to installation damage, that is, the greater its potential for survivability. The ability of a geosynthetic to survive installation damage is difficult to quantify using design equations, but one can do it based on the past experience. While selecting the geotextiles, one can follow the M288-00 geotextile specifications by the American Association of State Highway and Transportation Officials (AASHTO), as described in Sec. 4.11 (see Chapter 4).

Longevity addresses how the geosynthetic properties change over the life of the structure. All geosynthetics are likely to be exposed to weathering during storage and on the construction site before installation. The resistance to weathering is important for the performance of the selected geosynthetic. The weathering of geosynthetics is mainly initiated by the climatic influences through the action of solar radiation, heat, moisture and wetting. In service life, most of the geosynthetics will be covered by soil, while those that remain exposed during their entire life will need a far greater degree of resistance. Unless the geosynthetics are to be covered on the day of installation, all geosynthetics should be subjected to an accelerated weathering test. The principle of the accelerated weathering test is to expose the specimens to simulated solar ultraviolet (UV) radiation for different radiant exposures with cycles of temperature and moisture. The strength retained by the geosynthetic at the end of testing, together with the specific application of the geosynthetic, will define the length of the time that the geosynthetic may be exposed on site. Extended artificial weathering tests are required for geosynthetics which are to be exposed for longer durations. If the geosynthetics are to be used for reinforcement applications, an appropriate partial safety factor should be applied to allow for the reduction in strength.

Generally, as the ambient temperature is increased, the strength, creep and durability characteristics of geosynthetics deteriorate. In fact, heat exposure causes a change in its chemical structure resulting in changes in its physical properties and sometimes in the appearance of a polymer. Geosynthetics are likely to encounter high temperature only in paving applications, where they come in contact with hot asphaltic materials. This application favours the use of PP grids in preference to PE grids because of their temperature resistance being greater. High temperature extremes should always be avoided. A geosynthetic tested for resistance to oxidation (temperature stability) in accordance with ENV ISO 13438 should have the minimum percentage retained strength of 50%.

Geosynthetics may degrade when exposed to the UV component (wavelength shorter than 400 nm) of sunlight. UV light stimulates oxidation by which the molecular chains are cut off. If this process starts, the molecular chains degrade continuously and the original molecular structure changes, resulting in a substantial reduction of the mechanical resistance and also in the geosynthetic becoming brittle. In most applications, as geosynthetics are exposed to UV light for only a limited time during storage, transport, and installation and are subsequently protected by a layer of soil, UV degradation is not a major cause of concern, provided sensible placement procedures are followed.

Generally, those geosynthetics that are white or grey in colour are likely to be the most vulnerable to UV degradation. Carbon black and other stabilizers are added to many polymers during the production process to provide long-term resistance to UV-induced degradation. To achieve this, carbon black and other stabilizers should be dispersed and distributed uniformly throughout the as-manufactured geosynthetic material. The uniformity of carbon black dispersion can be checked by the microscopic evaluation.

The study of the long-term performance of geosyntheics in sunlight can be carried out either by exposing the geosynthetics to natural radiation from outdoor exposure or by artificial radiation such as carbon (or xenon) arc lighting in a laboratory. An outdoor exposure test evaluates geosynthetics under site-specific atmospheric conditions generally over an 18-month period. Exposure shall begin so as to ensure that the geosynthetic is exposed during the maximum intensity of UV light of the year. A degradation curve in the form of a graph of percent tensile strength retained, percent strain at failure, or modulus versus exposure time or all of these may be developed for the geosynthetic being evaluated. The durability is generally assessed by comparing the ultimate tensile strength and the elongation at ultimate tensile strength of exposed specimens with that of unexposed specimens. The artificial exposure test that may be completed even within a week has the advantage of not only accelerating testing by increasing mean irradiance level and temperature, and eliminating the cycles of night and day, winter and summer, but also controlling the exposure parameters. Outdoor exposure tests or artificial exposure tests at one location may not be applicable to a project site at another location. The UV degradation test results therefore must be analysed for practical applications keeping in view geographic location, radiation angle, temperature, humidity, rainfall, wind, air pollution, etc. associated with a particular construction site. DE and PP products perform worst, with the majority having a 50% strength loss in about 4–24 weeks exposure of UV radiations.

Geosynthetics may come into contact with chemicals/leachates that are not normally part of the soil environment. Site-specific tests must be performed to assess the chemical degradation of the geosynthetic, resulting in a reduction in molecular weight of polymer and in the deterioration of their engineering properties. Index tests are generally used in chemical-resistance studies. Geosynthetic specimens are exposed by immersion to liquids under specific conditions for a specified time (generally 15 days, unless otherwise specified). The durability is assessed by comparing the ultimate tensile strength and the elongation at ultimate tensile strength of exposed specimens with that of unexposed specimens. Chemical-resistance testing of geosynthetics should employ worst-case scenario conditions. This is necessary to ensure that when in actual use, the geosynthetics will not be subjected to conditions worse than those experienced in the testing laboratory. Accelerated tests should have generally an accepted relationship to real conditions. Geosynthetic composition should be considered in cases of complex chemical exposure (e.g. leachate) and burial in metal-rich soils. In the absence of actual test data, chemical resistance can be evaluated, at least initially, by comparing chemicals anticipated in the application with manufactures' published resistivity charts.

All polymeric materials have a tendency to absorb water over time. The absorbed water causes chain scission and reduction in the molecular weight of the polymer along with some swelling; this degradative chemical reaction is called *hydrolysis*. However, the effect of hydrolysis is probably not enough to cause significant changes in the mechanical or hydraulic properties of geosynthetics. A geosynthetic consisting solely of PET can be tested for resistance to hydrolysis in accordance with ISO-13439. The minimum percentage retained strength should be 50%. A geosynthetic that fulfils this requirement is estimated to have the following minimum retained strengths in saturated soil after 25 years:

- At 25°C: 95%
- At 30°C: 90%
- At 35°C: 80%

For geosynthetics, oxidation and hydrolysis are the most common forms of chemical degradation as these are processes that involve solvents. Generally, chemical degradation is accelerated by elevated temperatures because the activation energy for these processes is commonly high. The moderate temperatures associated with most installation environments is, therefore, not expected to promote excessive degradation within the usual service lifetimes of most civil engineering systems. Most of the geosynthetics may be considered as having sufficient durability for a minimum service life of 25 yeas provided that it is used in natural soils with a pH of between 4 and 9 and at a soil temperature less than 25°C. Prudent attention should always be paid to unique environments to assess their potential for causing polymer degradation. Resistance to specific chemical attacks (e.g. highly alkaline, pH > 9, or acidic, pH < 4, environments) should be investigated on a site-specific basis. For example, tests on geotextiles in contact with uncured concrete indicate that PP products are largely unaffected, whereas PET products can lose about 50% of their strength in two months of prolonged exposure (Wewerka, 1982). Since many geosynthetic users are not familiar with polymer chemistry, it would be better to assess geosynthetic performance on a functional basis and reserve the polymer chemistry for interpreting unsatisfactory test results or performing forensic studies, if necessary.

Macrobiological degradation is the attack and physical destruction of a geosynthetic by macroorganisms (e.g. insects, rodents and other higher life forms) leading to a reduction in physical properties. *Microbiological degradation* is the chemical attack of a polymer by enzymes or other chemicals exerted by microorganisms (e.g. bacteria, fungi, algae, yeast, etc.) resulting in a reduction of molecular weight and changes in physical properties. All geosynthetic resins are very high in molecular weight with relatively few chain endings for the initiation of biological degradation. Therefore, geosynthetics commonly manufactured from high molecular weight polymers are in general not affected by the biological elements. So far there has been no evidence of any biological degradation in geosynthetics. Only those based on natural fibres degrade, as is the intention.

Microbiological degradation cannot be accelerated beyond the selection of optimum soil conditions and temperature; if it is accelerated further, the microorganisms will be destroyed. It can be studied by conducting *soil burial tests* in which geosynthetic specimens are buried in a prepared microbially active soil bed that is placed in an incubator maintained at a temperature of 28°C and not less than 85% relative humidity for a specified time (generally 14 days, unless otherwise specified). The durability is assessed by comparing the ultimate tensile strength and the elongation of the exposed specimens with that of unexposed specimens. Note that there is no need to inoculate the soil with specific bacteria or fungi; all relevant species are assumed to be already present and those that benefit from the nutrients in the geosynthetic, if any, will multiply and accelerate the attack. The soil must be allowed to stabilize before the specimens are placed in it. A sample of untreated cotton is used to test the soil: if the tensile strength of the cotton strips is less than 25% of the original tensile strength after a seven-day exposure, the soil is regarded as biologically active. A good-quality horticultural compost should be sufficient for soil burial tests (Greenwood *et al.*, 1996). The biological test is generally not required for geosynthetics manufactured from virgin (not recycled) PE, PP, PET and polyamide (PA). It may be applied to other materials including natural fibre-based products, new materials, geocomposites, coated material and others, which are of doubtful quality.

Ageing is the alteration of the physical, chemical, and mechanical properties of geosynthetics caused by the combined effect of environmental conditions over time. It therefore includes both polymer degradation and reduced geosynthetic performance and is dependent on the specific application environment. The resistance of a geosynthetic to ageing is referred to as *durability* as defined earlier. Ageing and burial test procedures and results are becoming more critical as the long-term demands on geosynthetics increase. It should be noted that daily and seasonal variations occur with decreasing intensity as the distance from the ground surface increases. For example, the daily variation in atmospheric temperature and solar radiation is felt to a depth of half a metre. Since higher temperatures increase the rates of ageing and creep of polymers disproportionally, their effect on the geosynthetic behaviour may have to be considered for material installed close to the surface. Ageing is an area that requires much research work. The knowledge and understanding of long-term, in-service behaviour of geosynthetics are vital to the continued growth of this geosynthetic industry and to the science of geosynthetic design.

Most geosynthetics do not suffer from problems with brittle behaviour. However, certain geosynthetic materials may be subject to brittle behaviour as a result of *environmental stress cracking* (ESC), which is the embrittlement of polymers caused by the combination of mechanical stress and environmental conditions. Semi-crystalline polymers such as PE can be more susceptible to environmental stress cracking. Drawn PET or PP fibres, or the drawn ribs of extruded geogrids, are comparatively resistant to ESC. Susceptibility to ESC can be measured by immersing notched samples under load in a bath of liquid and can be accelerated by raising the load or changing the environmental conditions. It is then necessary to carry out tests on a longer term to establish the degree of acceleration.

It must be noted that there are many ways to perform a given test in the laboratory, depending on the case to be designed. The recommended way is the one that best simulates the actual performance of the geosynthetic at the site. Usually, a laboratory test simulates the field situation at only one point of the geosynthetic. When the whole field situation can be simulated in a laboratory test, the test results can be applied to the field situation either directly or using minor mathematical adjustments to deal with the difference in scale between the laboratory and the field. In this case, the test is a model test and an analogical method of design is used. This method of design is the simplest one but it can be rarely used; so other methods such as the analytical method (based on mathematical theories and the basic parameters of geosynthetics) and empirical methods (based on experience and sometimes, systematic testing including full-scale tests) are needed (Giroud, 1980).

3.6 Test and allowable properties

There are presently a large number of geosynthetic products available commercially, each having different properties, but their inclusion in this chapter is beyond the scope of the book. For obtaining the specific values of the various properties of geosynthetics, the users should consult the respective manufacturers or suppliers (see Appendix C for some useful websites). Some representative properties of typical commercially available geosynthetics are listed in Table 3.3. A comparison of the properties of woven and nonwoven geotextiles having the same area density is also given in Table 3.4.

In Chapter 2, it is mentioned that a geosynthetic performs one or more functions in a specific field application. A particular function of the geosynthetic is required to be evaluated using some of its properties. Table 3.5 provides a list of important properties related to the basic functions of geosynthetics. These properties are sometimes referred to as *functional properties*. It should be noted that the data on soil–geosynthetic interface characteristics are necessary for the reinforcement and separation when the geosynthetic is used in a situation where a differential movement can take place between the geosynthetic and adjacent material (soil/geosynthetic), which may endanger the stability of the structure. The data on tensile creep may be required to give an indication of the resistance to sustained loading, when the geosynthetic fulfils a reinforcement function. Data on static puncture strength are necessary for the filtration and separation functions when the site loading conditions are such that there is a potential risk of static puncture of the geosynthetic.

Table 3.3 General range of some specific properties of commercially available geosynthetics (based on the information compiled by Lawson and Kempton (1995))

Types of geosynthetics	Tensile strength (kN/m)	Extension at max. load (%)	Apparent opening size (mm)	Water flow rate (volume permeability) (litres/m^2/s)	Mass per unit area (g/m^2)
Geotextiles					
Nonwovens					
Heat-bonded	3–25	20–60	0.02–0.35	10–200	60–350
Needle-punched	7–90	30–80	0.03–0.20	30–300	100–3000
Resin-bonded	5–30	25–50	0.01–0.25	20–100	130–800
Wovens					
Monofilament	20–80	20–35	0.07–4.0	80–2000	150–300
Multifilament	40–1200	10–30	0.05–0.90	20–80	250–1500
Flat tape	8–90	15–25	0.10–0.30	5–25	90–250
Knitteds					
Weft	2–5	300–600	0.20–2.0	60–2000	150–300
Warp	20–800	12–30	0.40–1.5	80–300	250–1000
Stitch-bondeds	30–1000	10–30	0.07–0.50	50–100	250–1000
Geogrids					
Extruded	10–200	20–30	15–150	NA	200–1100
Textile-based					
Knitted	20–400	3–20	20–50	NA	150–1300
Woven	20–250	3–20	20–50	NA	150–1100
Bonded cross-laid strips	30–200	3–15	50–150	NA	400–800
Geomembranes					
Natural					
Reinforced (made from bitumen and nonwoven geotextile)	20–60	30–60	0	0	1000–3000
Plastomeric (made from plastomers such as HDPE, LDPE, PP, or PVC)					
Unreinforced	10–50	50–200	0	0	400–3500
Reinforced	30–60	15–30	0	0	600–1200
Elastomeric (made from elastomers, i.e. rubbers of various types)					
Reinforced	30–60	15–30	0	0	500–1500
Geocomposites					
Geosynthetic clay liners	10–20	10–30	0	0	5000–8000
Linked structures (geostrip-based)[1]	100–1500	3–15	NA	NA	400–4500

Notes
NA is not applicable.

1 Geostrips are geocomposites having tensile strength in the range 20–200 kN and extension at max. load in the range 3–15%. Geobars are geocomposites having tensile strength in the range 20–1000 kN, if reinforced internally and in the range 20–300 kN if reinforced externally, and extension at max. load in the range 3–15% for both cases.

Table 3.4 Comparison of some properties of woven and nonwoven geotextiles having the same area density

Property	Woven	Nonwoven
Fibre arrangement	Orthogonal	Random
Breaking strength	Higher	Lower
Breaking elongation	Lower	Higher
Initial modulus	Higher	Lower
Tear resistance	Lower	Higher
Openings	Can be regular	Irregular
Filtration	Single layer	Multi-layer
Porosity	35–45%	55–93%
In-plane flow	Lower	Higher
Edge	May ravel	Does not ravel

Table 3.5 Important properties of geosynthetics related to their basic functions

Geosynthetic functions	Geosynthetic properties
Reinforcement	Strength, stiffness, soil–geosynthetic interface characteristics (frictional and interlocking characteristics), creep, stress relaxation, durability
Separation	Characteristic opening size, strength, soil–geosynthetic interface characteristics (frictional and interlocking characteristics), durability
Filtration	Characteristic opening size, permittivity, clogging, puncture strength, durability
Drainage (fluid transmission)	Characteristic opening size, transmissivity, clogging, durability
Fluid barrier	Permittivity, strength, durability, abrasion resistance
Protection	Puncture strength, burst strength, stiffness, abrasion resistance, durability

Geosynthetics almost always encounter soil and environmental conditions that would be expected to cause reductions in their performance. Their properties can be changed unfavourably by several means such as ageing, mechanical damage, creep, hydrolysis (reaction with water), chemical and biological attack, etc., as described in the previous section. These factors have to be taken into account when geosynthetics are selected. For instance, a reduction factor has to be taken into account in the calculation of the decline of strength caused by these factors.

If the test methods for determining the geosynthetic properties are not site specific and completely field simulated, before using the test functional property in calculation of the design factor of safety according to Equation (2.1), it must be modified to an allowable property taking into account of all unfavourable conditions up to the end of the design life as follows:

$$\text{Allowable functional property} = \frac{\text{Test functional property}}{f_1 \times f_2 \times f_3 \times \cdots}, \quad (3.14)$$

where f_1, f_2, f_3, etc. are the various *reduction factors* (a.k.a. *partial factors of safety*) required to account for differences between the test and the site-specific conditions. These reduction factors reflect appropriate degradation processes and are equal to or greater than one.

For example, the laboratory-generated tensile strength is usually an ultimate value, which must be reduced before being used in design. This can be carried out using the following equation:

$$T_{\text{allow}} = T_{\text{ult}} \left[\frac{1}{f_{\text{ID}} \times f_{\text{CR}} \times f_{\text{CD}} \times f_{\text{BD}}} \right], \quad (3.15)$$

where T_{allow} is allowable tensile strength to be used in Equation (2.1) for final design purposes, T_{ult} is ultimate tensile strength from test, f_{ID} is the reduction factor for installation damage (1.1–3.0 for geotextiles, 1.1–1.6 for geogrids), f_{CR} is the reduction factor for creep (1.0–4.0 for geotextiles, 1.5–3.0 for geogrids), f_{CD} is the reduction factor for chemical degradation (1.0–2.0 for geotextiles, 1.0–1.6 for geogrids), and f_{BD} is the reduction factor for biological degradation (1.0–1.3 for geotextiles, 1.0–1.2 for geogrids).

While dealing with flow-related problems through or within a geosynthetic, several reduction factors are required to be considered suitable, as mentioned in the following expression for allowable permittivity (ψ_{allow}):

$$\psi_{\text{allow}} = \psi_{\text{ult}} \left[\frac{1}{f_{\text{CB}} \times f_{\text{CR}} \times f_{\text{IN}} \times f_{\text{CC}} \times f_{\text{BC}}} \right] \quad (3.16)$$

where, ψ_{ult} is the ultimate permittivity from test; f_{CB} is the reduction factor for soil clogging, blinding, and blocking (2.0–10 for geotextiles); f_{CR} is the reduction factor for creep reduction of void volume (1.0–3.0 for geotextiles; 1.0–2.0 for geonets); f_{IN} is the reduction factor for intrusion of adjacent materials into the void volume of geotextile (1.0–1.2 for geotextiles, 1.0–2.0 for geonets); f_{CC} is the reduction factor for chemical clogging (1.0–1.5 for geotextiles, 1.0–2.0 for geonets); and f_{BC} is the reduction factor for biological clogging (1.0–10.0 for geotextiles, 1.0–2.0 for geonets).

It is important to underline that the values of reduction factors are highly dependent on the area of application and the prevailing site conditions. For example, in Equation (3.16), the reduction factor for biological clogging can be higher for turbidity and/or microorganism contents greater than 5000 mg/l. The low end of the range for creep reduction factors refer to applications which have relatively short service lifetimes and/or situations where creep deformations are not critical to the overall system performance. Thus, the designer must use the engineering judgement appropriately based on the available information while selecting the reduction factors.

ILLUSTRATIVE EXAMPLE 3.4

If the ultimate tensile strength of a geogrid from an index-type test is 80 kN/m, then determine the allowable tensile strength to be used in the design of a geotextile-reinforced retaining wall.

SOLUTION

Values of reduction factors are decided based on the site-specific situation. Guidelines given in the local codes of practice, if available, should be considered while deciding these factors. For the present problem consider the following values of reduction factors:

$$f_{ID} = 1.1, \ f_{CR} = 2.0, \ f_{CD} = 1.2, \text{ and } f_{BD} = 1.1.$$

Now, from Equation (3.15),

$$T_{allow} = T_{ult} \left[\frac{1}{f_{ID} \times f_{CR} \times f_{CD} \times f_{BD}} \right]$$

$$= 80 \left[\frac{1}{1.1 \times 2.0 \times 1.2 \times 1.1} \right] \text{kN/m} = 27.5 \text{ kN/m} \qquad \textbf{Answer}$$

3.7 Description of geosynthetics

Manufacturers' literature generally provide the product information and the relevant properties of geosynthetics. If these property values are being used for design, modification must be made, as described in the previous section. Geosynthetics, in general, are commercially described as follows:

1. polymer type
2. type of element (e.g. fibre, yarn, strand, rib), if applicable
3. manufacturing process, if essential
4. type of geosynthetic
5. mass per unit area and/or thickness, if applicable
6. additional information/property in relation to specific field applications.

For example, PP staple filament needle-punched nonwoven, 400 g/m²; PET extruded uniaxial geogrid, with 20 mm by 10 mm openings; HDPE roughened sheet geomembrane, 2.0 mm thick, etc. are a few descriptions of geosynthetics. Before unrolling a roll of geosynthetic at the job site, it must be properly identified.

If geosynthetics are used as a paving fabric, some or all of the following commonly specify them:

1. mass per unit area
2. grab tensile strength in the weakest principal direction
3. elongation
4. bitumen retention
5. fabric storage
6. heat resistance.

It should be noted that properties used in the specification of geosynthetics are established from index tests or from performance tests. As was already discussed, index tests are used by manufacturers for quality control and by installers for product comparison, material specifications and construction quality assurance. Index tests describe the general strength and hydraulic and durability properties of the geosynthetic. General properties are used to distinguish between polymer type and mass per unit area. Performance tests are used by designers to establish, where necessary, design parameters under site-specific conditions using soil samples taken from the site.

Self-evaluation questions

(Select the most appropriate answers to the multiple-choice questions from 1 to 24)

1. The most useful geosynthetic physical property which is closely related to engineering performance is

   (a) Thickness.
   (b) Mass per unit area.
   (c) Strength.
   (d) Stiffness.

2. The base polymer of a geosynthetic can be identified by determining

   (a) Mass per unit area.
   (b) Strength.
   (c) Specific gravity.
   (d) None of the above.

3. The thickness of a geotextile is measured at a specified normal compressive stress, generally equal to

   (a) 2.0 kPa for 5 s.
   (b) 2.0 kPa for 10 s.
   (c) 20.0 kPa for 5 s.
   (d) None of the above.

4. 1 mil is equal to

   (a) 0.1 in.
   (b) 0.01 in.
   (c) 0.001 in.
   (d) None of the above

5. The compressibility is relatively high for

   (a) Woven geotextiles.
   (b) Needle-punched nonwoven geotextiles.
   (c) Thermally bonded geotextiles.
   (d) Knitted geotextiles.

6. The gauge length of the geosynthetic specimen for the wide-width tensile strength test is

    (a) 10 mm.
    (b) 100 mm.
    (c) 200 mm.
    (d) None of the above.

7. If the strength of a geotextile in a technical report is written as 100/40 kN/m, then its strength in the cross machine direction will be

    (a) 100 kN/m.
    (b) 40 kN/m.
    (c) 60 kN/m.
    (d) None of the above.

8. Which one of the following depicts the deformation required to develop a given stress in the geosynthetic?

    (a) Strength.
    (b) Modulus.
    (c) Compressibility.
    (d) None of the above.

9. The woven geotextiles have generally

    (a) High tensile strength.
    (b) High modulus.
    (c) Low elongation.
    (d) All of the above.

10. Typical monofilament woven geotextiles used in construction have a strength in the range of

    (a) 8–90 kN/m.
    (b) 20–80 kN/m.
    (c) 40–1200 kN/m.
    (d) None of these.

11. The ability of a geosynthetic to withstand localized stresses generated by penetrating objects under quasi-static conditions is called its

    (a) Tensile strength.
    (b) Tearing strength.
    (c) Bursting strength.
    (d) Puncture strength.

12. The preferred measure of in-plane water flow capacity of a geotextile is

    (a) Permeability.
    (b) Transmissivity.
    (c) Permittivity.
    (d) Volume rate of flow.

Properties and their evaluation 101

30. What are the tensile properties of a geosynthetic? Explain the test results of the wide-width tensile test and discuss their limitations.
31. What is aspect ratio? What is its significance?
32. Discuss the influence of geotextile specimen width and mass per unit area on the tensile strength.
33. What is the purpose of conducting the multi-axial tensile strength test on geosynthetics?
34. What is the Minimum Average Roll Value of a geosynthetic property? How is it related to Minimum, Average and Maximum values?
35. Draw a typical load–strain curve for geotextiles. Differentiate between the offset modulus and the secant modulus using this typical curve.
36. In geosynthetic testing,

    (a) what is an index test?
    (b) what is a performance test?

37. List the survivability properties of geosynthetics. Give some examples of geosynthetic applications where one or all of these properties will require an essential check.
38. Suggest some field situations showing puncturing and bursting of geosynthetics.
39. Do you observe any limitations in the currently available dynamic puncture strength test? Explain.
40. What are the currently used test methods to evaluate the soil–geosynthetic interface characteristics? Explain the basic principles of these methods by means of neat sketches.
41. Discuss the typical results from the direct shear test on geotextiles.
42. Direct shear test was conducted in the laboratory to study the compacted fly ash – nonwoven geotextile interface characteristics. The following data were obtained:

Normal stress (kPa)	Shear strength (kPa)
100	42
200	68
300	95

Plot the Mohr failure envelope and obtain the fly ash – geotextile interface characteristics, that is angle of interface shear resistance and adhesion.

43. What is the pullout capacity formula? What are your comments on the accuracy of this formula?
44. Calculate the porosity of the geotextile with the following properties:

Thickness, $\Delta x = 1.8$ mm
Mass per unit area, $m = 150$ g/m^2
Density of polymer solid, $\rho_s = 910$ kg/m^3.

45. For geotextiles the pore size is measured directly, rather than using particle size as an estimate of pore size, as is done with soils. Is there any specific reason for this? If yes, explain.
46. List the various test methods developed for measuring the size of openings in geotextiles. Compare these methods by giving relevant details.

102 Properties and their evaluation

47. Draw the pore size distribution curves for typical woven and nonwoven geotextiles. Do you observe any specific differences in the two curves? If yes, list the differences.
48. Define the following terms: equivalent opening size (EOS), apparent opening size (AOS), and filtration opening size (FOS).
49. Differentiate between permittivity and transmissivity.
50. It is preferable to determine and report the permittivity and transmissivity values of geotextiles rather than their coefficients of in-plane and cross-plane permeability respectively. Explain the reasons for this.
51. Why is permittivity used in filtration and transmissivity used in drainage, rather than just their respective coefficients of permeability?
52. What are the long-term normal stress and environmental implications for flow rate capability of geosynthetics?
53. How does the confining pressure affect the thickness, permittivity and transmissivity of needle-punched nonwoven geotextiles?
54. Calculate the transmissivity of a geonet using the following laboratory-based data:

    Flow rate per unit width, $q = 0.72 \times 10^{-4}$ m²/s
    Hydraulic gradient, $i = 0.05$.

55. In a laboratory constant head cross-plane permeability test on a 50-mm diameter geotextile specimen, the following parameters were measured:

    Nominal thickness, $\Delta x = 2.1$ mm
    Flow rate of water normal to the plane of the geotextile, $Q_n = 0.317$ l/s
    Head loss across the geotextile, $\Delta h = 300$ mm.

    Calculate the permittivity and the cross-plane coefficient of permeability of the geotextile.

56. In a laboratory constant head in-plane permeability test on a 300-mm length (flow direction) by a 200-mm width geotextile specimen, the following parameters were measured:

    Nominal thickness, $\Delta x = 2.6$ mm
    Flow rate of water in the plane of the geotextile, $Q_p = 68$ cm³/min
    Head loss in the plane of the geotextile, $\Delta h = 150$ mm.

    Calculate the transmissivity and the in-plane coefficient of permeability of the geotextile.

57. In a laboratory determination of transmissivity by radial method on a geotextile specimen (outer radius = 150 mm, inner radius = 25 mm), the following parameters were measured:

    Nominal thickness, $\Delta x = 2.6$ mm
    Flow rate of water in the plane of the geotextile, $Q_p = 1620$ cm³/min
    Head loss in the plane of the geotextile, $\Delta h = 300$ mm.

    Calculate the transmissivity and the in-plane coefficient of permeability of the geotextile.

58. State Fick's first law of diffusion. How is this useful for geomembranes?
59. Do you think that creep is an important factor in the design and performance of some geosynthetic-reinforced structures? If yes, provide the list of such geosynthetic-reinforced structures along with a proper justification in support of your answer.
60. What is the role of geosynthetic creep in the selection of the allowable load to be used in designs? Explain.

Application	Purpose	Function
Railway tracks	To prevent ballast contamination	Separation
	To dispose of water to side drains	Filtration, drainage
	To prevent contamination in railroad refuelling areas, to prevent upward groundwater movement in a railroad cut	Fluid barrier, protection
	To reinforce track systems and distribute loads	Reinforcement
Slopes	To protect soil slope against erosion along with slope armour	Filtration
	To protect earthen slopes against erosion while vegetation is being established	Vegetative reinforcement, surface stabilization
	To prevent erosion and scouring using bags, tubes, and mattresses filled with soil	Containment, screening
	To prevent soil slope against movement/sliding	Reinforcement
Landfills	To prevent leachate from infiltrating into soil	Fluid barrier, protection
	To drain leachate	Filtration, drainage
Earth dams	To reduce seepage through the dam embankment, to provide upstream face infiltration cut-off	Fluid barrier, protection
	To prevent internal erosion/piping	Filtration, protection
	To drain seepage water	Drainage, filtration
Containment ponds, reservoirs, and canals	To reduce leakage/seepage of water/liquid into ground	Fluid barrier, protection
	To minimize the migration of sediments, to prevent the transportation of solid particles suspended in water	Screening
	To prevent erosion of the earthen surfaces while vegetation is being established	Surface stabilization, vegetative reinforcement
	To prevent erosion and scouring of earthen surfaces using bags, tubes, and mattresses filled with soil	Containment, screening
Filters and drains	To protect the drainage medium, to provide drainage medium	Filtration, drainage, separation
Tunnels and underground structures	To prevent seepage	Fluid barrier, protection
	To provide drainage of seepage water	Drainage

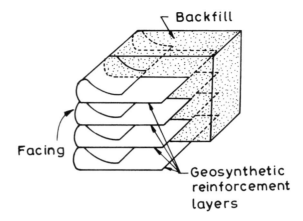

Figure 4.1 Schematic diagram of a geosynthetic-reinforced soil retaining wall.

percentage should not exceed 15%. Particles in the granular backfill material should generally be smaller than 19 mm sieve size. If particles larger than 19 mm sieve size are present in the backfill, then the geosynthetic strength reduction due to installation damage must be considered in the design.

A geosynthetic is mainly used to function as a reinforcement. It resists the lateral earth pressure and thus maintains the stability of the backfill. Its presence also causes reduction in the load-carrying requirements of the wall-facing elements resulting in material and time saving. Filtration and drainage are secondary functions to be served by the geosynthetic in retaining walls.

Woven geotextiles and geogrids with a high modulus of elasticity are generally used as soil reinforcing elements in geosynthetic-reinforced retaining walls. The facing can dictate the type of geosynthetic reinforcement. Because of the permanent reinforcement function, high demands are made upon the durability of the geosynthetic. The force is transmitted to the geotextile layers through friction between their surfaces and soil, and to the geogrid layers through passive soil resistance on grid transverse members as well as through friction between the soil and their horizontal surfaces. It is to be noted that the long-term load transfer is greatly governed by durability and creep characteristics of geosynthetics.

The performance of a geosynthetic-reinforced wall is highly dependent on the type of facing elements used and the care with which it is designed and constructed. Facing elements can be installed as the wall is being constructed or after the wall is built. Geosynthetic wraps, segmental, or modular concrete blocks (MCBs), full-height precast concrete panels, welded wire panels, gabion baskets, and treated timber panels are the facing elements, which are installed as the wall is being constructed. Geosynthetic layers are attached directly to these facing elements. Figure 4.2 shows the schematic diagrams of geosynthetic-reinforced retaining walls with different facing elements, which are commonly used in practice.

The wrap around wall face tends to exhibit relatively large deformation at the wall face and significant settlement at the crest adjacent to the wall face. It is also not aesthetically appealing, since it gives an impression of a relatively low-quality structure. However, it is

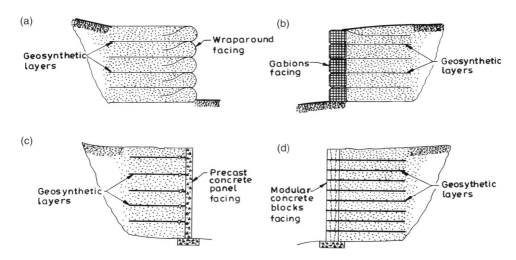

Figure 4.2 Side views of geosynthetic-reinforced retaining walls: (a) with wraparound geosynthetic facing; (b) with gabion facing; (c) with full-height precast concrete panel facing; (d) with segmental or MCBs facing.

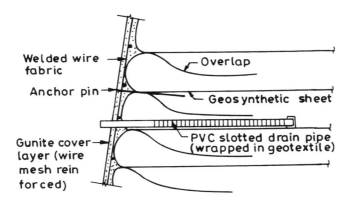

Figure 4.3 Protection of geotextile wraparound facing.

the most economical facing and it was used on many early retaining walls. The wraparound facings are usually sprayed with bitumen emulsion, concrete mortar, or gunite (material similar to mortar)/shotcrete in lifts to produce a thickness in the order of 150–200 mm (Fig. 4.3). A wire mesh anchored to the geotextile wraparound facing may be necessary to keep the coating on the face of the wall. This coating provides protection against ultraviolet (UV) light exposure, potential vandalism and possible fire. If facing elements are required to be installed at the end of wall construction, then shotcrete, cast-in-place concrete panels, precast concrete panels and timber panels can be attached to steel bars placed or driven between the layers of geosynthetic wrapped wall face.

Geogrids along with filter layer (nonwoven geotextile or conventional granular blanket), can also be used for wraparound facings (Fig. 4.4). With proper UV light stabilizer geogrids

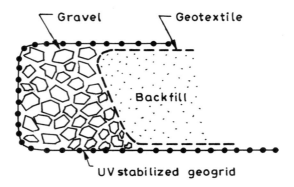

Figure 4.4 Geogrid wraparound facing.

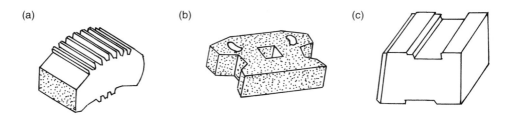

Figure 4.5 Examples of MCB units used in the UK: (a) porcupine; (b) keystone; (c) geoblock (after Dikran and Rimoldi, 1996).

can be left uncovered for a number of years, even for a design life of 50 years or more, provided they are heavy and stiff (Wrigley, 1987).

The modular concrete blocks may have some kind of keys or inserts, which provide a mechanical interlock with the layer above. They provide flexibility with respect to the layout of curves and corners. They can tolerate larger differential settlements than conventional structures. Modular concrete blocks are manufactured from concrete and produced in different sizes, textures and colours; therefore, they provide a varied choice to engineer (Fig. 4.5). Typically all the blocks shown in Figure 4.5 are 250–450 mm in length, 250–500 mm in width and 150–200 mm in height. The mass of each block varies typically from 25 to 48 kg.

In a permanent geosynthetic-reinforced retaining wall (or steep-sided embankment), the geosynthetic load remains constant throughout the life of the structure, and therefore it is an example of a time-independent reinforcement application (Fig. 4.6). In this case, the creep strain may be very high and, therefore, the factor of safety (FS) should not be compromised.

Geosynthetic-reinforced retaining walls are generally an economical alternative to conventional gravity or cantilevered retaining walls, especially for higher retaining walls in fill sections, as found in a large number of retaining wall projects completed successfully worldwide in the past. They can usually be sited on or near the ground surface, which avoids excavation and replacement, costly deep foundation construction and use of ground improvement techniques. In the case of a very weak foundation soil, a geosynthetic-reinforced base can be economically provided for the reinforced wall (Fig. 4.7). Even greater

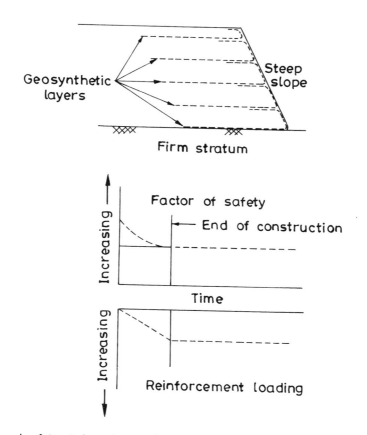

Figure 4.6 Example of time-independent reinforcement application (after Paulson, 1987).

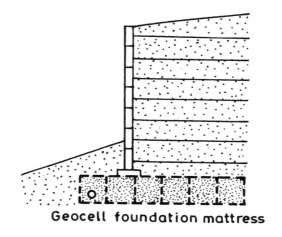

Figure 4.7 Reinforced wall with a reinforced base.

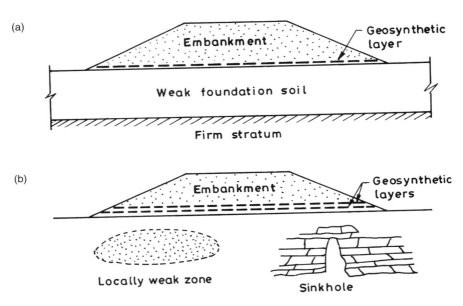

Figure 4.8 Embankment over weak foundation soils: (a) embankment on uniform weak foundation soil; (b) embankment on locally weak foundation soil with lenses of clay or peat, or with sinkholes (after Bonaparte and Christopher, 1987).

economy can be achieved through the use of low-quality backfill that may be available near the construction sites. Being relatively more flexible, the geosynthetic-reinforced retaining walls are very suitable for sites with poor foundation soils and for seismically active areas.

4.3 Embankments

The construction of embankments over weak/soft foundation soils is a challenge for geotechnical engineers. In the conventional method of construction, the soft soil is replaced by a suitable soil or it is improved (by preloading, dynamic consolidation, lime/cement mixing or grouting) prior to the placement of the embankment. Other options such as staged construction with sand drains, the use of stabilizing berms and piled foundations are also available for application. These options can be either time consuming, expensive, or both. The alternate option is to place a geosynthetic (geotextile, geogrid, or geocomposite) layer over the soft foundation soil and construct the embankment directly over it (Fig. 4.8(a)). More than one geosynthetic layer may be required, if the foundation soil has voids or weak zones caused by sinkholes, thawing ice, old streams, or weak pockets of silt, clay or peat (Fig. 4.8(b)). In such situations, the geosynthetic layer is often called a *basal geosynthetic layer*. In some cases, the most effective and economic solution may be some combination of a conventional ground improvement and/or construction alternative together with a geosynthetic layer. For example, taking into account the strength gain that occurs with staged embankment construction, lower strength and therefore lower cost geosynthetic can be utilized.

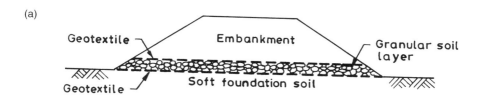

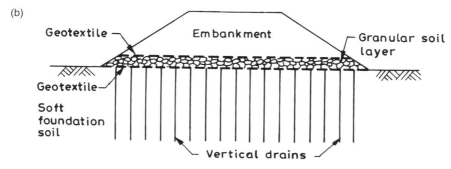

Figure 4.9 Embankment over weak foundation soil: (a) with basal drainage layer; (b) with vertical drains and basal drainage layer.

The geosynthetic as the basal layer in the embankment over soft foundation soil can serve one of the following basic functions or a combination thereof:

1. reinforcement
2. drainage
3. separation/filtration.

The reinforcement function usually aims at a temporary increase in the FS of embankment, which is associated with a faster rate of construction or the use of steeper slopes that would not be possible in the absence of reinforcement. The drainage function is associated with the increase in the rate of consolidation to have a more stable embankment or staged construction. In fact, the geosynthetic allows for free drainage of the foundation soils to reduce pore pressure buildup below the embankment (Fig. 4.9(a)). The consolidation of soft foundation soil can be further accelerated by installing vertical drains along with the basal drainage blanket (Fig. 4.9(b)). The separation function helps in preventing the mixing of the embankment material and the soft foundation soil, thus reducing the consumption of embankment material.

The use of a geosynthetic basal layer is generally attractive for low ratios between foundation soil thickness and embankment base width (say, less than 0.7). For thick foundation soils, the contribution of the reinforcement can be less significant (Palmeira, 2002).

Geosynthetics used to provide reinforcement function include woven geotextiles and/or geogrids. The following factors may be of major concern when choosing the basal geosynthetic to function as a reinforcement:

- tensile strength and stiffness
- soil–reinforcement bond characteristics

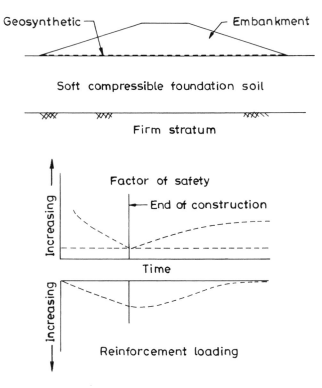

Figure 4.10 Example of time-dependent reinforcement application (after Paulson, 1987).

- creep characteristics
- geosynthetic resistance to mechanical damage
- durability.

In most cases, the geosynthetic reinforcement is required beneath an embankment only during embankment construction and for a short period afterwards, because the consolidation of the soft foundation soil results in an increase in the load-bearing capacity of the foundation soil in due course of time. When a basal geosynthetic is used beneath a permanent embankment, the strain becomes fairly constant, once most of the settlement has taken place. In such a situation, there may be loss of tensile stress experienced by the geosynthetic with time (Fig. 4.10). The phenomenon of the decrease in stress, at constant strain, with time is called *stress relaxation*, which is closely related to creep. Fortunately, during this period the underlying soil is consolidating and increasing in strength. The subsoil is therefore able to offer greater resistance to failure as time passes. The factor of safety should not be compromised if the rate at which the geosynthetic loses its stress is greater than the rate of strength gain occurring in the foundation soil.

If the consolidation of the foundation soil is required to be accelerated for a consequent gain in strength, nonwoven geotextiles may be recommended. Where settlement criteria require high strength and high modulus geosynthetic, geocomposites may be used to

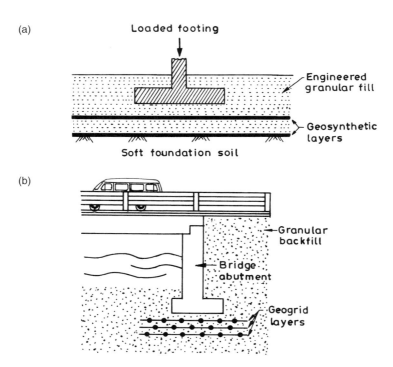

Figure 4.11 Reinforced foundation soils supporting footings of structures.

provide the drainage function. It should be noted that at some very soft soil sites, especially where there is no vegetative layer, a geogrid layer, if laid, may require a lightweight nonwoven geotextile layer as separator/filter to prevent contamination of the first lift, especially if it is an open-graded soil. The geotextile layer is not required if a sand layer is placed as the first lift, which meets soil filtration criteria.

4.4 Shallow foundations

The geosynthetic-reinforced foundation soils are being used to support footings of many structures including warehouses, oil drilling platforms, platforms of heavy industrial equipments, parking areas, and bridge abutments. In usual construction practice, one or more layers of geosynthetic (geotextile, geogrid, geocell, or geocomposite) are placed inside a controlled granular fill beneath the footings (Fig. 4.11). Such reinforced foundation soils provide improved load-bearing capacity and reduced settlements by distributing the imposed loads over a wider area of weak subsoil. In the conventional construction techniques without any use of the reinforcement, a thick granular layer is needed which may be costly or may not be possible, especially in the sites of limited availability of good-quality granular materials.

The geosynthetics, in conjunction with foundation soils, may be considered to perform mainly reinforcement and separation functions. The reinforcement function of geosynthetics can be observed in terms of their several roles, as discussed in Sec. 2.2. Geosynthetics

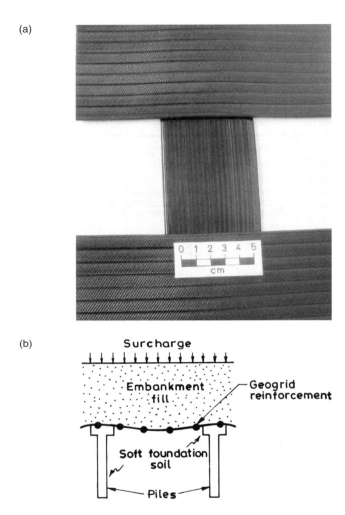

Figure 4.12 'Paralink' geogrid: (a) pictorial view; (b) use over piles.

(particularly, geotextiles, but perhaps also geogrids) also improve the performance of the reinforced soil system by acting as a separator between the soft foundation soil and the granular fill. In many situations, the separation can be an important function compared to the reinforcement function as discussed in Sec. 2.2. In general, the improved performance of a geosynthetic-reinforced foundation soil can be attributed to an increase in shear strength of the foundation soil from the inclusion of the geosynthetic layer(s). The soil–geosynthetic system forms a composite material that inhibits development of the soil-failure wedge beneath shallow spread footings. Geosynthetic products like 'Paralink' as shown in Figure 4.12 (a) can be very effective for use over soft foundation soils as well as over voids and piles (Fig. 4.12(b)).

The ideal reinforcing pattern has geosynthetic layers placed horizontally below the footing, which becomes progressively steeper farther from the footing (Fig. 4.13(a)). It means that

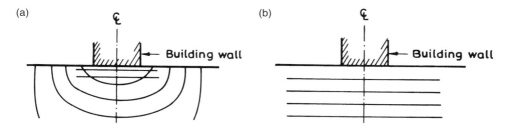

Figure 4.13 Arrangement of reinforcement layers beneath a footing: (a) ideal arrangement (after Basset and Last, 1978); (b) practical arrangement.

the reinforcement should be placed in the direction of the major principal strain. However, for practical simplicity, geosynthetic sheets are often laid horizontally as shown in Figure 4.13(b).

4.5 Roads

Roads often have to be constructed across weak and compressible soil subgrades. It is therefore common practice to distribute the traffic loads in order to decrease the stresses on the soil subgrade. This is generally done by placing a granular layer over the soil subgrade. The granular layer should present good mechanical properties and enough thickness. The long-term interaction between a fine soil subgrade and the granular layer, under dynamic loads, is likely to cause pumping erosion of the soil subgrade and penetration of the granular particles into the soil subgrade, giving rise to permanent deflections and eventually to failure. At present, geosynthetics are being used to solve many such problems.

Based on the type of pavement surfacing provided, roads can be classified as (i) unpaved roads and (ii) paved roads. If roads are not provided with permanent hard surfacing (i.e. asphaltic/bituminous or cement concrete pavement), they are called unpaved roads. Such roads have stone aggregate layers, placed directly above soil subgrades, and they are at most surfaced with sandy gravels for reasonable ridability; thus the granular layer serves as a base course and a wearing course at the same time. If permanent hard pavement layers are made available to unpaved roads, to be called paved roads, their behaviour under traffic loading changes significantly. It can be noted that unpaved roads can be utilized as temporary roads or permanent roads, whereas paved roads are, in most cases, utilized as permanent roads which usually remain in use for 10 years or more.

4.5.1 Unpaved roads

Geosynthetics, especially geotexiles and geogrids, have been used extensively in unpaved roads to make their construction economical by reducing the thickness of the granular layer as well as to improve their engineering performance and to extend their life. A geosynthetic layer is generally placed at the interface of the granular layer and the soil subgrade (Fig. 4.14). Reinforcement and separation are two major functions served by the geosynthetic layer (see Table 4.2). As discussed in Sec. 2.2, if the soil subgrade is soft, that is, the California Bearing Ratio (CBR) of the soil subgrade is low, say its unsoaked value is less

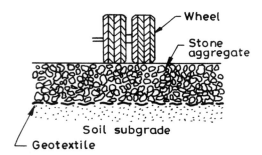

Figure 4.14 A typical cross-section of geosynthetic-reinforced unpaved road.

Table 4.2 Primary function of geosynthetic layer in unpaved road construction based on field CBR value

Soil subgrade description	CBR		Primary function of the geosynthetic	Cost justification for use of the geosynthetic
	Unsoaked	Soaked		
Soft	Less than 3	Less than 1	Reinforcement	Significantly less granular material utilization
Medium	3–8	1–3	Stabilization (an interrelated group of separation, filtration, and reinforcement functions)	Less granular material utilization and longer lifetime
Firm	Greater than 8	Greater than 3	Separation	Much longer lifetime

than 3 (or soaked value is less than 1), then reinforcement will be the primary function because of adequate tensile strength mobilization in the geosynthetic through large deformation, that is, deep ruts (say, greater than 75 mm) in the soil subgrade. Geosynthetics, used with soil subgrades with an unsoaked CBR higher than 8 (or soaked CBR higher than 3), will have negligible amount of reinforcement occurring, and in such cases the primary function will uniquely be separation. For soils with intermediate unsoaked CBR values between 3 and 8 (or soaked CBR values between 1 and 3), there will be an interrelated group of separation, filtration, and reinforcement functions, may be called *stabilization* function of the geosynthetic. Geosynthetics, especially geotextiles and some geocomposites, may also provide performance benefits from their filtration and drainage functions by allowing excess pore water pressure, caused by traffic loads in the soil subgrade, to dissipate into the granular base course and in the case of poor-quality granular materials, through the geosynthetic plane itself.

By providing a geosynthetic layer, improvement in the performance of an unpaved road is generally observed in either of the following two:

1. for a given thickness of granular layer, the traffic can be increased;
2. for the same traffic, the thickness of the granular layer can be reduced, in comparison with the required thickness when no geosynthetic is used.

The introduction of a geotextile layer can typically save one-third of the granular layer thickness of the roadway over moderate to weak soils. Giroud et al. (1984) reported reduction of about 30–50% of thickness of the aggregate layer with the inclusion of geogrids. Improvement in the performance of unpaved roads can also be observed in the form of reduction in permanent (i.e. non-elastic) deformations to the order of 25–50% with the use of geosynthetics, as reported by several workers in the past (De Garidel and Javor, 1986; Milligan et al., 1986; Chaddock, 1988; Chan et al., 1989; Hirano et al., 1990).

4.5.2 Paved roads

Pavements are civil engineering structures used for the purpose of operating wheeled vehicles safely and economically. Paved roadways that include the carriageways and the shoulders have been constructed for more than a century. Their basic design methods and construction techniques have undergone some changes, but the development of geosynthetics in the past four decades has provided the strategies for enhancing the overall performance of the paved roadways. Governments in most of the countries devote unprecedented time and resources to roadway construction, maintenance and repair. Efforts are also being made to apply newfound technology to old pavement problems.

Geosynthetic layer at the soil subgrade level

Geosynthetic layers are used in paved roads usually at the interface of the granular base course and the soft soil subgrade during the initial stage of their construction and may be called *unpaved age*, as a stabilizer lift, to allow construction equipment access to weak soil subgrade sites, and to make possible proper compaction of the first few granular soil lifts. In case of thicker granular bases, the geosynthetic layer may be placed within the granular layer, preferably near midlevel, to achieve optimum effect. The presence of a geosynthetic layer at the interface of the granular base course and the soft soil subgrade improves the overall performance of paved roads, with their long operating life, because of its *separation, filtration, drainage,* and *reinforcement* functions (Holtz et al., 1997; Shukla, 2005).

During construction as well as during the operating life of paved roads, contamination of the granular base course by fines from the underlying soft soil subgrade leads to promoting pavement distress in the form of structural deficiencies (loss of vehicular load-carrying capacity) or functional deficiencies (development of conditions such as rough riding surface, cracked riding surface, excessive rutting, potholes, etc. causing discomfort) that result in early failure of the roadway (Perkins et al., 2002). This is mainly because of the reduction of the effective granular base thickness, by contamination, to a value less than the design value already adopted in practice. This problem may cease to exist in the presence of a geosynthetic layer at the interface of granular base course and the soft soil subgrade because of its role as a separator and/or a filter (Fig. 4.15).

Geosynthetics, especially bitumen-impregnated geotextiles, are used to improve the paved roads, as a separator and/or a fluid barrier, by providing capillary breaks to reduce frost action in frost-susceptible soils (fine-grained soils – silts, clays, and related mixed soils). The paved roads can also be improved by providing the membrane-encapsulated soil layers (MESL) as a moisture-tight barrier beneath the wearing course with an aim to reduce the effects of seasonal water content changes in soils (Fig. 4.16). If good-quality granular

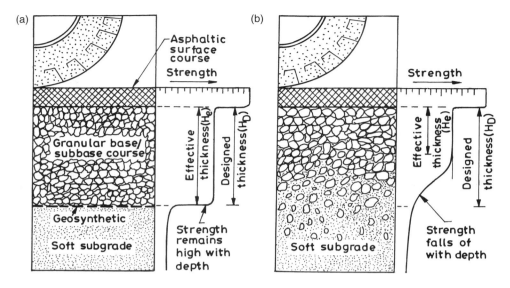

Figure 4.15 Concept of geosynthetic separation in paved roadways (modified from Rankilor, 1981).

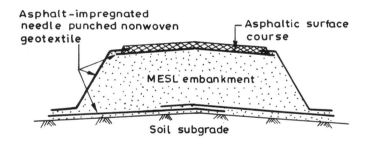

Figure 4.16 Concept of membrane-encapsulated soil layer (MESL) as a base/subbase course in paved roadways.

materials are not available for base/subbase courses, then the concept of MESL can be used to construct base/subbase courses of paved roadways even using locally available poor-quality soils. Commercially available thin-film geotextile composites (Fig. 4.17) are also used as moisture barriers in roadway construction to prevent or minimize moisture changes in pavement subgrades.

Pavement distress can also be caused by inadequate lateral drainage through granular base course. It has been observed that adequate drainage of a pavement extends its life by up to 2–3 times that of a similar pavement having inadequate drainage (Cedergren, 1987). A geosynthetic layer, especially a thick geotextile or a drainage geocomposite, can act as a drainage medium to intercept and carry water in its plane to side drains on either side of the pavement.

The use of a geosynthetic layer also helps in enhancing the structural characteristics and in controlling the rutting of the paved roadway through its reinforcement function. It is to be

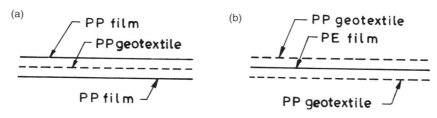

Figure 4.17 Thin-film geotextile composites.

noted that the principal reinforcing mechanism of the geosynthetic in paved roads is its *confinement effect*, not its *membrane effect*, which is applicable to unpaved roads allowing large rutting. The lateral confinement provided by the geosynthetic layer resists the tendency of the granular base courses to move out under the traffic loads imposed on the asphaltic or cement concrete wearing surface. In the case of paved roads on firm subgrade soils, pre-stressing the geosynthetic by external means can significantly increase lateral confinement to granular base course. It also significantly reduces the total and differential settlements of the reinforced soil system under applied loads (Shukla and Chandra, 1994b). It is to be noted that prestressing the geosynthetic can be an effective technique to adequately improve the behaviour of geosynthetic-reinforced paved roads in general situations, if it is made possible to adopt the prestressing process in field in an economical manner.

Geosynthetic layer at the overlay base level

Commonly a paved road becomes a candidate for maintenance when its surface shows significant cracks and potholes. Cracks in the pavement surface cause numerous problems, including

- riding discomfort for the users;
- reduction of safety;
- infiltration of water and subsequent reduction of the load-bearing capacity of the subgrade;
- pumping of soil particles through the crack;
- progressive degradation of the road structure in the vicinity of the cracks due to stress concentrations.

The construction of bituminous/asphalt overlays is the most common way to renovate both flexible and rigid pavements. Most overlays are done predominantly to provide a waterproofing and pavement crack retarding treatment. A minimum thickness of the asphalt concrete overlay may be required to provide an additional support to a structurally deficient pavement. An asphalt overlay is at least 25 mm thick and is placed on top of the distressed pavement. Overlays are economically practical, convenient and effective. The cracks under the overlay rapidly propagate through to the new surface. This phenomenon is called *reflective cracking*, which is a major drawback of asphalt overlays. Because asphalt overlays are otherwise an excellent option, research and development has focused on preventing reflective cracking.

Reflective cracks in an asphalt overlay are basically a continuation of the discontinuities in the underlying damaged pavement. When an overlay is placed over a crack, the crack

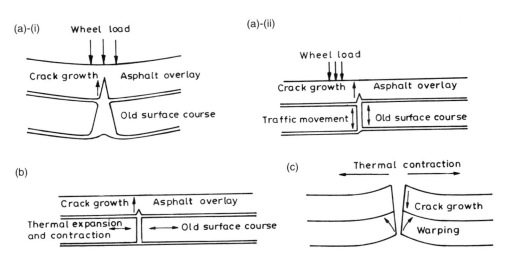

Figure 4.18 Mechanisms of crack formation and propagation in asphalt overlay: (a) traffic induced – (i) repeated bending, (ii) shear effect; (b) thermally induced; (c) surface initiated.

grows up to the new surface. The causes of crack formation and enlargement in asphalt overlays are numerous, but the mechanisms involved may be categorized as *traffic induced, thermally induced* and *surface initiated* (Fig. 4.18). Surface cracking in overlays can occur from traffic induced fatigue as a result of repeated bending condition in the pavement structure or shear effect causing the pavement on one side of a crack (in the old layer) to move vertically relative to the other side of the crack during traffic movement. High axle loads or increased traffic can further increase the stresses and strains in the pavement that lead to surface cracking. In the case of an asphalt overlay on top of a concrete pavement, cracks may be reflected to the overlay as the concrete slabs expand and contract under varying temperatures. The expansion and contraction of the overlays and upper asphalt layers can lead to tension within the surfacing which can also lead to surface cracking. The stresses are at their maximum at the pavement surface where the temperature variation is the greatest. In this case, the cracks are initiated at the surface and propagate downwards. It should be noted that the term 'reflective cracking' is often used to describe all these types of cracking.

Methods for controlling reflective cracking and extending the life of overlays consider the importance and effectiveness of overlay thickness and proper asphalt mixture specification. Asphalt mixes have been improved and even modified by adding a variety of materials. In the past a number of potential solutions have also been evaluated including unbound granular base 'cushion courses' and wire mesh reinforcement. All have been found either marginally effective or extremely costly.

The most basic way to slow down the reflective cracking is to increase the overlay thickness. In general, as the overlay thickness increases, its resistance to reflective cracks increases. However, the upper limit of overlay thickness is highly governed by the expense of the asphalt and the increase in height of the road structure.

Asphalt additives do not stop reflective cracking, but do tend to slow down the development of cracks and convert a large crack in the old pavement into multiple small cracks in the

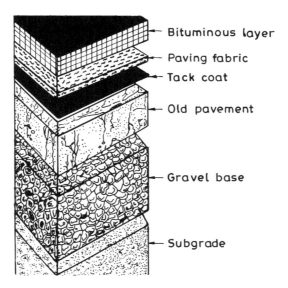

Figure 4.19 Typical cross-section of a paved roadway with a paving fabric interlayer.

overlay. Mixing glass fibres, metal fibres, or polymers in asphalt prior to paving creates *modified or optimized asphalt*, which is not always specified because it is much more expensive than unimproved asphalt and the relationship between investment and improvement has not been established.

The crack resistance of the overlay can also be enhanced via interlayer systems. An interlayer is a layer between the old pavement and new overlay, or within the overlay, to create an overlay system. The benefits of a geosynthetic interlayer include

- waterproofing the pavement;
- delaying the appearance of reflective cracks;
- lengthening the useful life of the overlay;
- added resistance to fatigue cracking;
- saving up to 50 mm of overlay thickness.

A geosynthetic layer, especially a geotextile layer, is used beneath asphalt overlays, ranging in thickness from 25 to 100 mm, of asphalt concrete (AC) and Portland cement concrete (PCC) paved roads. The geotextile layer is generally combined with asphalt sealant or tack coat to form a membrane interlayer system known as a *paving fabric interlayer*. Figure 4.19 shows the layer arrangement in paved roads with paving fabric interlayer. When properly installed, a geotextile layer beneath the asphalt overlay mainly functions as the following (Holtz *et al.*, 1997; Shukla and Yin, 2004):

- fluid barrier (if impregnated with bitumen, that is asphalt cement), protecting the underlying layers from degradation due to infiltration of road-surface moisture;
- cushion, that is, stress-relieving layer for the overlays, retarding and controlling some common types of cracking, including reflective cracking.

124 Application areas

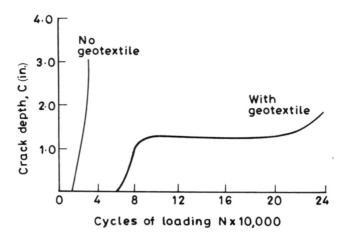

Figure 4.20 Fatigue response of asphalt overlay (after IFAI, 1992).

A paving fabric, in general, is not used to replace any structural deficiencies in the existing pavement. However, the above functions combine to extend the service life of overlays and the roadways with reduced maintenance cost and increased pavement serviceability.

The pavements typically allow 30–60% of precipitation to infiltrate and weaken the road structure. The fluid barrier function of the bitumen-impregnated geotextile may be of considerable benefit if the subgrade strength is highly moisture sensitive. In fact, excess moisture in the subgrade is the primary cause of premature road failures. Heavy vehicles can cause extensive damage to roads, especially when the soil subgrade is wet and weakened. The pore water pressure can also force the soil fines into the voids in the sub-base/base layer, weakening them if a geotextile is not used as a separator/filter. Therefore, efforts should be made to keep the soil subgrade at fairly constant and low moisture content by stopping moisture infiltration into the pavement and providing proper pavement drainage.

A stress-relieving interlayer retards the development of reflective cracks in the overlay by absorbing the stresses induced by underlying cracking in the old pavement. The stress is absorbed by allowing slight movements within the paving fabric interlayer inside the pavement without distressing the asphalt concrete overlay significantly. In fact, the addition of a stress-relieving interlayer reduces the shear stiffness between the old pavement and the new overlay, creating a buffer zone (or break layer) that gives the overlay a degree of independence from movements in the old pavement. Pavements with paving fabric interlayers also experience much less internal crack developing stress than those without. This is why the fatigue life of a pavement with a paving fabric interlayer is many times that of a pavement without it, as shown in Figure 4.20. A stress-relieving interlayer also waterproofs the pavement, so when cracking does occur in the overlay, water cannot worsen the situation.

Geotextiles generally have performed best when used for load-related fatigue distress, for example, closely spaced alligator cracks. Fatigue cracks, mainly caused by too many flexures of the pavement system, should be less than 3 mm wide for best results. Geotextiles used as a paving fabric interlayer to retard thermally induced fatigue cracking, caused by actual expansion and contraction of underlying layers, mostly within the overlay, have, in general, been found to be ineffective. For obtaining the best results on existing cracked pavements, the geotextile layer is laid over the entire pavement surface or over the crack, spanning it 15–60 cm on each side, after placement of an asphalt levelling course followed by an application of asphalt tack coat; and then the asphalt overlay is placed above (Fig. 4.19). This construction technique is adopted keeping in view that much of the deterioration that occurs in overlays is the result of unrepaired distress in the existing pavement prior to the overlay.

The selection of a geosynthetic for use in asphalt overlays is complicated by the variable deterioration conditions of the existing roadway systems. The deterioration may range from simple alligator cracking of the pavement surface to significant potholes caused by failure of the underlying subgrade. It is important to note that an overlay system as well as a paving fabric interlayer will fail if the deficiencies already present in the existing pavements are not corrected prior to the placement of overlay and/or paving fabric.

The selected paving grade geosynthetic must have the ability to absorb and retain the bituminous tack coat sprayed on the surface of the old pavement and to effectively form a permanent fluid barrier and cushion layer. The most common paving grade geosynthetics are lightweight needle-punched nonwoven geotextiles, with a mass per unit area of 120–200 g/m². Woven geotextiles are ineffective paving fabrics, because they have no interior plane to hold asphalt tack coat and so do not form an impermeable membrane. They also do not perform well as a stress-relieving layer to help reduce cracking.

Tests should be performed to determine the bitumen (asphalt cement) retention of paving fabrics for their effective application. In the most commonly used test procedure, after taking weights individually, test specimens are submerged in the bitumen at a specified temperature, generally 135°C for 30 min. Specimens are then hung to drain in the oven at 135°C for 30 min from one end and also 30 min from the other end to obtain a uniform saturation of the fabric. Upon completion of specimen submersion in bitumen and draining, the individual specimens are weighed and bitumen retention, R_B, is calculated as follows (ASTM D6140-00):

$$R_B = \frac{W_{sat} - W_f}{\gamma_B A_f}, \qquad (4.1)$$

where W_{sat} is the weight of the saturated test specimen, in kg; W_f is the weight of paving fabric, in kg; A_f is the area of fabric test specimens, in m²; and γ_B is the unit weight of bitumen/asphalt cement at 21°C, in kg/l. The average bitumen retention of specimens are calculated and reported in l/m².

Paving fabrics precoated with modified bitumen are also available commercially in the form of strips. These products perform the same functions of waterproofing and stress relief as the field impregnated paving fabrics; however, they are more expensive. Their applications are economical if only limited areas of the pavement need a paving fabric interlayer system. For waterproofing and covering the potholes, the precoated paving fabrics are good.

Heavy-duty composites of geosynthetic and bituminous membrane are used, especially over cracks and joints of cement concrete pavements that are overlaid with asphalt concrete. Geogrids and geogrid–geotextile composites are also commercially available

126   Application areas

Figure 4.21 Asphalt reinforcement geogrid.

for overlay applications to function as *reinforcement interlayer* for holding the crack, if any, together and dissipating the crack propagation stress along its length. It has been reported that the reinforcement geogrid, as shown in Figure 4.21, if used beneath the overlay, can reduce the crack propagation by a factor of up to 10 when traffic induced fatigue is the failure mechanism (Terram Ltd, UK). The study conducted by Ling and Liu (2001) shows that the geogrid reinforcement increases the stiffness and load-bearing capacity of the asphalt concrete pavement. Under dynamic loading, the life of the asphalt concrete layer is prolonged in the presence of geosynthetic reinforcement. The stiffness of the geogrid and its interlocking with the asphalt concrete contributes to the restraining effect.

It should be noted that choosing proper application sites for the paving geosynthetic is a function of the existing pavement's structural integrity and crack types – not its surface condition. For successful performance, proper installation must occur on a pavement without significant vertical or horizontal differential movement between cracks or joints and without local deflection under design loading (Marienfeld and Smiley, 1994).

Geosynthetics are used in airfield pavements and parking lots for their enhanced perfromance in the same way as described for road pavements. One of the basic differences lies in the fact that airfield pavements and parking lots are wider than road pavements. In wide pavements with free draining bases, the bases must be tied into an effective edge-drain or underdrain system to help drain the water quickly. Another simple solution is to keep the water from entering the pavement base from the start using a paving fabric interlayer. When properly installed, the paving fabric interlayer keeps the water out of the road base for maximum pavement life. It is very useful for wide pavements, especially airfields and parking lots, where the path to underdrains or edge drains can be at a large distance. The use of paving fabric interlayer is based on the fact that it is much easier to handle surface water than water in the pavement base.

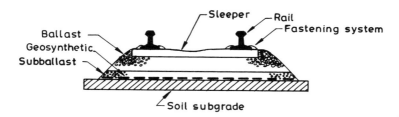

Figure 4.22 Components of the railway track structure.

4.6 Railway tracks

Railway tracks serve as a stable guideway to trains with appropriate vertical and horizontal alignment. To achieve this role each component of the track system (see Fig. 4.22) must perform its specific functions satisfactorily in response to the traffic loads and environmental factors imposed on the system.

Geosynthetics play an important role in achieving higher efficiency and better performance of modern-day railway track structures. They are nowadays used to correct some track support problems. Acceptance and use of geotextiles for track stabilization is now common practice in the USA, Canada and Europe. Geotextiles are also being used in high maintenance locations such as turnouts, rail crossings, switches and highway crossings. One of the most important areas served by geotextiles is beneath the mainline track for stabilization of marginal or poor subgrade, which can suffer from severe mud-pumping and subsidence.

In normal static ground conditions, such as in standard drainage, the cohesive nature of clay and silty clay soils allows the soil particles to bridge over the fine sand size pores, while allowing water to filter through. Under the dynamic conditions caused by pulsating train loading, the action known to engineers as pumping occurs beneath the ballast. This is the phenomenon where clay and silt soil particles are washed upwards into the ballast under pressure from train loading. The subgrade mud-pumping and the load-bearing capacity failure beneath railway tracks are problems that can be handled by the use of geotextiles, geogrids, and/or geomembranes at the ballast–subgrade interface (see Fig. 4.22). The design difficulty lies in the choice of the most suitable geosynthetic.

It is important to underline that not all fines originate from the ground below. As ballast ages, the stone deteriorates through abrasive movement and weathering, producing silty fines which reduce the performance of ballast until it needs cleaning or replacing.

Four principal functions are provided when a properly designed geosynthetic is installed within the track structure. These are

- separation, in new railway tracks, between soil subgrade and new ballast;
- separation, in rehabilitated railway tracks, between old contaminated ballast and new clean ballast;
- filtration of soil pore water rising from the soil subgrade beneath the geosynthetic, due to rising water conditions or the dynamic pumping action of the wheel loadings, across the plane of the geosynthetic;
- lateral confinement-type reinforcement in order to contain the overlying ballast stone;
- lateral drainage of water entering from above or below the geosynthetic within its plane leading to side drainage ditches.

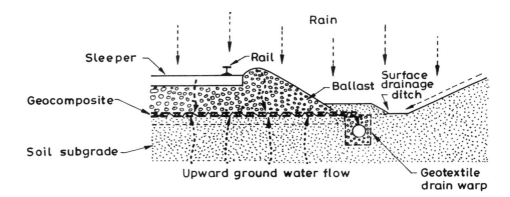

Figure 4.23 A typical drainage system in railway tracks (after Jay, 2002).

The separation function of the geotextile in railway tracks is to prevent the ballast, which is both expensive and difficult to replace, from being pushed down into the soil subgrade and effectively lost. Similarly, the geotextile needs to prevent the soil subgrade working its way up into the ballast, contaminating it, and causing loss of ballast effectiveness.

The drainage has been found to be the most critical aspect for achieving long-term stability in the railway track structure. Excess moisture in the track is found to reduce the subgrade strength and provide easy access for soil fines to foul the open ballast. The drainage function of the geotextile is to allow any ground water within the subgrade to escape upwards and through the geotextile towards the side drains (Fig. 4.23). If water is trapped beneath a geotextile, it may weaken the soil foundation to a very significant extent – which is why a geotextile is made permeable. A well-engineered geotextile allows groundwater to escape upwards easily, involving the reduction in excess pore water pressures generated from repeated applied axle loads of a passing train. At the same time, during rainfall, the downward flow of water is encouraged by the geotextile to be shed into trackside drains. If the water table is not available at perhaps 600 mm or more below the formation, then the upward release of water pressure offered by a permeable geotextile is not required. Instead, an impermeable membrane can be used, because it keeps the rain water from reaching the soil subgrade, as well as separating the soil.

Railway track specifications seem to favour relatively heavy nonwoven needle-punched geotextiles because of their high flexibility and in-plane permeability (transmissivity) characteristics. The logic behind high flexibility is apparent, since geotextiles must deform around relatively large ballast stone and not fail or form a potential slip plane. In-plane drainage itself is not a dominant function, because any geotextile that acts as an effective separator and filter would preserve the integrity of the drainage of the ballast. It should be noted that geotextiles, generally installed at a 300-mm depth below the base of the ties, are likely to be damaged during any subsequent rehabilitation so that the geotextile life only needs to extend the full rehabilitation cycle (Raymond, 1999).

The extension of existing railway routes or construction of new rail routes, to take higher axle loads as well as higher volume and speed of traffic, requires that the load-carrying system should be strengthened. In particular, the load-carrying capacity of the railway subballast must be increased. A subgrade protective layer between the subgrade and the subballast can bring about an increase in load-carrying capacity and reduce settlement. A high-strength geogrid works effectively as a subgrade protective layer, even at small deformations, because they can absorb high tensile forces at low strain. The geogrid layer has a stiffening effect on the track structure, reducing the rate of track settlement to that approaching a firm foundation. The elastic deflections are also reduced, smoothing out variable track quality. If the construction of railway track project has to be completed in a very limited time-span, then in the present-day track construction technology, use of geogrid layers is the best option.

Excess water may create a saturated state in ballast and subballast and cause significant increases in track maintenance costs. Because each source of water requires different drainage methods, the sources must be identified in order to determine effective drainage solutions. The use of geotextile-wrapped trench drains (a.k.a. French drains) and fin drains can provide rapid and cost-effective solutions for the requirements of subsurface and side drainage in railway tracks. The details of the drainage system are described in the following section.

It should be noted that the function of a geosynthetic beneath a railway track is fundamentally different from that beneath a road (described in Sec. 4.5). The following essential differences must be kept in mind while designing the railway track structure (Tan, 2002):

- The ballast used to support the sleeper is very coarse, uniform and angular.
- The regular repeated loading from the axles can set up resonant oscillations in the subgrade making wet subgrade with fine soils very susceptible to mud-pumping.
- The rail track system produces long-distance waves of both positive and negative pressure into the ground ahead of the train itself.

4.7 Filters and drains

The role of groundwater flow and good drainage in the stability of pavements, foundations, retaining walls, slopes, and waste-containment systems is gaining attention from engineers, practitioners, and researchers alike. That is why geosynthetics are being increasingly employed either as filters, in the form of geotextiles (nonwovens and lightweight wovens), in conjunction with granular materials and/or pipes (Fig. 4.24(a)), or as both filters and drains in the form of geocomposites (Fig. 4.24(b)). Filters also form an essential part of many types of hydraulic structures. Thus, there are several application areas for filters and drains including buried drains as pavement edge drains/underdrains, seepage water transmission systems in pavement base course layers and railway tracks, abutments and retaining wall drainage systems, slope drainage, erosion control systems, landfill leachate collection systems, drains to accelerate consolidation of soft foundation soils, drainage blanket to dissipate the excess pore pressure beneath embankments and within the dams and silt fences/barriers.

A filter consists of any porous material that has openings small enough to prevent movement of soil into the drain and that is sufficiently pervious to offer little resistance to seepage. When a geosynthetic is used as a filter in drainage applications, it prevents upstream soils from entering adjacent granular layers or subsurface drains. When properly designed, the geosynthetic filter promotes the unimpeded flow of water by preventing the unacceptable movement of fines into the drain, which can reduce the performance of the drain. Geosynthetic filters are being used successfully to replace conventional graded granular filters in several drainage applications. In fact, filter structures can be realized by using

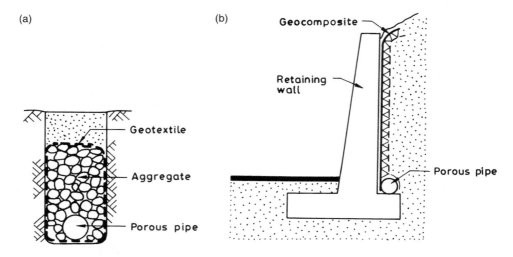

Figure 4.24 (a) A use of geotextile filter; (b) a use of drainage geocomposite.

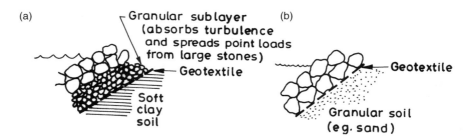

Figure 4.25 Filter layers using geotextile.

granular materials (i.e. crushed stone) or geotextiles or a combination of these materials (Fig. 4.25). The choice between the graded granular filter or geotextile filter depends on several factors. In general, geotextile filters provide easier and more economical placement/installation, and continuity of the filter medium is assured whether the construction is below or above water level. In addition, quality control can be ensured more easily for geotextile-filter systems. Table 4.3 provides a comparison of granular and geotextile filters.

When using riprap–geotextile filter, it is recommended that a layer of aggregate be placed between the geotextile and the riprap, for the following reasons (Giroud, 1992):

- to prevent damage of the geotextile by the large rocks;
- to prevent geotextile degradation by light passing between large rocks;
- to apply a uniform pressure on the geotextile, thereby ensuring close contact between the geotextile filter and the sloping ground, which is necessary to ensure proper filtration;
- to prevent geotextile movement between the rocks because of wave action, thereby ensuring permanent contact between the geotextile filter and the sloping ground, which is also necessary to ensure proper filtration.

Table 4.3 Comparison of granular and geotextile filters (modified from Pilarczyk, 2000)

Objective	Granular filter	Geotextile filter
Similarities		
Complex structure and distribution of open area		
Sensitive in respect to changing of permeability		
Differences		
Determination of characteristic opening size	By particle size analysis	By pore size analysis
Thickness (filtration length)	Long	Very short
Porosity	25–40%	75–95%
Compactibility	Low	High for needle-punched nonwoven geotextiles
Uniformity	Natural variation in grading and density	Greater uniformity due to industrial processing and control
Transmissivity	Independent of stresses	Often dependent on exerted stresses
Internal stability	Can be unstable	Stable
Durability	High	High but not yet properly defined
Placement and execution	Proper quality control necessary, more excavation required	Quick and relatively easy placement, less excavation required

In sediment control applications like silt fences/barriers used to remove soil from runoff, the filter performance of geosynthetics are evaluated in terms of the *filtering efficiency (FE)*. This term is defined as the per cent of sediment removed from sediment-laden water by a geosynthetic over a specified period of time.

It is a misconception that the geosynthetic can replace a granular filter completely. A granular filter serves also other functions related to its thickness and weight. It can often be needed to damp (i.e. to reduce) the hydraulic loadings (internal gradients) to an acceptable level at the soil interface, after which a geotextile can be applied to fulfil the filtration function.

Geotextiles with high in-plane drainage ability and several geocomposites are nowadays commercially available for use as a drain itself, thus replacing the traditional granular drains. The drainage geocomposites consist of drainage cores of extruded and fluted sheets, three-dimensional meshes and mats, random fibres and geonets, which are covered by a geotextile on one or both sides to act as a filter. The cores are usually produced using polyethylene, polypropylene, or polyamide (nylon). Geocomposite drains may be prefabricated or fabricated on site. They offer readily available material with known filtration and hydraulic flow properties, easy installation and thereby construction economies, and protection of any waterproofing applied to the structure's exterior (Hunt, 1982).

Vertical strip drains (also called prefabricated vertical band drains (PVD) or wick drains) are geocomposites used for land reclamation or for stabilization of soft ground. They accelerate the consolidation process by reducing the time required for the dissipation of excess pore water pressure. The efficiency of the drains is partly controlled by the transmissivity, that is discharge capacity that can be measured, using the drain tester, to check their short-term and long-term performance. The discharge capacity of drains is affected by several factors such as confining

132 Application areas

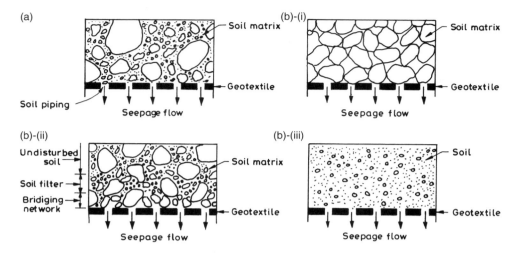

Figure 4.26 (a) Idealized soil/geotextile interface conditions immediately following geotextile installation; (b) idealized interface conditions at equilibrium between three different soil types and geotextile filter – (i) single sized soil and geotextile filter, (ii) well-graded soil and geotextile filter, and (iii) cohesive soil and geotextile filter (Courtesy Terram Ltd, UK).

pressure, hydraulic gradient, length of specimen, stiffness of filter and the duration of loading. The experimental study, conducted in the laboratory by Broms *et al.* (1994), suggests that the effect of the length of the drains and the duration of loading on the discharge capacity of the drain is small, whereas the stiffness of the filter of drain can have a considerable effect. The discharge capacity of the drain decreases with decreasing stiffness of the filter.

Presently, drainage geocomposites are designed for structures requiring vertical drainage such as bridge abutments, building walls, and retaining walls. The composite normally consists of a spacer sandwiched between two geotextiles. This construction combines in a single flexible sheet.

The primary function of the geosynthetic in subsurface drainage applications is filtration. The successful use of a geotextile in a filtration application is dependent on a thorough knowledge of the soil to be retained. The essential properties to be determined are particle size distribution, permeability, plasticity index and dispersiveness. It is crucial to adequately characterize the soil to be retained in order to ensure its compatibility with the chosen geotextile. In certain applications, such as the use of geotextile filters below waste deposits, the nature of the leachate is of crucial importance, because the bacterial growth process may render the geotextile impermeable. Therefore, leachate parameters such as total suspended solids, chemical oxygen demand and biological oxygen demand may be required to be determined (Fourie, 1998).

When a geotextile is placed adjacent to a base soil (the soil to be filtered), a discontinuity arises between the original soil structure and the structure of the geotextile, as discussed in Sec. 2.2. This discontinuity allows some soil particles, particularly particles closest to the geotextile filter and having diameters smaller than the filter opening size, to migrate through the geotextile under the influence of seepage flows. This condition is shown in an idealized manner in Figure 4.26(a). For a geotextile to act as a filter, it is essential that a condition of

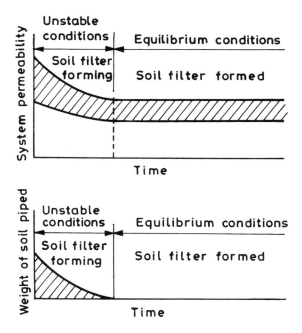

Figure 4.27 Overall requirements for optimal filter performance (after Lawson, 1986).

equilibrium is established at the soil–geotextile interface as soon as possible, after installation, to prevent soil particles from being piped indefinitely through the geotextile; if this were to happen, the drain would eventually become blocked. As fines are washed out from the base soil, the coarser particles located at the filter interface will maintain their positions and a natural filtration zone will be formed immediately above the soil–geotextile interface. These larger particles will, in turn, stop smaller particles, which then stop even finer particles. As a consequence, coarse particles at the filter interface cause a filtration phenomenon within the soil itself, stopping the migration of fines. At equilibrium, which may take normally between 1 and 4 months to occur in practice, the soil adjacent to the filter becomes more permeable.

The structure, or stratification, of the soil immediately adjacent to the geotextile at the onset of equilibrium conditions dictates the filtering efficiency of the system. The stratification is dependent on the type of soil being filtered, the size and frequency of the pores of geotextile, and the magnitude of the seepage forces present. Figure 4.26(b) shows typical stratification occurring with three different soil types – single sized soil, well-graded soil and cohesive soil. When the soil is well graded, considerable rearrangement of the soil takes place. At equilibrium, three zones may be identified: the undisturbed soil, a 'soil filter' layer which consists of progressively smaller particles as the distance from the geotextile increases and a bridging layer which is a porous, open structure. Once the stratification process is complete, it is actually the soil filter layer which actively filters the soil. If the geotextile is chosen correctly, it is possible for the soil filter layer to be more permeable than the undisturbed soil. The function of the geotextile is to ensure that the soil remains in an undisturbed state without any soil piping as shown in Figure 4.27. It should be noted that the

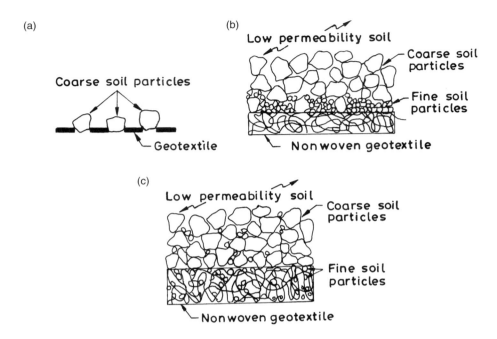

Figure 4.28 Schematic views of (a) blocking; (b) blinding; (c) clogging mechanisms (after Palmeira and Fannin, 2002).

time taken to reach 'constant system permeability conditions' should coincide with the time taken for 'zero soil piping' to be effected. It is also to be noted that for most soils (we can refer to them as stable soils; more discussions can be found in Sec. 5.8), it is not necessary for a geosynthetic filter to screen out all the particles in the soil. Instead, a geosynthetic filter needs only to restrain the coarse fraction of the various particle sizes present.

In filter applications, the design must be prepared so as to avoid, throughout the design life, the following three phenomena causing decrease of the permeability of the geotextile filter in course of time:

1   blocking
2   blinding
3   clogging.

When a geotextile is selected to retain particles of low concentrated suspensions or whenever there is a lack of direct contact between the soil and the geotextile, coarse particles of a size equal to or larger than the pore sizes of the geotextile may migrate and locate themselves permanently at the entrance of the pores of the geotextile, as shown in Figure 4.28(a). This phenomenon that develops at the soil–geotextile interface resulting in the decrease of geotextile permeability is called *blocking*. *Blinding* is a phenomenon similar to blocking and is used to describe the mechanism occurring when coarse particles retained by the geotextile, or geotextile fibres, intercept fine particles migrating from the soil in such a way that a low-permeable layer (often called *soil cake*) is formed very quickly at the interface with the geotextile, thereby reducing the hydraulic conductivity of the system (Fig. 4.28(b)). The phenomenon of accumulation of soil particles within the openings (voids) of a geotextile, thereby reducing its hydraulic conductivity, is called *clogging* (Fig. 4.28(c)). This phenomenon may result in a complete shut off of water flow through the geotextile filter.

It should be noted that clogging, in general, takes place very slowly. It should also be noted that blinding of the filter is far more detrimental than clogging. Geotextiles with a tortuous surface in contact with the soil, such as needle-punched nonwoven geotextiles, do not favour the development of a continuous cake of the fine soil particles, whereas geotextile filters with a smooth surface may favour the development of such a cake (Giroud, 1994). Furthermore, geotextiles with a tortuous surface do not favour the mechanism of blocking, because they do not have individual openings.

4.8 Slopes

4.8.1 Erosion control

The problem of soil movement due to erosive forces by moving water and/or wind as well as by seeping water is called *soil erosion*. Gravity is also one of the prime agents of soil erosion, particularly on steep slopes. Soil erosion is associated with negative economic and environmental consequences in many areas such as agriculture, river and coastal engineering, highway engineering, slope engineering and some more sections of civil engineering. Construction sites with unvegetated steep slopes are prime targets for soil erosion.

Soil erosion by moving water is caused by two mechanisms: (1) detachment of particles due to raindrop impact and (2) movement of particles from surface water flow (Wu and Austin, 1992). The dislodged particles carry with them seeds and soil nutrients. Natural growth of vegetation on the exposed soil slope surface is thus hindered. High velocity runoff can cause not only surface soil movement downslope, but their scouring effects can cause total undermining of slopes. Rain erosion can act upon a land surface of any degree of slope; however, the severity of rain erosion increases with increasing slope steepness and slope length (Ingold, 2002).

The exposed denuded slopes become increasingly vulnerable to erosion agents and are ultimately destabilized. To control erosion is to curb or restrain (not to stop completely) the gradual or sudden wearing away of soils by wind and moving water. The goal of any erosion control project should be to stabilize soils and manage erosion in an economical manner. Since surface water flow cannot be eliminated, the most feasible solution to erosion problems is slope protection. The slope protection serves two functions: (1) it slows down the surface water flow and (2) it holds soil particles, grass or seedlings in place. If an element (natural or synthetic) is incorporated into the soil to prevent the detachment and transportation of soil particles, then the slope would be able to withstand greater forces.

The solutions of soil erosion problems typically involve the use of basic erosion control techniques such as soil cover and soil retention. The use of revetments is very common in civil engineering practice for erosion control (Fig. 4.29). A cover layer (called armour) of a revetment can be permeable or impermeable. An open cover layer substantially reduces the uplift pressures, which can be induced in the sublayers and provides protection against the external loads. Riprap, blocks and block mats, grouted stones, gabions and mattresses, and concrete and asphalt slabs are most commonly used as revetment armours.

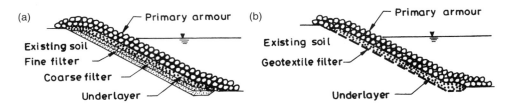

Figure 4.29 Revetment systems: (a) conventional revetment system consisting wholly of granular materials; (b) revetment system containing a geotextile filter.

Riprap consists of stone dumped in place on a filter blanket (4.29(a)) or a prepared slope to form a well-graded mass with a minimum of voids. The stone used for riprap is hard, dense, durable, angular in shape, resistant to weathering and to water action and free from overburden, spoil, shale and organic material. The riprap material can also be placed on gravel bedding layer and/or a woven monofilament or nonwoven geotextile filter (Fig. 4.29(b)).

Concrete block systems consist of prefabricated concrete panels of various geometries, which may be attached to and laid upon a properly designed woven monofilament or nonwoven geotextile filter. Gabions and mattresses are compartmented rectangular containers made of galvanized steel hexagonal wire mesh or rectangular plastic mesh and filled with hand-sized stones. Compared with rigid structures, the advantages of gabions include flexibility, durability, strength, permeability and economy. The growth of native plants is promoted as gabions collect sediment in the stone fill. A high percentage of installations are underlaid by woven monofilament and nonwoven geotextiles to reduce hydrostatic pressure, facilitate sediment capture and prevent washout from behind the structure (Theisen, 1992).

The basic function of revetments is to protect the slopes (coastal shorelines, river/stream banks, canal slopes, hill slopes, and embankment slopes) against hydraulic loadings (forces by water waves and currents as well as seepage forces). The resistance of the erosion control is derived mainly from friction, weight of different elements, interlocking and mechanical strength. As a result of the difference in strength properties, critical loading conditions are also different. Maximum velocities and impacts will be determinants for grass mats and riprap, as they cause displacement of the material. Uplift pressures and impacts, however, are of prime importance for paved revetments and slabs, as they tend to lift the revetment. As these phenomena vary both in space and time, critical loading conditions vary both with respect to position along the slope and the time during the passage of water wave, particularly along the coastal shorelines. Cement concrete and bituminous blocks will mainly respond to uplift forces as maximum loads are distributed more evenly over a layer area, thus causing a higher resistance against uplift, compared with loose block revetment (Pilarczyk, 2000).

Fundamentally, the function of armouring systems is to protect and hold in place the filtering system, which is nowadays almost always a geotextile. The first line of defence against soil erosion is the geotextile, since it is in intimate contact with the soil. If the block systems do not have enough mass and frictional characteristics, this intimate contact between the soil and filter may be lost, which usually results in a rapid degradation of the slope beneath the armouring system.

Scarcity of land and often limited financial resources are forcing engineers to become more innovative and to utilize new products. Geosynthetics have already earned recognition in the

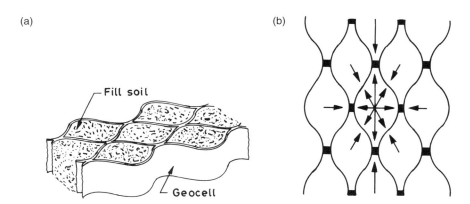

Figure 4.30 (a) Physical confinement of soil in a geocell; (b) confinement forces generated by the resistance of the cell walls and by the passive resistance of the soil in the adjacent cells.

area of erosion control. Geotextiles and geonets are nowadays used as a replacement of graded granular filters typically used beneath revetment armour to keep erodible soil in place and have been found very effective in erosion control. The basic objective in using a geotextile filter is to effectively protect the subsoil from being washed away by the hydraulic loads.

Geosynthetic nets and meshes available in various forms have proven to be successful in both temporary and long-term erosion control projects. These products protect the soil surface from water and wind erosion while accelerating vegetative development. Nettings or meshes may contain UV stabilizers for controlled degradation. Perhaps most advantageous to the environment, these meshes and nettings may ultimately become biodegradable.

Three-dimensional erosion control geosynthetic mats and geocells that are nowadays commercially available with various dimensions can be used in permanent erosion control systems. As was mentioned in Sec. 1.2, geocells are three-dimensional honeycomb structures that have a unique cellular confinement system formed by a series of self-containing cells up to 20 cm deep. They have the ability to physically confine the soil placed inside the cells (Fig. 4.30). They retain soil, moisture and seed, and thus create situations for the growth of vegetative mats on slopes where vegetation may be difficult to establish. The vegetative mats provide reinforcement and the system's cells increase the natural resistance of these mats to erosive forces and protect the root zone from soil loss. At the same time, the cellular confinement system facilitates slope drainage.

Long-term nondegradable (permanent) rolled erosion control three-dimensional products are also manufactured commercially as a *turf reinforcement mat* (*TRM*). Turf reinforcement is in fact a method or system by which the natural ability of vegetation to protect soil from erosion is enhanced through the use of geosynthetic materials. A flexible three-dimensional matrix retains seeds and soil, stimulates seed germination, accelerates seedling development, and, most importantly, synergistically meshes with developing plant roots and shoots to permanently anchor the matting to the soil surface. TRMs provide more than twice the erosion protection of unreinforced vegetation. In fact; these systems are capable of withstanding short-term high velocity flows without erosion; thus they are most suitable where heavy runoff or channel scouring is anticipated. The higher resistance to flow has resulted

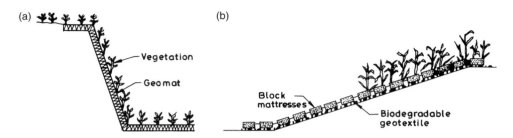

Figure 4.31 Erosion control using geosynthetic mat and geotextile along with vegetation.

in the widespread practice of turf reinforcement as an alternative to riprap, concrete and other armour systems in the protection of open channels, drainage ditches, detention basins and steepened slopes.

A grassed clay dike revetment is also one of the types of revetments used mostly in agricultural applications aimed at preventing the erosion of a dike by hydraulic forces. The development of a strong grass revetment is a matter of time. Also its success depends on the good maintenance.

The most common and natural element used for erosion control is vegetation. Roots of the grasses protect the slope surface from erosion. The deeper roots of plants, shrubs and trees tend to reinforce and stabilize the deeper soils. The application of vegetation as bank protection is preferred rather than the application of conventional materials such as riprap, concrete blocks, etc. If necessary, vegetation and appropriate geosynthetics (geomats, geonets, geocells, etc.) can be applied in combination (Fig. 4.31). The selection of vegetation must be done on the basis of soil and climatic conditions of the specific area of application. The vegetation will on the one hand stabilize the body of the channel, consolidate the soil mass of the slope and bed and reduce erosion. On the other hand, the presence of vegetation will result in extra turbulence and retardation of flow. Geotextiles and other perforated geosynthetics and open blocks provide additional strength to the root mat and can reduce much of the direct mechanical disturbance to plants and soil.

In erosion control applications, where vegetation is considered to be the long-term solution from an environmental point of view, the short-term erosion control is technically performed excellently in a diverse set of environments and soil conditions by jute products (*geojutes*). The low cost (despite significant costs of transportation) and the inherent variability of soil application well accommodate a natural fibre product. An additional advantage of these biodegradable products is that on biodegradation, they improve the quality of the soil for quick vegetation growth. However, a geojute has drawbacks: its open weave construction leaves soil exposed, the organic material tends to shrink and swell under changing moisture, and it is extremely flammable. It should be noted that temporary erosion protection is important, but the long-term goal of any vegetated erosion control technique is to provide a permanent erosion protection through permanent vegetation and/or subsequent root reinforcement.

Geotextiles are also used in toe and bed protection, which consists of the armouring of the beach or bottom surface in front of a structure to prevent scouring and undercutting by water waves and currents (Fig. 4.32). The stability of toe is essential, because its failure will generally lead to failure of the entire structure.

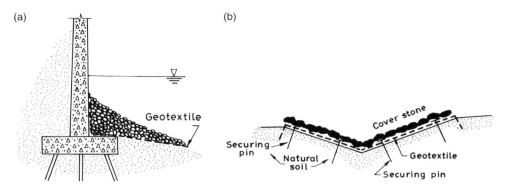

Figure 4.32 (a) Scour protection for abutment; (b) erosion control of ditch.

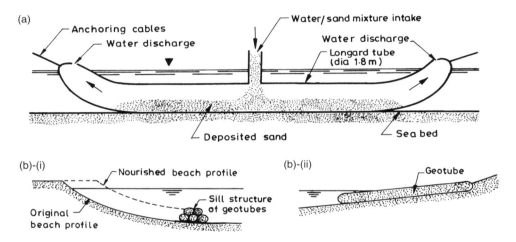

Figure 4.33 Geocontainer: (a) filling procedure of a geotube; (b) erosion control applications (after Pilarczyk, 2000).

In many cases, geotextile is used to wrap a fill material (sand, gravel, asphalt or mortar), creating *geobags, geotubes* or *geomats*, known collectively as *geocontainers*, which are used in hydraulic and coastal engineering (Fig. 4.33). The geotextile cover has to act as a filter towards the fill, which is permeable, as is the container. Sometimes the container also is used as a filter element with respect to the subsoil, that is, when used as a scour fill or a scour prevention layer. The volume of actually used geocontainers varies from 100 to 1000 m^3. Geocontainers are suitable for slope, toe and bed protection but the main application is construction of groynes, perched beaches and offshore breakwaters. They can also be used to store and isolate contaminated materials obtained from harbour dredging and/or as bunds for reclamation works. Geocontainers offer the advantages of simplicity in placement and constructability, cost effectiveness and minimum impact on environment. Thus, they can be a good alternative for traditional materials and erosion control systems and hence they deserve to be applied large scale.

140 Application areas

Table 4.4 Short-term erosion control systems (after Theisen and Richardson, 1998)

Type	Description	Flow velocity range, fps
Hay/straw/hydraulic mulches	Typically machine applied over newly seeded sites	1–3
BOP (Biaxially oriented process nets)	Polypropylene or polyethylene nets used to anchor loose fibre mulches, such as straw, or as a component of erosion control blankets	2–5[1]
ECM (woven erosion control meshes)	Twisted fibres of polypropylene, jute, or coir, woven into a dimensionally stable blanket. Excellent for bioengineering or sod reinforcement	3–6[1]
ECB (erosion control blankets)	BOP stitched or glued on one or both sides of a biodegradable fibre blanket composed of straw, wood, excelsior, coconut, etc.	3–6[1]

Note
1 Depending on vegetation, composition, and density.

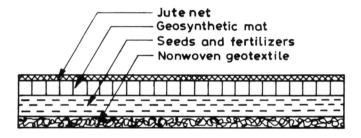

Figure 4.34 Geosynthetic mulching mat (after Ahn et al., 2002).

In the recent past, numerous geosynthetic erosion control systems, including rolled erosion control products (RECPs) (geosynthetic products manufactured or fabricated into rolls), have been developed to provide an aesthetically appealing and maintenance-free option. Some of these geosynthetic systems are intended to provide temporary protection to the top soil cover and seeds against raindrop impact and sheet erosion until vegetation can grow, after which they biodegrade or photodegrade. Others are intended to remain intact and to control erosion even in the absence of vegetation. These systems are usually preformed mats, meshes, blankets and cells. For these systems, degradation of the geosynthetic material constitutes failure of the system. Theisen and Richardson (1998) present an excellent and comprehensive overview of the various existing short-term and long-term erosion control systems, as summarized in Tables 4.4 and 4.5, respectively. These systems may offer cost advantages and improved aesthetics over more traditional designs.

Based on the experimental study, Ahn et al. (2002) pointed out that the mulching mats can be used effectively to provide good plant growth and adequately stabilized soil slopes. The mulching mat can be made of geosynthetics and jute nets with a needle-punched structure (Fig. 4.34). It should be manufactured to have the following properties:

1   It should be biodegradable.
2   It should be very light and portable.

Table 4.5 Long-term erosion control systems (after Theisen and Richardson, 1998)

Type	Description	Flow velocity, fps[1]
Soft armouring systems		
GCS (vegetated geocellular containment)	Polymeric honeycomb-shaped three-dimensional cell systems filled with soil and vegetation	4–6[2]
FRS (UV-stabilized fibre roving systems)	Strands of polypropylene or fiberglass fibre blown onto the ground surface, then anchored in place using emulsified asphalt	6–9
TRM (turf reinforcement mats)	A three-dimensional matrix of polypropylene, polyethylene, or nylon fibres or yarns, mechanically stitched woven or thermally bonded. Designed to be seeded and then filled with soil.	10–25
CBS (vegetated concrete block systems)	Articulate or hand-placed concrete blocks filled with soil, then vegetated	10–25[2]
Hard armouring systems		
GCS (geocellular containment systems)	Polymeric honeycomb cells filled with gravel or concrete	6–25[2]
FFR (fabric-formed revetments)	Geotextiles filled with grout or slurry	15–25
CBS (concrete block systems)	Articulating or hand-placed concrete blocks	15–25
Gabions	Rock-filled wire baskets	15–25
Riprap	Quarried rock of sufficient density	6–30 (depends on mean diameter)

Notes
1 Some systems with greater mass and/or ground cover may exceed these limits.
2 Depending on infill material.

3   It should be capable of holding water but remain substantially unaffected.
4   It should be capable of holding seeds and fertilizer to prevent them from being washed away by rain and wind.

A mulching mat with the above properties has several advantages. First, since seeds and fertilizers are tied into the mat, the seeds are not washed or blown away from the soil slope. Second, the mulching mat prevents evaporation from the soil slope and helps plants to grow more successfully. Third, the mulching mat maintains ground temperature and reduces damage.

The choice and suitability of a particular geosynthetic system for controlling soil erosion depends on whether the system is intended to provide long-term or short-term protection, the degree of protection it can provide under different climatic, topographic and physiographic conditions and the cost-protection efficiency measure (Rustom and Weggel, 1993). Selection of an appropriate rolled erosion control products to protect disturbed soil slopes depends on many factors, including expected project life, down-slope length, soil type, vegetative class, local climatic conditions, slope angle, slope orientation, drainage patterns

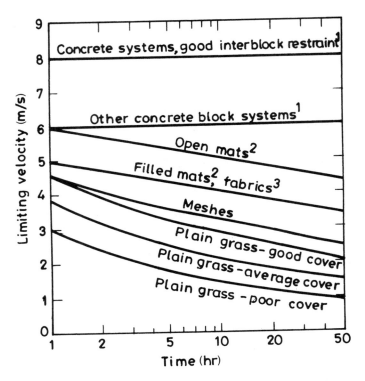

Figure 4.35 Limiting velocities for erosion resistance (after Hewlett et al., 1987).

Notes
1 Minimum superficial mass 135 kg/m^2;
2 minimum nominal thickness 20 mm;
3 installed with 20 mm of soil surface or in conjunction with a surface mesh; all reinforced grass values assume well-established good grass cover.

and available experience. Most manufacturers of rolled erosion control products provide extensive case histories, field and laboratory test data and design software to make the job easier.

Figure 4.35 indicates prototype data on the stability of revetments subject to water current attack. It is noted that concrete systems are excellent erosion control measures, whereas grass mats work most effectively if water current continues for a short period of time.

It is important to note the results of experimental studies, conducted by Cancelli et al. (1990), to separate out the performance of several erosion control products based on rain splash and runoff in controlling the soil erosion on a 1V:2H slope of silty fine sand. It was reported that under rain splash, jute returned by far the highest efficiency with a soil loss of approximately 3 g/l of rainfall (Fig. 4.36(a)); however, under runoff the jute returned by far the worst result (Fig. 4.36(b)). These results suggest that a geocomposite based on jute and synthetic products will have better efficiency in terms of soil loss under a typical combination of rain splash and runoff.

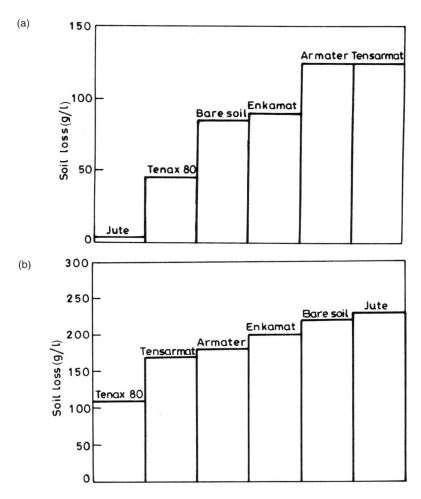

Figure 4.36 Erodibility of erosion control products: (a) based on rain splash; (b) based on runoff (after Cancelli et al., 1990).

4.8.2 Stabilization

Slopes can be natural or man-made (cut slopes or embankment slopes). Several natural and man-made factors, which have been identified as the cause of instability to slopes, are well known to the civil engineering community (Shukla, 1997). Many of the problems of the stability of natural slopes (a.k.a. hillside) are radically different from those of man-made slopes (a.k.a. artificial slopes) mainly in terms of the nature of soil materials involved, the environmental conditions, location of groundwater level, and stress history. In man-made slopes, there are also essential differences between cuts and embankments. The latter are structures which are (or at least can be) built with relatively well-controlled materials. In cuts, however, this

Figure 4.37 A severe landslide causing inconvenience to traffic movement (after Shukla and Baishya, 1998).

possibility does not exist. The failures of slopes, called landslides, may result in loss of property and lives and create inconvenience in several forms to our normal activities (Fig. 4.37).

Several slope stabilization methods are available to improve the stability of unstable slopes (Broms and Wong, 1990; Abramson *et al.*, 2002). The slope stabilization methods generally reduce driving forces, increase resisting forces, or both. The advent of geosynthetic reinforcement materials has brought a new dimension of efficiency to stabilize the unstable and failed slopes by constructing various forms of structures such as reinforced slopes, retaining walls, etc. mainly due to their corrosive resistance and long-term stability. In recent years geosynthetic-reinforced slopes have provided innovative and cost-effective solutions to slope stabilization problems, particularly after a slope failure has occurred or if a steeper than safe unreinforced slope is desirable. They provide a wide array of design advantages as mentioned below (Simac, 1992):

- reduce land requirement to facilitate a change in grade;
- provide additional usable area at toe or crest of slope;
- use available on-site soil to balance earthwork quantities;
- eliminate import costs of select fill or export costs of unsuitable fill;
- meet steep changes in grade, without the expense of retaining walls;
- eliminate concrete face treatments, when not required for surficial stability or erosion control;
- provide a natural vegetated face treatment for environmentally sensitive areas;
- provide noise abatement for high traffic areas and minimize vandalism;
- offer a design that is easily adjustable for surcharge loadings from buildings and vehicles.

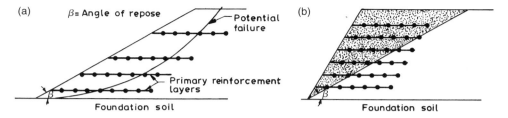

Figure 4.38 Role of reinforcement in slopes: (a) increase factor of safety; (b) stabilize steepened portion of slope (after Simac, 1992).

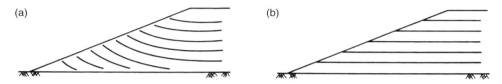

Figure 4.39 Reinforcement orientations: (a) idealized; (b) practical (after Ingold, 1982a).

Construction of reinforced slopes may highlight some of the above advantages in the following applications:

- repair of failed slopes;
- construction of new embankments;
- widening of existing embankments;
- construction of alternatives to retaining walls.

Reinforced slopes are basically compacted fill embankments that incorporate geosynthetic tensile reinforcement arranged in horizontal planes. The tensile reinforcement holds the soil mass together across any critical failure plane to ensure stability of the slope. Facing treatments ranging from vegetation to armour systems are applied to prevent ravelling and sloughing of the face.

Figure 4.38 illustrates the two basic applications of slope reinforcement for stability enhancement in relation to the slope angle, which also represents the angle of repose, defined to be the steepest slope angle that may be built without reinforcement, that is, FS equal to 1. Geosynthetic tensile reinforcement may be used to improve the stability of slopes (= or <β, angle of repose) that are at or slightly greater than FS of 1 (Fig. 4.38(a)). This would be typical of a landslide repair, where grades are established but the soil has failed. Alternatively, more frequent tensile reinforcement may be incorporated into a slope (>β) (Fig. 4.38(b)) that cannot be otherwise built or stand on its own. This creates a sloped earth retaining structure for steep changes in grade that previously required a wall.

The tensile reinforcement should, to be effective, be placed in the direction of tensile normal strains, ideally in the direction and along the line of action of the principal tensile strain. Figure 4.39(a) shows the ideal reinforcement layout. As can be seen, although the horizontal layers of reinforcement would be correctly aligned under the crest of the

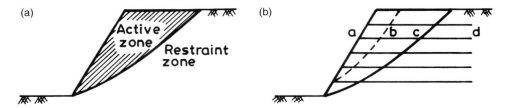

Figure 4.40 Modes of slope reinforcement failure (after Ingold, 1982a).

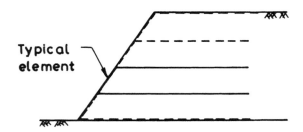

Figure 4.41 Encapsulating reinforcement (after Ingold, 1982a).

slope, they would have inappropriate inclinations under the batter, especially at the toe. Even though an idealized reinforcement layout might be determined it would be impractical if it took the form shown in Figure 4.39(a). Consequently the geosynthetics are usually placed in horizontal layers within the slope as shown in Figure 4.39(b).

Figure 4.40(a) shows an active zone of the soil slope where instability will occur and the restraint zone in which the soil will remain stable. The required function of any reinforcing system would be to maintain the integrity of the active zone and effectively anchor this to the restraint zone, to maintain overall integrity of the soil slope. This function may be achieved by the introduction of a series of horizontal reinforcements or restraining members as indicated in Figure 4.40(b). This arrangement of reinforcement is associated with three prime modes of failure, namely, *tensile failure of the reinforcement, pullout from the restraint zone* or *pullout from the active zone*. Using horizontal reinforcement, it would be difficult to guard against the latter mode of failure. There may be a problem of obtaining adequate bond lengths. This can be illustrated by reference to Figure 4.40(b), which shows a bond length ac for the entire active zone. This bond length may be adequate to generate the required restoring force for the active zone as a rigid mass; however, the active zone contains infinite prospective failure surfaces. Many of these may be close to the face of the batter as typified by the broken line in Figure 4.40(b) where the bond length would be reduced to length ab and as such be inadequate to restrain the more superficial slips. This reaffirms the soundness of using encapsulating reinforcement or facing elements where a positive restraining effect can be administered at the very surface of the slope by the application of normal stresses (Fig. 4.41). The shallow or surficial soil slope failure (Fig. 4.42(a)) can also be prevented by installing shorter, more closely spaced, surficial reinforcement layers in addition to primary reinforcement layers (Fig. 4.42(b)). A second purpose of surficial reinforcement is to provide lateral resistance during compaction of the soil.

In the past, limited experimental studies were conducted to understand the behaviour of reinforced soil slopes. Das *et al.* (1996) presented the results of bearing capacity tests for a

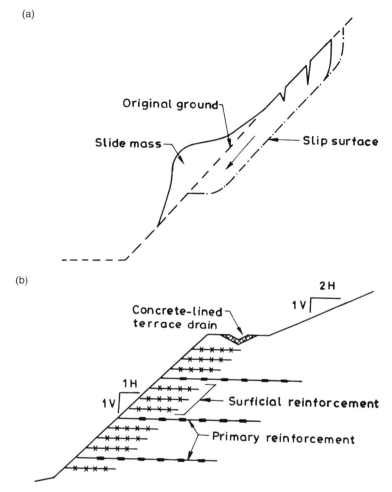

Figure 4.42 (a) Typical surficial soil slope failure; (b) typical cross section of reinforced soil slope (after Collin, 1996).

model strip foundation resting on a biaxial geogrid-reinforced clay slope. The geometric parameters of the test model are shown in Figure 4.43. Based on the study, the following conclusions can be drawn:

1. The first layer of geogrid should typically be located at a depth of 0.4B (B = width of footing) below the footing for maximum increase in the ultimate bearing capacity derived for reinforcement.
2. The maximum depth of reinforcement, which contributes to the bearing capacity improvement, is about 1.72B.

Geotextiles, both woven and nonwoven, and geogrids are being increasingly used for reinforcing steep slopes. Geotextiles, especially nonwoven, exhibit considerable strain

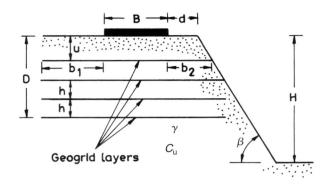

Figure 4.43 Geometric parameters for a surface strip foundation on geogrid-reinforced clay slope (after Das et al., 1996).

before breaking. Also a nonwoven geotextile is much less stiff than the ground. Hence the deformation of a geotextile-reinforced soil slope is dominated not by geotextile but by the soil slope. Due to large extensibility of nonwoven geotextiles, relatively low stresses are induced in them. Their functions, however, are to provide adequate deformability and to redistribute the forces from areas of high stresses to areas of low stresses, thus avoiding the crushing of the soil material. Further, the nonwoven geotextiles facilitate better drainage and help prevent the build up of pore pressures, causing reduction in shear strength.

4.9 Containment facilities

Containment facilities in various forms are being constructed to meet the varying needs of the society. These containment facilities can be categorized in the following three types (Giroud and Bonaparte, 1989):

1. facilities containing solids such as landfills, waste piles and ore leach pads;
2. facilities containing liquids such as dams, canals, reservoirs (to which a variety of names are given such as ponds, lagoons, surface impoundments and liquid impoundments);
3. facilities containing mostly liquids at the beginning of operations and mostly solids at the end such as settling ponds, evaporation ponds and sludge ponds.

Geosynthetics are used in the construction of the above containment facilities to perform various functions: fluid barrier, drainage, filtration, separation, protection and reinforcement. Out of these functions, fluid barrier is generally the required function of the geosynthetic in almost all the containment facilities. In other words, in most of the containment facilities fluid barrier is the primary function of the geosynthetic. The present section describes some of the containment applications with more attention to the fluid barrier function of the geosynthetics.

4.9.1 Landfills

Our activities create several types of waste such as municipal solid waste (MSW), industrial waste, and hazardous waste. We should always attempt to minimize the amount of waste by

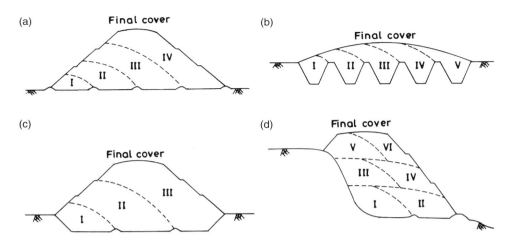

Figure 4.44 Types of solid waste landfill geometry: (a) area fill; (b) trench fill; (c) above and below ground fill; (d) valley fill (after Repetto, 1995).

designing and implementing programmes focussed on waste *reuse, recycling* and *reduction*, and may be called *RRR-concept*. The remaining waste has to be disposed off by suitable disposal methods such as incineration, deep well injection, surface impoundments, composting, and shallow/deep burial in soil and rock. Incineration is not a viable method of disposal for a wide variety of wastes, and furthermore, it may lead to air pollution. It also creates an ash residue that still must be landfilled. In fact, the need for landfilling of solid wastes will continue indefinitely for a number of reasons.

An engineered landfill is a controlled method of waste disposal. It is not an open dump. It has a carefully designed and constructed envelope that encapsulates the waste and that prevents the escape of *leachate* (the mobile portion of the solid waste as the contaminated water) into the environment. Leachate is generated from liquid squeezed out of the waste itself (*primary leachate*) and by water that infiltrates into the landfill and percolates through the waste (*secondary leachate*). It consists of a carrier liquid (solvent) and dissolved substances (solutes).

There are basically two major types of landfill: the *MSW landfill* (also called *sanitary landfill*) to keep the commercial and household solid wastes and the *hazardous waste landfill* for the deposition of the hazardous waste materials. MSW landfill is the most common type of landfill. The geometrical configurations of this landfill commonly include *area fill, trench fill, above and below ground fill*, and *valley (or canyon) fill*, as shown in Figure 4.44. The area fill type of landfill is used in areas with high groundwater table or where the ground is unsuitable for excavation. The trench fill is generally used only for small waste quantities. The depth of trench excavation normally depends on the depth of the natural clay layer and the groundwater table. Above and below type of landfill is like a combination of the area fill and the trench fill. If the solid waste is kept between hills, that is in the valley, it is called a valley fill type of landfill.

The process of selecting a landfill site is complex and sometimes costly; however, a proper siting can be economical to the extent it contributes to the reduction in design and/or construction costs, as well as in long-term expenses with operation and maintenance. Factors that must be considered in evaluating potential sites for the long-term disposal of solid waste include haul distance, location restrictions, available land area, site access, soil

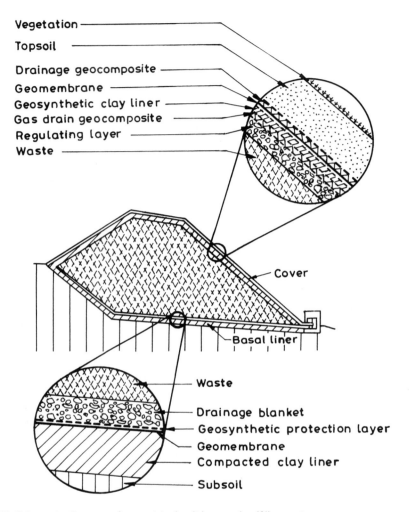

Figure 4.45 Schematic diagram of a municipal solid waste landfill containment system.

conditions, topography, climatological conditions, surface water hydrology, geologic and hydrogeologic conditions, local environmental conditions and potential end uses of the completed site (Tchobanoglous *et al.*, 1993). The geology of the site is an important barrier to the migration of harmful substances. The ground should have a low hydraulic conductivity and a high capacity for the adsorption of toxic material; it must be sufficiently stable and should not undergo excessive settlements under the load of the landfill body.

The engineered landfills, particularly the MSW landfills, consist primarily of the following elements or systems (Fig. 4.45):

1 *Liner system:* This system consists of multiple barrier and drainage layers and is placed on the bottom and lateral slopes of a landfill to act as a barrier system against the leachate transport, thus preventing contamination of the surrounding soil and groundwater.

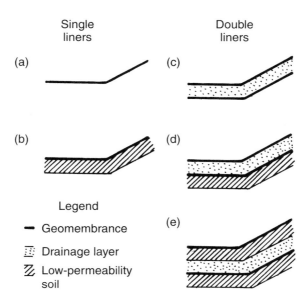

Figure 4.46 Some examples of lining systems: (a) single geomembrane liner; (b) single composite liner; (c) double geomembrane liner; (d) double liner with geomembrane top (or primary) liner and composite bottom (or secondary) liner; (e) double composite liner (after Giroud and Bonaparte, 1989).

2 *Leachate collection and removal system*: This is used to collect the leachate produced in a landfill and to drain it to a wastewater treatment plant for treatment and disposal. The materials used to construct this system are high-permeability materials including the following:

- soils such as sands and gravels, often combined with pipes;
- geosynthetic drainage materials such as thick needle-punched nonwoven geotextiles, geonets, geomats and geocomposites.

3 *Gas collection and control system*: This is used to collect the gases (generally methane and carbon dioxide) that are generated during decomposition of the organic components of the solid waste. One can use the landfill gases to produce a useful form of energy.

4 *Final cover (top cap) system*: This system consists of barrier and drainage layers to minimize water infiltration into the landfill so that the amount of leachate generated after closure can be reduced.

It should be noted that out of four major components of a landfill, the liner system is the single most important component. The barrier in a landfill liner or cover system may consist of a compacted clay liner (CCL), geomembrane (GMB), geosynthetic clay liner (GCL), and/or a combination thereof. Figure 4.46 shows some examples of lining systems. If the combination of a geomembrane layer and an underlying layer of low-permeability soil (clayey soil) placed in good hydraulic contact (often called *intimate contact*) is used as a liner, then this combined system is called a *composite liner*. It is important to note that the terms 'liner' and 'lining system' are not synonymous. The liner refers to only the low-permeability barrier that impedes the flow of fluids towards the ground; whereas the lining system includes the combination of the low-permeability barriers and drainage layers in a containment facility.

Double composite liners with both primary and secondary leachate collection systems are essential for hazardous waste landfills. A leakage if any, through a primary liner, can be collected and properly disposed off in the case of a double lined system. In case of single lined systems, downstream monitoring wells are required for post-construction leak detection. The cost of design and construction of such monitoring wells may exceed the cost of additional liner/leak detection layers of a double lined system. With a leak detection layer, not only the quantity of liquid be monitored over time, but it can also be remediated in place, because the secondary liner is in place to prevent leachate from leaving the site.

Note that most of the major types of geosynthetics, namely geomembranes, geotextiles, geonets, geosynthetic clay liners, and geogrids, are used in landfill engineering to perform various functions: drainage, filtration, protection and reinforcement. In landfills, the geotextiles replace conventional granular soil layers resulting in decreased weight and reduction in landfill settlement. They also prevent puncture damage to the geomembrane liner by acting as a cushion in addition to functioning as a filter in leachate collection systems. Thus, geotextiles can be used to augment and/or replace conventional protective soil layers for geomembrane liners. Even with a geotextile, the minimum protective soil layer should be 300 mm thick. The geotextile will have better puncture protection, if it has a higher mass per unit area.

In case of sanitary landfills, it is always recommended to provide a cover for working faces at the end of the working day to control disease vectors, odours, fires, etc. eliminating the threat to human health and environment. A 15-cm soil layer is traditionally used as the daily cover material. Geosynthetics are more effectively used as an alternative daily cover material in reusable or nonreusable forms. The reusable geosynthetic is placed over the working face at the end of the day and retrieved prior to the start of the next operating day. Tyres, sandbags, or ballast soils are placed along the edges to anchor the geosynthetic cover.

4.9.2 Ponds, reservoirs, and canals

Liquid containment and conveyance facilities, such as ponds, reservoirs and canals, are required in several areas including hydraulic, irrigation and environmental engineering. Unlined ponds, reservoirs, and canals can lose 20–50% of their water to seepage. Traditionally, soil, cement, concrete, masonry or other stiff materials have been used for lining ponds, reservoirs and canals. The effectiveness and longevity of such materials are generally limited due to cracking, settlement and erosion. Sometimes the traditional materials may be unavailable or unsuitable due to construction site limitations, and they may also be costly.

Flexible geosynthetic lining materials, such as geomembranes, have been gaining popularity as the most cost-effective lining solution alone or in combination with conventional lining material for a number of applications, including irrigation and potable water. Figure 4.47 shows typical schematics of liquid containment and conveyance facilities (ponds, reservoirs, and canals) involving application of geosynthetics in addition to conventional materials.

Geosynthetic liner/barrier materials can be classified as GMBs, GCLs, thin-film geotextile composites or asphalt cement-impregnated geotextiles. The selection of lining material

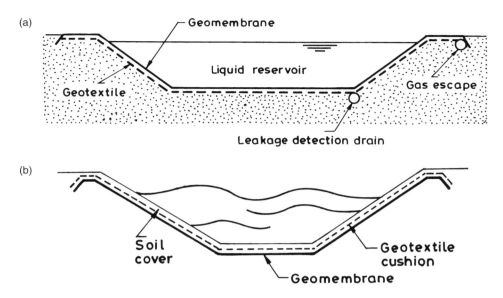

Figure 4.47 Typical cross-sections: (a) liquid pond/reservoir; (b) canal.

is governed by the location and environmental factors. Placement, handling and soil covering operations can also affect geosynthetic selection. When GMBs are used as lining material, geotextiles can be used with GMBs for their protection against puncture by the granular protective layer, which may also be required to prevent UV- and infrared-induced ageing of geosynthetics, as well as any effects of vandalism and burrowing animals. A geotextile, if used below the GMB liner, can function as a protection layer as well as a drainage medium for the rapid removal of leaked water, if any. For economical reasons, the GMB liner may be left uncovered.

4.9.3 Earth dams

Earth dams are water impounding massive structures and are normally constructed using locally available soils and rocks. One of the principal advantages of earth dams is that their construction is very economical compared to the construction costs of concrete dams. Apart from the conventional materials used in the earth dam, geosynthetics are being employed in recent times for new dam constructions and for the rehabilitation of the older dams. Properly designed and correctly installed geosynthetics, in an earth dam, contribute to increase in its safety which corresponds to a positive environmental impact on dam structures (Singh and Shukla, 2002). The reasons for which geosynthetics are used extensively in earth dam construction and rehabilitation are the following:

- The use of geosynthetics in earth dams may serve several functions: water barrier, drainage, filtration, protection and reinforcement.
- The geosynthetics are soft and flexible – therefore, they can endure some elasto-plastic deformations resulting from the subsidence, expansion, landslide and seepage of soil.

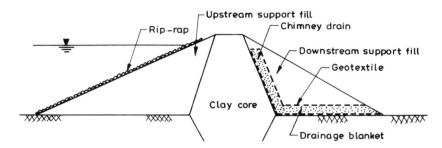

Figure 4.48 A typical cross section of the earth dam with geotextile filters.

- The geosynthetics (geotextiles and geogrids) possess certain mechanical strength, which is favourable as dam-filling materials.
- The permeability of geomembranes is much lower than that of clay or concrete.

The long-term performance of various components of an earth dam is critical to the performance of the dam as a whole. If a geotextile is to be used as a filter, careful assessment of the properties, extensive testing and monitoring are required to ensure its suitability. The locations in earth dams where geotextile filters may be used are in the downstream chimney drain and in the downstream drainage blanket (see Fig. 4.48). If the dam is subjected to rapid drawdown, then drainage systems using geosynthetics may also be installed on the upstream side of the core. In the past, geotextile filters, mostly nonwovens, have been used for the construction or the rehabilitation of numerous embankment dams (i.e. earth or earth and rockfill dams) in various parts of the world.

The chimney drain concept can also be used for rehabilitation purposes. In case of embankment dams, which exhibit seepage through their downstream slope, the construction of a drainage system in the downstream zone is required. A geocomposite drain, placed on the entire downstream slope or only on the lower portion of it and covered with backfill, also performs well. The geocomposite drain must be connected with the new toe of the Dam with outlet pipes or with a drainage blanket. This technique has been used at *Reeves Lake Dam* in USA which is a 13-m high dam repaired in 1990 by placing geocomposite drain (including a PE geonet core between two PP thermobonded nonwoven geotextile filters) on the downstream slope (Wilson, 1992).

To reduce the rate of seepage through dam embankment, a geomembrane sheet may be installed on the upstream face of the embankment. The geomembrane sheet acts as a barrier to water flow. A thick geotextile must be placed on one or both sides of the geomembrane, to protect it from potential damage by adjacent materials, typically the granular layer underneath and the external cover layer. The lower geotextile is generally bonded to the geomembrane in factory, while the upper geotextile is independently placed between the geomembrane and the cast-in-place concrete cover layer (see Fig. 4.49).

Geomembranes can be used for the lining of concrete or masonry dams. In this application, a thick needle-punched nonwoven geotextile is used as a cushion and drainage layer between the geomembrane and the dam. The geotextile is connected to a collector pipe at the toe of the dam. Because of the geotextile layer, the concrete, generally saturated with water, is allowed to drain, thereby slowing the mechanisms of concrete deterioration caused by the presence of water. This method can also be used to control seepage through the wall.

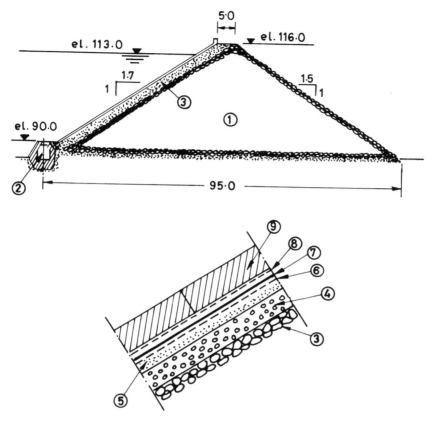

Figure 4.49 Codole rockfill dam, constructed by France in 1983 with the use of geomembrane barrier and geotextile protective layers (after ICOLD, 1991).

Notes
1 Rockfill (up to 1 m size).
2 Inspection and drainage gallery.
3 Sand and gravel layer (2 m thick, 25–120 mm grain size).
4 Gravel layer (0.15 m thick, 25–50 mm grain size).
5 Cold premix layer (50 mm thick, 6–12 mm grain size).
6 Geotextile (mass per unit area = 400 g/m^2) bonded to geomembrane.
7 PVC geomembrane (thickness = 2 mm).
8 Geotextile (mass per unit area = 400 g/m^2).
9 Concrete slabs (0.14 m thick, 4.5 m × 5 m size).

Geosynthetics are also used in dams as a reinforcement. If steep slopes are required, then geogrids or woven geotextiles are generally used for reinforcement purposes. For rehabilitation purposes, like increasing the height of the dam itself to increase the storage capacity or the free board, the use of geosynthetic is ideal and intensive research is recommended in this direction. The first application of geosynthetic reinforcement in dam construction occurred in 1976, when Maraval Dam, which is 8 m high, was constructed in Pierrefeu, France (Giroud, 1992).

Geosynthetics are also used to control surficial erosion of earth dams, both for new construction and rehabilitation purposes. The upstream slope of dams can be eroded by wave action and the downstream slope by runoff from precipitation or overtopping in case of flood. In all these cases, solutions incorporating geosynthetics have been used. In the case of erosion caused by rain water, the entire downstream slope and the upper portion of the upstream slope is protected using typical techniques adopted for river bank revetments, as a riprap, in which geotextiles perform as a filter or, in other solutions (see ICOLD, 1993), as soil–cement blankets, concrete slabs, bituminous concrete layers and so on, in which the geosynthetics could be incorporated with a separation or even a reinforcement function. The products commonly used to control surficial erosion due to atmospheric agents are mainly geomats and geocells. Sometimes biotechnical mats are also adopted, particularly when biodegradation is desirable as in the case of a temporary role to be played during the vegetation growth. It is common practice to solve the problems induced by erosion due to rainfall and consequent runoff as described in Sec. 4.8.1.

The most challenging application of geosynthetics is related to the protection against overtopping, which represents a very crucial aspect of dam engineering. Many failures of embankment dams have been induced in the past by overtopping. Articulate concrete blocks linked by cables, and resting on a geotextile, can be used in order to protect the crest and the downstream slope against overtopping. For this, woven geotextiles are adopted mainly to perform as a filter material. However, the opening size of the geotextile can be selected not only to satisfy the filter criteria but also to allow penetration by grass roots. In fact, the articulate blocks are covered by grassed topsoil layer to give it a natural appearance and additional anchorage for articulate concrete blocks. This system was used in the Blue Ridge Parkways dams in the USA (see Fig. 4.50). Geosynthetics, associated with earth materials and vegetation, can thus form a stable solution to resist overtopping phenomena in earth embankment dams. Gabions and mattresses can also be used to protect the upstream and downstream faces of earth dams.

Use of geosynthetics is associated to a reduction of natural earth materials which are to be exploited and placed on the dam sites. One should note that this shows a positive environmental impact.

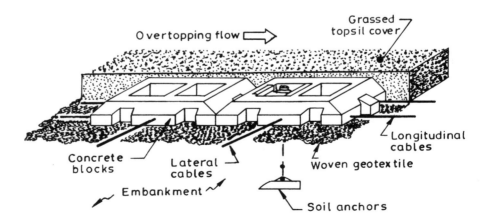

Figure 4.50 Detail of articulate concrete blocks, with a geotextile filter and a grassed topsoil cover, for the downstream face protection against overtopping of Blue Ridge Parkways dams in the USA, 8.5–12 m high (after Wooten et al., 1990).

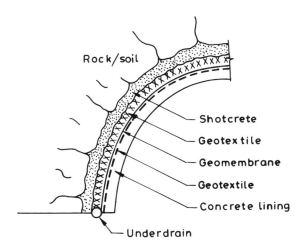

Figure 4.51 Cross-section of a tunnel vault showing the general arrangement of the lining system.

4.10 Tunnels

Tunnels are used for various purposes in civil engineering, including traffic movement and fluid flow. Waterproof tunnels are required at some sections of the highway and railway alignments. A crack-free concrete lining is needed for a waterproof tunnel. Geotextiles and geomembranes are commonly used in modern-day tunnel technology to construct waterproof tunnels.

Figure 4.51 shows the cross-section of a tunnel vault with the general arrangement of the lining system. The shotcrete lining placed over the excavated surface provides a smooth surface for the geosynthetics. In addition the rock surface is supported by the shotcrete immediately after excavation so that the radially acting forces can be accepted adhesively (Wagner and Hinkel, 1987). The nonwoven geotextile (generally needle-punched) acts as a drainage layer and as protection for a waterproofing geomembrane. It also acts as a cushion (stress-relieving layer) to significantly reduce the formation of cracks in the inner concrete lining by allowing free shrinkage deformation of the concrete during the setting process.

It should be noted that geomembrane sheet sealing with a protective nonwoven geotextile drainage layer has predominated over the conventional sealing methods such as asphalt membranes or spray applied glass fibre-reinforced plastic or bitumen-latex based products. The geosynthetic system not only meets the demands of the rapid tunnelling rates but also the demands for rough construction treatment.

4.11 Installation survivability requirements

When geosynthetics are used in a specific application, or in the solution of a particular engineering problem, it is for the designer to determine what properties are required. The role of a designer should be to specify the properties required to have by the geosynthetic to solve the specific problem rather than starting with a geosynthetic of predetermined properties and defining the problem which this geosynthetic might solve. However, the recommended geosynthetics should always satisfy the installation survivability requirements.

While selecting the geosynthetics, particularly geotextiles, for some applications, one can follow the M288-00 geotextile specifications laid down by the American Association of State Highway and Transportation Officials (AASHTO) to meet the installation survivability requirements. These specifications cover geotextiles for use in subsurface filtration, separation, stabilization (an interrelated group of separation, filtration and reinforcement functions), erosion control (filtration), temporary silt fence and paving fabrics. Table 4.6 provides strength properties for three geotextile classes (Class 1, Class 2 and Class 3) that are required for survivability under typical installation conditions for different functions. Class 1 is recommended for use in more severe or harsh installation conditions where there is a great potential for geotextile damage. Class 2 can be used as default classification in the absence of site-specific information. One can use Class 3 for mild survivability conditions.

Table 4.7 indicates the classes required for some functions or applications; for example, for filtration applications the geotextile should meet the Class 2 specifications. The geotextile should conform to the properties of Table 4.6 based on the geotextile class mentioned in Table 4.7 and also of Table 4.8, 4.9, 4.10 or 4.11 for the indicated application. Property requirements for temporary silt fence and paving fabrics are given in Table 4.12 and 4.13 respectively. The geosynthetic silt fence functions as a vertical, permeable interceptor designed to remove suspended soil from overland water flow. The function of a temporary silt fence is to filter and allow settlement of soil particles from sediment-laden water. The purpose is to prevent the eroded soil from being transported off the construction site by water runoff. In Tables 4.6 to 4.13, all property values, with the exception of the apparent opening size (AOS), represent minimum average roll values (MARV) in the weakest principal direction. Values of AOS represent maximum average roll values. The geotextile properties for each class are dependent upon geotextile elongation. It must be noted that these guidelines are based on geotextile survivability from installation stresses, and, therefore, these should be used as a base point only. Specific design and site conditions often require individual geotextile properties and construction recommendations to be modified to ensure that the guidelines are consistent with the project needs.

ILLUSTRATIVE EXAMPLE 4.1
Using the AASHTO M288–00 specifications, recommend the properties of a woven geotextile (elongation, $\varepsilon < 50\%$) required for its application as a separator between soil subgrade (with CBR = 5) and the granular base course under typical installation survivability conditions.

SOLUTION
From Tables 4.6 and 4.9, the recommended properties for the woven geotextile as a separator are the following:

Permittivity ≥ 0.02 s^{-1}
AOS ≤ 0.60 mm
Grab strength ≥ 1100 N
Sewn seam strength ≥ 990 N
Tear strength ≥ 400 N
Puncture strength ≥ 400 N
UV stability $\geq$ (50% of 1100 N = 550 N) after 500 h of exposure.   **Answer**

Table 4.6 AASHTO M288-00 geotextile strength property requirements

Property	ASTM test method	Units	Geotextile class[1,2]					
			Class 1		Class 2		Class 3	
			Elongation <50%[3]	Elongation ≥50%[3]	Elongation <50%[3]	Elongation ≥50%[3]	Elongation <50%[3]	Elongation ≥50%[3]
Grab strength	D4632	N	1400	900	1100	700	800	500
Sewn seam strength[4]	D4632	N	1260	810	990	630	720	450
Tear strength	D4533	N	500	350	400[5]	250	300	180
Puncture strength	D4833	N	500	350	400	250	300	180
Permittivity	D4991	s^{-1}	Minimum property values for permittivity, AOS, and UV stability are based on geotextile application. Refer to Table 4.8 for subsurface drainage, Table 4.9 for separation, Table 4.10 for stabilization and Table 4.11 for permanent erosion control.					
Apparent opening size (AOS)	D4751	mm						
Ultraviolet (UV) stability	D4355	%						

Notes
1 Required geotextile class is designated in Tables 4.8, 4.9, 4.10 or 4.11 for the indicated application. The severity of installation conditions for the application generally dictates the geotextile class. Class 1 is specified for more severe or harsh installation conditions where there is a greater potential for geotextile damage, Classes 2 and 3 are specified for less severe conditions.
2 All numeric values represent Minimum Average Roll Values (MARV) in the weaker principal direction.
3 As measured in accordance with ASTM D4632.
4 When sewn seams are required.
5 The required MARV tear strength for woven monofilament geotextiles is 250 N.

Table 4.7 AASHTO M288-00 default geotextile class

Application	Default geotextile class
Filtration applications in subsurface drainage	2
Separation of soil subgrades (soaked CBR $\geq$ 3) or shear strength equal to or greater than approximately 90 kPa	2
Stabilization of soft subgrades (soaked 1 < CBR < 3; or shear strength between approximately 30 kPa and 90 kPa)	1
Permanent erosion control, for example geotextiles beneath rock riprap	2 for woven monofilament, 1 for all other geotextiles

Table 4.8 AASHTO M288-00 subsurface filtration (called 'Drainage' in the specification) geotextile property requirements

Property	ASTM test method	Units	Requirements (% in situ soil passing 0.075 mm[1])		
			<15	15 to 50	>50
Strength			Class 2 from Table 4.6[2]		
Permittivity[3,4]	D4491	s^{-1}	0.5	0.2	0.1
Apparent opening size[3,4]	D4751	mm	0.43 max. avg. roll value	0.25 max. avg. roll value	0.22[5] max. avg. roll value
Ultraviolet (UV) stability (retained strength)	D4355	%	50% retained after 500 h of exposure		

Notes
1 Based on grain-size analysis of in situ soil in accordance with AASHTO T88.
2 Default geotextile selection. The engineer may specify a Class 3 geotextile from Table 4.6 for trench drain applications based on one or more of the following:
   a The engineer has found Class 3 geotextiles to have sufficient survivability based on field experience.
   b The engineer has found Class 3 geotextiles to have sufficient survivability based on laboratory testing and visual inspection of a geotextile sample removed from a field test section constructed under anticipated field conditions.
   c Subsurface drain depth is less than 2 m, drain aggregate diameter is less than 30 mm and compaction requirement is less than 95% of AASHTO T99.
3 These default filtration property values are based on the predominant particle sizes of in situ soil. In addition to the default permittivity value, the engineer may require geotextile permeability and/or performance testing based on engineering design for drainage systems in problematic soil environments.
4 Site-specific geotextile design should be performed especially if one or more of the following problematic soil environments are encountered: unstable or highly erodible soils such as non-cohesive silts, gap-graded soils, alternating sand/silt laminated soils, dispersive clays and/or rock flour.
5 For cohesive soils with a plasticity index greater than 7, the geotextile maximum average roll value for apparent opening size is 0.30 mm.

Table 4.9 AASHTO M288-00 separation geotextile property requirements

Property	ASTM test method	Units	Requirements
Strength			Class 2 from Table 4.6[1]
Permittivity	D4491	s^{-1}	0.02[2]
Apparent opening size	D4751	mm	0.60 max. avg. roll value
Ultraviolet (UV) stability (retained strength)	D4355	%	50% retained after 500 h of exposure

Notes

The separation requirements will be applicable to the use of geotextile at the subgrade level if the soaked CBR $\geq$ 3 or shear strength is equal to or greater than 90 KPa. These are appropriate for unsaturated subgrade soils.

1 Default geotextile selection: the engineer may specify a Class 3 geotextile from Table 4.6 based on one or more of the following:
  a The engineer has found Class 3 geotextiles to have sufficient survivability based on field experience.
  b The engineer has found Class 3 geotextiles to have sufficient survivability based on laboratory testing and visual inspection of a geotextile sample removed from a field test section constructed under anticipated field conditions.
  c Aggregate cover thickness of the first lift over the geotextile exceeds 300 mm and the aggregate diameter is less than 50 mm.
  d Aggregate cover thickness of the first lift over the geotextile exceeds 150 mm, aggregate diameter is less than 30 mm and construction equipment pressure is less than 550 kPa.
2 Default value: permittivity of geotextile should be greater than that of the soil ($\psi_g > \psi_s$). The engineer may also require the permeability of the geotextile to be greater than that of the soil ($k_g > k_s$).

Table 4.10 AASHTO M288-00 stabilization geotextile property requirements

Property	ASTM test method	Units	Requirements
Strength			Class 1 from Table 4.6[1]
Permittivity	D4491	s^{-1}	0.05[2]
Apparent opening size	D4751	mm	0.43 max. avg. roll value
Ultraviolet (UV) stability (retained strength)	D4355	%	50% retained after 500 h of exposure

Notes

The stabilization requirements will be applicable to the use of a geotextile layer at the subgrade level to provide the coincident functions of separation, filtration and reinforcement if the subgrade soil is in wet, saturated conditions due to a high groundwater table or due to prolonged periods of wet weather. Stabilization is appropriate if the subgrade soils are having soaked 1 < CBR < 3 or shear strength approximately between 30 kPa and 90 kPa.

1 Default geotextile selection: the engineer may specify a Class 2 or 3 geotextile from Table 4.6 based on one or more of the following:
  a The engineer has found the class of geotextile to have sufficient survivability based on field experience.
  b The engineer has found the class of geotextile to have sufficient survivability based on laboratory testing and visual inspection of a geotextile sample removed from a field test section constructed under anticipated field conditions.
2 Default value: permittivity of geotextile should be greater than that of the soil ($\psi_g > \psi_s$). The engineer may also require the permeability of the geotextile to be greater than that of the soil ($k_g > k_s$).

Table 4.11 AASHTO M288-00 permanent erosion control geotextile property requirements

Property	ASTM test method	Units	Requirements (% in situ soil passing 0.075 mm[1])		
			<15	15 to 50	>50
Strength			For woven monofilament geotextiles, Class 2 from Table 4.6[2]		
			For all other geotextiles, Class 1 from Table 4.6[2,3]		
Permittivity[1,4]	D4491	s^{-1}	0.7	0.2	0.1
Apparent opening size[3,4]	D4751	mm	0.43 max. avg. roll value	0.25 max. avg. roll value	0.22[5] max. avg. roll value
Ultraviolet (UV) stability (retained strength)	D4355	%	50% retained after 500 h of exposure		

Notes

The erosion control requirements will be applicable to the use of a geotextile layer between energy absorbing armour system and the in situ soil to prevent soil loss resulting in excessive scour and to prevent hydraulic uplift pressures causing instability of the permanent erosion control system. This specification does not apply to other types of geosynthetic soil erosion control materials such as the turf reinforcement mats.

1 Based on grain-size analysis of in situ soil in accordance with AASHTO T88.
2 As a general guideline, the default geotextile selection is appropriate for conditions of equal or less severity than either of the following:
  a Armour layer stone weights do not exceed 100 kg, stone drop height is less than 1m, and no aggregate bedding layer is required.
  b Armour layer stone weighs more than 100 kg, stone drop height is less than 1m, and the geotextile is protected by a 150-mm thick aggregate bedding layer designed to be compatible with the armour layer. More severe applications require an assessment of geotextile survivability based on a field trial section and may require a geotextile with higher strength properties.
3 The engineer may specify a Class 2 geotextile from Table 4.6 based on one or more of the following:
  a The engineer has found Class 2 geotextiles to have sufficient survivability based on field experience.
  b The engineer has found Class 2 geotextiles to have sufficient survivability based on laboratory testing and visual inspection of a geotextile sample removed from a field test section constructed under anticipated field conditions.
  c Armour layer stone weights less than 100 kg, stone drop height is less than 1m, and the geotextile is protected by a 150-mm thick aggregate bedding layer designed to be compatible with the armour layer.
  d Armour layer stone weights do not exceed 100 kg and stone is placed with a zero drop height.
4 These default filtration property values are based on the predominant particle sizes of in situ soil. In addition to the default permittivity value, the engineer may require geotextile permeability and/or performance testing based on engineering design for drainage systems in problematic soil environments.
5 a Site-specific geotextile design should be performed especially if one or more of the following problematic soil environments are encountered: unstable or highly erodible soils such as non-cohesive silts, gap-graded soils, alternating sand/silt laminated soils, dispersive clays, and/or rock flour.
  b For cohesive soils with a plasticity index greater than 7, geotextile maximum average roll value for apparent opening size is 0.30 mm.

Table 4.12 AASHTO M288-00 temporary silt fence geotextile property requirements

Property	ASTM test method	Units	Requirements		
			Supported silt fence[1]	Unsupported silt fence	
				Geotextile elongation $\geq 50\%$[2]	Geotextile elongation $<50\%$[2]
Minimum post spacing		m	1.2	1.2	2
Grab strength	D4632	N			
Machine direction			400	550	550
Cross-machine direction			400	450	450
Permittivity[3]	D4491	s^{-1}	0.05	0.05	0.05
Apparent opening size	D4751	mm	0.60 max. avg. roll value	0.60 max. avg. roll value	0.60 max. avg. roll value
Ultraviolet (UV) stability (retained strength)	D4355	%	70% retained after 500 h of exposure		

Notes

These requirements will be applicable to the use of a geotextile as a vertical, permeable interceptor designed to remove suspended soil from overland water flow. In fact, the function of a temporary silt fence is to filter and allow settlement of soil particles from sediment-laden water. The purpose is generally to prevent the eroded soil from being transported off the construction site by water runoff.

1 Silt fence support shall consist of 14-gauge steel wire with a mesh spacing of 150 mm by 150 mm or prefabricated polymeric mesh of equivalent strength.
2 As measured in accordance with ASTM D4632.
3 These default filtration property values are based on empirical evidence with a variety of sediments. For environmentally sensitive areas, a review of previous experience and/or site or regionally specific geotextile tests should be performed by the agency to confirm suitability of these requirements.

Table 4.13 AASHTO M288-00 paving fabric geotextile property requirements[1]

Property	ASTM test methods	Units	Requirements
Grab strength	D4632	N	450
Ultimate elongation	D4632	%	≥ 50
Mass per unit area	D5261	g/m^2	140
Asphalt retention	D6140	l/m^2	Notes 2 and 3
Melting point	D276	°C	150

Notes

These requirements will be applicable to the use of a paving fabric, saturated with asphalt cement, between pavement layers. The function of the paving fabric is to act as a waterproofing and stress-relieving membrane within the pavement structure. This specification is not intended to describe fabric membrane systems specifically designed for pavement joints and localized (spot) repairs.

1 All numeric values represent MARV in the weaker principal direction.
2 Asphalt required to saturate paving fabric only. Asphalt retention must be provided in manufacturer certification. Value does not indicate the asphalt application rate required for construction.
3 Product asphalt retention property must meet the MARV value provided by the manufacturer certification.

Illustrative Example 4.2

Using the AASHTO M288-00 specifications, recommend the properties of a nonwoven geotextile (elongation, $\varepsilon > 50\%$) required for its permanent erosion control application adjacent to a soil with 70% passing the 0.075 mm sieve under harsh installation survivability conditions.

Solution

From Tables 4.11 and 4.6, the recommended properties for the woven geotextile as a separator are the following:

Permittivity ≥ 0.1 s^{-1}
AOS ≤ 0.22 mm
Grab strength ≥ 900 N
Sewn seam strength ≥ 810 N
Tear strength ≥ 350 N
Puncture strength ≥ 350 N
UV stability $\geq$ (50% of 1100 N = 550 N) after 500 h of exposure   **Answer**

Self-evaluation questions

(Select the most appropriate answers to the multiple-choice questions from 1 to 22)

1. Which one of the following geosynthetics can be used as a reinforcement in reinforced soil retaining wall?

   (a) Nonwoven geotextile.
   (b) Woven geotextile.
   (c) Geonet.
   (d) Geomembrane.

2. A geosynthetic-reinforced foundation soil provides

   (a) Improved bearing capacity.
   (b) Reduced settlements.
   (c) Both (a) and (b).
   (d) None of the above.

3. A nonwoven geotextile at the base of an embankment on soft foundation soil

   (a) Acts principally as a reinforcement layer.
   (b) Acts principally as a separator.
   (c) Causes compaction of the ground.
   (d) Accelerates consolidation and subsequent gain in strength.

4. The use of a geosynthetic basal layer is generally attractive, if the ratio between the foundation soil thickness and the embankment base width is

   (a) Less than 0.7.
   (b) Greater than 0.7.
   (c) Extremely high.
   (d) None of the above.

5. The major functions served by the geotextile in unpaved roads are

   (a) Separation and filtration.
   (b) Separation and reinforcement.
   (c) Reinforcement and filtration.
   (d) Filtration and drainage.

6. If the soil subgrade is soft, that is, the CBR of the soil subgrade is low, say its unsoaked value is less than 3.0, then the geotextile layer at the subgrade level in unpaved roads will primarily function as a

   (a) Separator.
   (b) Reinforcement.
   (c) Filter.
   (d) Drainage medium.

7. In case of paved roads, the principal mechanism responsible for the reinforcement function of the geotextile is its

   (a) Shear stress reduction effect.
   (b) Membrane effect.
   (c) Confinement effect.
   (d) Interlocking effect.

8. When properly installed, a geotextile layer beneath the asphalt overlay mainly functions as

   (a) Reinforcement.
   (b) Fluid barrier.
   (c) Cushion.
   (d) Both (b) and (c).

9. The most common paving grade geosynthetics are lightweight needle-punched nonwoven geotextiles with a mass per unit area of

   (a) 60–120 g/m^2.
   (b) 120–200 g/m^2.
   (c) 200–400 g/m^2.
   (d) None of the above.

10. Paving geogrids can function as

    (a) Reflective crack retarder.
    (b) Effective fluid barrier.
    (c) Both (a) and (b).
    (d) None of these.

11. Most railway track specifications call for

    (a) Thin needle-punched nonwoven geotextiles.
    (b) Thin thermally bonded nonwoven geotextiles.
    (c) Thick needle-punched nonwoven geotextiles.
    (d) Thick thermally bonded nonwoven geotextiles.

12. Which one of the following dictates the filtering efficiency of the filter system at the onset of equilibrium conditions?

    (a) Structure of the geotextile filter.
    (b) Structure of the soil immediately adjacent to the geotextile filter.
    (c) Quality of water being filtered.
    (d) None of the above.

13. A drainage geocomposite behind a concrete retaining wall is beneficial mainly because it

    (a) Reduces the plasticity of the backfill.
    (b) Increases the durability of the concrete.
    (c) Improves the compaction requirements for the backfill.
    (d) Reduces the lateral pressures on the wall.

14. Which one of the following will have the highest efficiency in terms of soil loss under rain splash?

    (a) A jute product.
    (b) A synthetic product.
    (c) A geocomposite based on jute and synthetic products.
    (d) None of the above.

15. Within a soil slope, geosynthetic sheets are usually placed in

    (a) Inclined planes towards the slope face.
    (b) Inclined planes away from the slope face.
    (c) Vertical planes.
    (d) Horizontal planes.

16. Which one of the following is the most important component of a landfill?

    (a) A cover system.
    (b) A liner system.
    (c) A leachate collection and removal system.
    (d) A gas collection and control system.

17. The barrier in a liner or cover system of the landfill may consist of

    (a) A compacted clay liner (CCL).
    (b) A geomembrane (GMB) sheet.
    (c) A geosynthetic clay liner (GCL).
    (d) All of the above.

18. In case of sanitary landfills, the thickness of daily soil cover is generally

    (a) 5 cm.
    (b) 15 cm.
    (c) 30 cm.
    (d) None of the above.

19. The main purpose of using geotextiles in canals and rivers is to

    (a) Increase load-bearing capacity.
    (b) Distribute load.

(c) Replace or improve the traditional filters.
(d) Relieve pore water pressures.

20. In the construction and rehabilitation of earth dams, a geotextile filter is never used in
    (a) Downstream drainage blanket.
    (b) Downstream chimney drain.
    (c) Upstream side of the core.
    (d) All of the above.

21. In applications of geotextiles as a separator, their permittivity should be equal to or greater than
    (a) 0.01.
    (b) 0.002.
    (c) 0.02.
    (d) None of the above.

22. The grab tensile strength of a paving fabric geotextile should generally be equal to or greater than
    (a) 150 N.
    (b) 250 N.
    (c) 350 N.
    (d) 450 N.

23. What is the main aim of using geosynthetics in civil engineering projects?
24. What is the fastest growing application area in geosynthetic engineering?
25. What is the expected lifespan of geosynthetics?
26. Describe the basic components of a geosynthetic-reinforced retaining wall.
27. Provide a list of various facing elements. Which one is the most economical?
28. How can you make a geotextile wraparound facing UV-resistant?
29. What do you mean by a basal geosynthetic layer?
30. List the factors that may be of major concern when choosing the basal geosynthetic to function as a reinforcement.
31. Describe the ideal reinforcement pattern below a shallow footing. What are the difficulties in adopting this pattern in real-life projects? Can you suggest an effective practical reinforcement pattern?
32. What are the benefits of using the geotextile layer or layers in the construction of unpaved roads?
33. Make a list of mechanical properties of a geotextile that are of greatest importance when using it as a separator in an unpaved road at the interface of a stone base and relatively soft foundation soil.
34. What is MESL? Explain its uses.
35. What do mean by unpaved age of a paved roadway?
36. Describe the concept of geosynthetic separation in paved roadways.
37. What are the different mechanisms of crack propagation in asphalt overlays? How can geosynthetics be beneficial in preventing such cracks?
38. For geotextiles used to reinforce paved roads on firm soil subgrades, the geotextile must somehow be prestressed. Can you suggest some methods for prestressing geotextiles for such an application?

168 Application areas

39. What are paving fabrics? Woven geotextiles are ineffective paving fabrics. Why?
40. How will you determine the bitumen retention of a paving fabric for effective application?
41. Draw a neat diagram to show the components of a railway track with the placement of the geosynthetic layer(s).
42. List the functions, in order of priority as per your judgement, that act when geotextiles are placed beneath railway track ballast in new railway track construction. Do you feel that the order of priority will change when geotextiles are used in remediation of existing railway tracks?
43. The function of a geosynthetic beneath a railway track is fundamentally different from that beneath a roadway. What are the essential differences to be kept in mind while designing a railway track structure?
44. List the advantages of a geosynthetic filter over the graded granular filter.
45. Why is it recommended to place a layer of aggregates between the geotextile and the riprap?
46. Define the 'filtering efficiency'. Explain its significance.
47. It is a misconception that the geosynthetic filter can replace the function of a granular filter completely. Do you agree with this statement? Justify your answer.
48. Describe the structure of a drainage geocomposite.
49. What is the effect of stiffness of the geosynthetic on the discharge capacity of the geosynthetic drain?
50. What are the essential properties of soil to be determined for the successful use of a geotextile in a filtration application?
51. Define the following phenomena: 'blocking', 'blinding' and 'clogging'.
52. Blinding of the geosynthetic filter is far more detrimental than its clogging. Why?
53. Why would there be a need for strength criteria for geosynthetics used in hydraulic applications?
54. What is the application of a turf reinforcement mat (TRM)?
55. Compare the roles of a geojute and a geotextile in an erosion control application.
56. What is a geocontainer? List its applications.
57. Present a comprehensive review of the various existing short-term and long-term erosion control systems.
58. What are the prime modes of geosynthetic failure in a slope stabilization application?
59. List the findings of the model test carried out by Das *et al.* (1996) on a geogrid-reinforced clay slope.
60. How does an engineered landfill differ from an open dump of wastes?
61. Regarding the siting of a lined landfill, what are the major features to be considered?
62. List the main elements or systems comprising a modern municipal solid waste landfill. Describe their functions.
63. What is the purpose and function of a landfill liner system?
64. What factors will you consider for the selection of a geomembrane liner for ponds, reservoirs and canals?
65. Why are geosynthetics used extensively in earth dam construction and rehabilitation?
66. Draw a cross section of the typical tunnel vault with the general arrangement of the lining system.
67. What are the functions served by a nonwoven geotextile layer installed adjacent to the geomembrane layer in a waterproof tunnel?

68. What are the three geotextile classes that are required for survivability under typical installation conditions for different functions, as recommended by the American Association of State Highway and Transportation Officials (AASHTO)?
69. Using the AASHTO M288-00 specifications, recommend the properties of a nonwoven geotextile (elongation, $\varepsilon < 50\%$) required for its application as a subsurface drain filter adjacent to a soil with 40% passing the 0.075 mm sieve under typical installation survivability conditions.
70. What are the geotextile property requirements as per AASHTO M288-00 specifications for its application as a paving fabric?

Chapter 5

Analysis and design concepts

5.1 Introduction

When a geosynthetic is used in a civil engineering application, it is intended to perform particular function(s) for a minimum expected time period, called the *design life*. Geosynthetics are designed to perform a function, or a combination of functions, within the soil–geosynthetic system. Such functions are expected to be performed over the design life of the soil–geosynthetic system, which is typically less than 5 years for short-term use, around 25 years for temporary use and 50 to 100 years or more for permanent use. The nature of the application system and the consequences of its failure may influence the design life (e.g. 70 years for a wall and 100 years for an abutment). Geosynthetics may have a short-term function although the system is permanent; for example an embankment over a weak foundation may require a geosynthetic reinforcement only while consolidation is occurring and until the weak foundation has gained sufficient strength to support the embankment load. Design life for a soil–geosynthetic system is set by the client or designer and is decided at the design stage.

The primary responsibility of a designer is to design a facility that fulfils the operational requirements of the owner/operator throughout its design life, complies with accepted design practices as per the relevant standards and meets or exceeds the minimum requirements of the permitting agency. The designer should be aware of the possible constructional and maintenance constraints. Also, social conditions, safety requirements and environmental impact may affect the eventual outcome of the design process. Based on these facts and main functional objectives of the given structure, a set of technical requirements should be assessed.

The present chapter deals with the basic concepts of the analytical approach and methods of design process for major civil engineering applications of geosynthetics. For more detailed design process, one should follow the relevant standards, codes of practices and design manuals some of which are mentioned in Appendix B.

5.2 Design methodologies

The design of a structure incorporating geosynthetics aims to ensure its strength, stability and serviceability over its intended life span. There are mainly four design methods for the geosynthetic-related structures or systems. These methods are described as follows:

1 *Design-by-experience*: This method is based on one's past experience or that of other's. This is recommended if the application is not driven by a basic function or has a nonrealistic test method.

2 *Design-by-cost-and-availability*: In this method, the maximum unit price of the geosynthetic is calculated by dividing the funds available by the area to be covered by the geosynthetic. The geosynthetic with the best quality is then selected within this unit price limit according to its availability. Being technically weak, this method is nowadays rarely recommended by the current standards of practice.

3 *Design-by-specification*: This method often consists of a property matrix where common application areas are listed along with minimum (or sometimes maximum) property values. Such a property matrix is usually prepared on the basis of local experiences and field conditions for routine applications by most of the governmental agencies and other large users of geosynthetics. For example, the AASHTO M288-00 specifications, as described in Chapter 4, provides the designer and field quality inspector with a very quick method of evaluating and designing geotextiles for common applications such as filters, separators, stabilizers and erosion control layers.

4 *Design-by-function*: This method is the preferred design approach for geosynthetics. The general approach of this method consists of the following steps:

   a Assessing the particular application, define the primary function of the geosynthetic, which can be reinforcement, separation, filtration, drainage, fluid barrier or protection.
   b Make the inventory of loads and constraints imposed by the application.
   c Define the design life of the geosynthetic.
   d Calculate, estimate or otherwise determine the required functional property of the geosynthetic (e.g. strength, permittivity, transmissivity, etc.) for the primary function.
   e Test for or otherwise obtain the allowable property (available property at the end of the design life) of the geosynthetic, as discussed in Sec. 3.6.
   f Calculate the factor of safety, FS, using Equation (2.1), reproduced as below:

   $$FS = \frac{\text{Allowable (or test) functional property}}{\text{Required (or design) functional property}} \qquad (5.1)$$

   g If this factor of safety is not acceptable, check into geosynthetics with more appropriate properties.
   h If acceptable, check if any other function of the geosynthetic is also critical, and repeat the above steps.
   i If several geosynthetics are found to meet the required factor of safety, select the geosynthetic on the basis of cost–benefit ratio, including the value of available experience and product documentation.

It should be noted that the design-by-function method bears heavily on identifying the primary function to be performed by the geosynthetic. For any given application, there will be one or more basic functions that the geosynthetic will be expected to perform during its design life. Accurate identification of the geosynthetic function as primary function(s) is essential. Hence, a special care is required while identifying the primary function(s).

All geosynthetic designs should begin with a criticality and severity assessment of the project conditions for a particular application. The designer should always keep in mind the geosynthetic failure mechanisms that result in unsatisfactory performance (Table 5.1). The properties of geosynthetics should be selected to protect against excessive reductions in performance under the specific soil and environmental conditions during the whole design

Analysis and design concepts 173

Table 5.1 Geosynthetic failure mechanisms

Function	Failure mode(s)	Possible cause(s)
Reinforcement	Large deformation of the soil-geosynthetic structure	Excessive tensile creep of the geosynthetic
	Reduced tensile resisting force	Excessive stress relaxation of the geosynthetic
Separation/ filtration	Piping of soils through the geosynthetic	Openings in the geosynthetic may be incompatible with retained soil. Openings might have been enlarged as a result of in situ stress or mechanical damage
Filtration	Clogging of the geosynthetic	Permittivity of the geosynthetic might have been reduced as a result of particle buildup on the surface of or within the geosynthetic. Openings might have been compressed as a result of long-term loading
Drainage	Reduced in-plane flow capacity	Excessive compression creep of the geosynthetic
Fluid barrier	Leakage through the geosynthetic	Openings may be available in the geosynthetic as a result of puncture or seam failure
Protection	Reduced resistance to puncture	Excessive compression creep of the geosynthetic

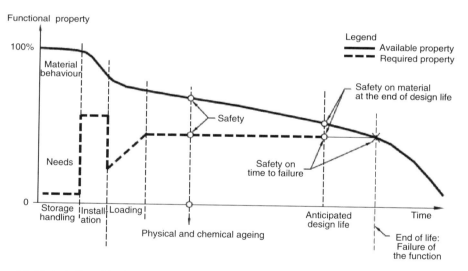

Figure 5.1 Typical allowable (or test) value and required (design) value of a functional property as a function of time (Reprinted, with permission, from HB 154-2002: *Technical Handbook: Geosynthetics – Guidelines on Durability*, copyright Standards Australia International Ltd, Sydney, NSW 2001).

life, as shown in Figure 5.1, and appropriate factors of safety must be utilized in designs incorporating geosynthetics. Note that the factor of safety is likely to decrease with time if geosynthetic properties are subject to degradation with time. Especially for most critical projects, conservative designs are recommended. Because of misconceptions with regard to the functioning of geosynthetics in various constructional and service stages of the project, it is possible that the designer formulates unnecessarily high requirements of geosynthetics. In fact, in most civil engineering applications, simple design rules are sufficient for a proper choice of geosynthetics. However, the designers should be aware of situations where a more sophisticated approach is necessary, and be able to explain to the client that the difference in approach depends on the situation such as type of application, loading conditions and design life.

The design-by-function approach described above is basically the traditional *working stress design approach* that aims to select allowable geosynthetic properties so that a nominated minimum total (or global) factor of safety is achieved. In geosynthetic applications, particularly reinforcement applications (e.g. geosynthetic-reinforced earth retaining walls), it is now common to use the *limit state design approach*, rather than the working stress design involving global safety factors. For the purpose of geosynthetic-reinforced soil design, a limit state is deemed to be reached when one of the following occurs:

1   Collapse, major damage or other similar forms of structural failure;
2   Deformations in excess of acceptable limits;
3   Other forms of distress or minor damage, which would render the structure unsightly, require unforeseen maintenance or shorten the expected life of the structure.

The condition defined in (1) is the ultimate limit state, and (2) and (3) are serviceability limit states. The practice in reinforced soil is to design against the ultimate limit state and check for the serviceability limit state. In reinforced soil design some of these limit states may be evaluated by conventional soil mechanics approaches (e.g. settlement). Margins of safety, against attaining the ultimate limit state, are provided by the use of partial material factors and partial load factors. Limit state design for reinforced soil employs four principal partial factors all of which assume prescribed numerical values of unity or greater. Two of these are load factors applied to dead loads (external dead load $-f_f$ and soil unit weight $-f_{fs}$) and to live loads (f_q). The principal material factor is f_m applied to geosynthetic reinforcement parameters, and f_{ms} applied to soil parameters. The fourth factor f_n is used to take into account the economic ramifications of failure. This factor is employed, in addition to the material factors, to produce a reduced design strength. Note that it is not feasible to uniquely define values for all these partial factors. Prescribed ranges of values are decided to take account of the type of geosynthetic application, the mode of loading and the selected design life. Partial factors are applied in a consistent manner to minimize the risk of attaining a limit state.

In limit state design of geosynthetic reinforcement applications, disturbing forces are increased by multiplying by prescribed load factors to produce design loads, whereas restoring forces (strength test values) are decreased by dividing by prescribed material factors to produce design strengths. There is deemed to be an adequate margin of safety against attaining the ultimate limit state if

Design strength (factored down strength)
$\geq$ Design load (stress due to factored loading)   (5.2)

In the case of drainage application of geosynthetics, this requirement can be expressed as

Design drainage capacity (factored down drainage capacity)
$\geq$ Design flow (factored up expected flow)   (5.3)

For assessing deformations or strains to determine compliance with the appropriate serviceability limit state, the prescribed numerical values of load factors are different from those used in assessing the ultimate limit state and usually assume a value of unity. In assessing magnitudes of total and differential settlements, all partial factors are set to a value of unity, except for those pertaining to the reinforcements (BS 8006-1995). With respect to serviceability limit state, the design requirement for a geosynthetic could be expressed as

$$\text{Allowable elongation} \geq \text{Elongation at serviceability loading} \tag{5.4}$$

In the generalized form, it can be said that the limit state design, considering all possible failure modes and all appropriate partial factors being applied, aims to produce a soil–geosynthetic system that satisfies the following principal equation (HB 154-2002):

$$\text{Design resistance effect} \geq \text{Design action effect} \tag{5.5}$$

for its all design elements.

It should be noted that this equation defines the fundamental principle of limit state design. In the case of internal stability, the design resistance effect may be generated in the soil and in reinforcement, whereas it is generated in the soil only in the case of external stability.

When the safety of man and environment is at great risk because of the failure of the geosynthetic used, or when a reliable method is not available to determine the requirements of the geosynthetic to be used, it becomes necessary to perform suitable practical tests. If the tests are being conducted in the laboratory, special attention must be required to get reliable data to be used for field applications.

The adoption of suitable design and construction method is essential not only to reduce design and construction costs, but also to minimize long-term operation, maintenance and monitoring expenses.

5.3 Retaining walls

The design of geosynthetic-reinforced retaining walls is quite well established. A number of design approaches have been proposed; however, the most commonly used design approach is based on limit equilibrium analysis. The analysis consists of three parts:

1. Internal stability analysis (a.k.a. 'local stability analysis' or 'tieback analysis'): An assumed Rankine failure surface is used, with consideration of possible failure modes of geosynthetic-reinforced soil mass, such as geosynthetic rupture, geosynthetic pullout, connection (and/or facing elements) failure (Fig. 5.2) and excessive geosynthetic creep. The analysis is mainly aimed at determining tension and pullout resistance in the geosynthetic reinforcement, length of reinforcement, and integrity of the facing elements.
2. External stability analysis (a.k.a. 'global stability analysis'): The overall stability of the geosynthetic-reinforced soil mass is checked including sliding, overturning, load-bearing capacity failure, and deep-seated slope failure (Fig. 5.3).
3. Analysis for the facing system, including its attachment to the reinforcement.

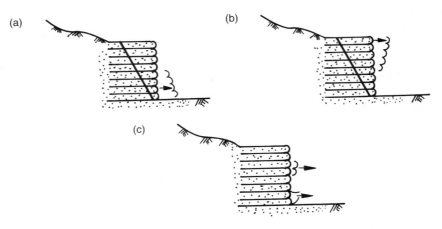

Figure 5.2 Internal failure modes of geosynthetic-reinforced soil retaining walls: (a) geosynthetic rupture; (b) geosynthetic pullout; (c) connection (and/or facing elements) failure.

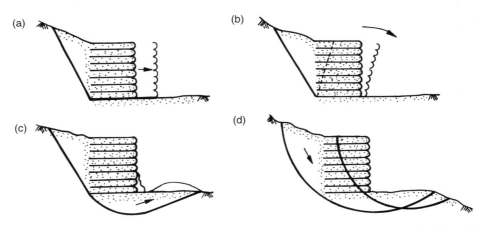

Figure 5.3 External failure modes of geosynthetic-reinforced soil retaining walls: (a) sliding; (b) overturning; (c) load-bearing capacity failure; (d) deep-seated slope failure.

Figure 5.4(a) shows a geotextile-reinforced retaining wall with a geotextile wraparound facing without any surcharge and live load. The backfill is a homogeneous granular soil. According to Rankine active earth pressure theory, the active earth pressure, σ_a, at any depth z is given by:

$$\sigma_a = K_a \sigma_v = K_a \gamma_b z \tag{5.6}$$

where, K_a is the Rankine earth pressure coefficient, γ_b is the unit weight of the granular backfill and

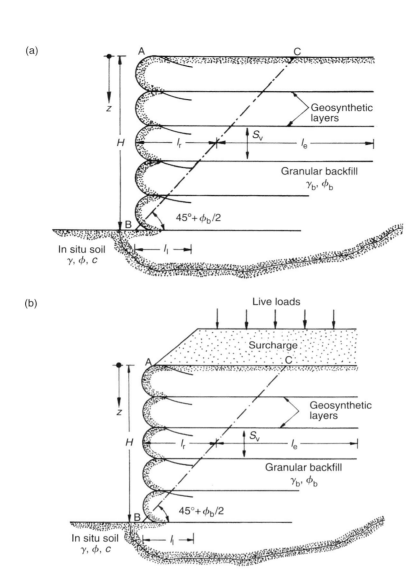
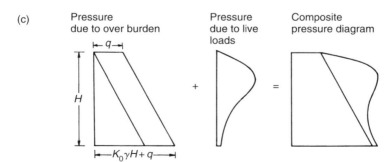

Figure 5.4 (a) Geosynthetic-reinforced retaining wall without surcharge and live load; (b) geosynthetic-reinforced retaining wall with surcharge and live load; (c) lateral earth pressure distribution.

The value of K_a can be estimated from

$$K_a = \tan^2(45° - \frac{\phi_b}{2}), \qquad (5.7)$$

where ϕ_b is the angle of shearing resistance of the granular backfill.

The factor of safety against the geotextile rupture at any depth z may be expressed as

$$FS_{(R)} = \frac{\sigma_G}{\sigma_a S_v}, \qquad (5.8)$$

where σ_G is the allowable geotextile strength in kN/m, and S_v is the vertical spacing of the geotextile layers at any depth z in metre. Since for retaining walls the geosynthetic reinforcement needs to provide stability throughout the life of the structure, the long-term sustained load test data, that is, the creep test data, should be used for design purpose.

The magnitude of the $FS_{(R)}$ is generally taken to be 1.3–1.5. Equation (5.8) can be rearranged as

$$S_v = \frac{\sigma_G}{\sigma_a FS_{(R)}} \qquad (5.9)$$

The geotextile layer at any depth, z, will fail by pullout (a.k.a bond failure) if the frictional resistance developed along its surfaces is less than the force to which it is being subjected. This type of failure occurs when the length of geotextile reinforcement is not sufficient to prevent its slippage with respect to the soil. The effective length, l_e, of a geotextile layer along which the frictional resistance is developed, may be conservatively taken as the length that extends beyond the limits of the Rankine active failure zone (ABC in Fig. 5.4(a)).

The factor of safety against the geosynthetic pullout at any depth z may be expressed as

$$FS_{(P)} = \frac{2l_e \sigma_v \tan \phi_r}{S_v \sigma_a}, \qquad (5.10)$$

where ϕ_r is the angle of shearing resistance of soil–geosynthetic interface and it is approximately equal to $2\phi_b/3$.

The magnitude of the $FS_{(P)}$ is generally taken to be 1.3–1.5. Using Equation (5.6), Equation (5.10) can be rearranged as:

$$l_e = \frac{S_v K_a [FS_{(P)}]}{2 \tan \phi_r}. \qquad (5.11)$$

The length, l_r, of geotextile layer within the Rankine failure zone can be calculated as:

$$l_r = \frac{H - z}{\tan(45° + \phi_b/2)}, \qquad (5.12)$$

where H is the height of the retaining wall.

The total length of the geotextile layer at any depth z is

$$l = l_e + l_r = \frac{S_v K_a [FS_{(P)}]}{2 \tan \phi_r} + \frac{H - z}{\tan(45° + \phi_b/2)}. \qquad (5.13)$$

It should be noted that mixed types of failures, that is, combinations of geotextile rupture and pullout failure, can also occur depending on geometry of the structure, external loads, etc. Usually in the lower parts of the retaining structure, the geotextile reinforcement is destroyed as rupture due to lack of strength and pulled out in upper parts due to insufficient resisting length.

For designing the facing system, it can be assumed that the stress at the face is equal to the maximum horizontal stress in geosynthetic-reinforced backfill. This assumption makes our design conservative because some stress reduction generally occurs near the face. In fact, the maximum stresses are usually located near the potential failure surface and then they decrease in both direction: towards the free end of the geotextile reinforcement and towards the facing. Values of the stress near the facing depend on its flexibility. In the case of rigid facing, the stresses near the facing and those at the potential failure surface do not differ significantly. In the case of flexible facing, the stress near the facing is lower than that at the potential failure surface (Sawicki, 2000). If the wraparound facing is to be provided, then the lap length can be determined using the following expression:

$$l_1 = \frac{S_v K_a [FS_{(P)}]}{4 \tan \phi_r}. \qquad (5.14)$$

The design procedure for geosynthetic-reinforced retaining walls with wraparound vertical face and without any surcharge is given in the following steps:

Step 1: Establish wall height (H).
Step 2: Determine the properties of granular backfill soil, such as unit weight (γ_b) and angle of shearing resistance (ϕ_b).
Step 3: Determine the properties of foundation soil, such as unit weight (γ) and shear strength parameters (c and ϕ).
Step 4: Determine the angle of shearing resistance of the soil–geosynthetic interface (ϕ_r).
Step 5: Estimate the Rankine earth pressure coefficient from Equation (5.7).
Step 6: Select a geotextile that has allowable fabric strength of σ_G.
Step 7: Determine the vertical spacing of the geotextile layers at various levels from Equation (5.9).
Step 8: Determine the length of geotextile layer, l, at various levels from Equation (5.13).
Step 9: Determine the lap length, l_1, at any depth z from Equation (5.14).
Step 10: Check the factors of safety against external stability including sliding, overturning, load-bearing capacity failure and deep-seated slope failure as carried out for conventional retaining wall designs assuming that the geotextile-reinforced soil mass acts as a rigid body in spite of the fact that it is really quite flexible. The minimum values of factors of safety against sliding, overturning, load bearing failure and deep-seated failure are generally taken to be 1.5, 2.0, 2.0 and 1.5, respectively.
Step 11: Check the requirements for backfill drainage and surface runoff control.
Step 12: Check both total and differential settlements of the retaining wall along the wall length. This can be carried out as per the conventional methods of settlement analysis.

For critical structures, especially permanent ones, efforts must be made to consider dead and live load surcharges (Fig. 5.4(b) and (c)) as well as the seismic loading in the design analysis as per the site location and situations. The readers can refer to the chapter contributed by Bathurst *et al.* (2002) in the book edited by Shukla (2002c) for more details on seismic aspects of geosynthetic-reinforced soil walls and slopes. Design factors of safety should be decided on the basis of local codes, if available, and on the basis of experience gained from safe and economical designs. Due to the flexibility of geosynthetic-reinforced retaining walls, the design factors of safety are generally kept lower than that normally used for rigid retaining structures. Unless the foundation is very strong, a minimum embedment depth must be provided as with most foundations.

The level of compaction of backfill influences the facing design. The connection stresses are caused by the settlement of the wall resulting from poor compaction of the backfill near the wall face. They can also be created by heavy compaction near the wall face. Therefore, an optimum level of compaction of granular fill is recommended near the wall face. The facing connection must be designed to resist lateral pressures, gravity forces, and seismic forces along with connection stresses, if there is any possibility of their occurrence.

When geosynthetic-reinforced retaining walls are chosen as a design alternative, readily available on-site soils are often figured into the design from the start. The suitability of soils as a backfill material can be decided on the basis of three key parameters, namely effective angle of shearing resistance, shear strength when compacted and saturated, and frost-heave potential. Table 5.2 provides some guidelines on the suitability of backfill materials using these parameters. It should be noted that as the quality of fill decreases, lower angles of shearing resistance are present, resulting in higher lateral earth pressures and flatter failure surfaces. Consequently, the amount of reinforcement strength and length increase. Fine-grained soils are recommended as a backfill material only when the following four additional design criteria are implemented (Wayne and Han, 1998):

1 Internal drainage must be designed and installed properly.
2 Only soils with low to moderate frost-heave potential should be considered.
3 The internal cohesive shear strength parameter, c, is conservatively ignored for long-term stability analysis.

Table 5.2 Retaining wall backfill (after NCMA, 1997)

Unified soil classification	Effective angle of shearing resistance (degrees)	Shear strength when compacted and saturated	Frost-heave potential	Comments
GW, GP	37–42	Excellent to good	Low	Recommended for backfill
GM, SW, SP	33–40	Excellent to good	Moderate	Recommended for backfill
GC, SM, SC, ML, CL	25–32	Good to fair	Moderate to high	Recommended for backfill with additional criteria
MH, CH, OH, OL	N/A	Poor	High	Generally not recommended for backfill
Pt	N/A	Poor	High	Not recommended for backfill

Analysis and design concepts 181

4   The final design is checked by a qualified geotechnical engineer to ensure that the use of cohesive soils does not result in unacceptable, time-dependent movement of the retaining wall.

ILLUSTRATIVE EXAMPLE   5.1
Consider:

Height of the retaining wall, $H = 8$ m
For the granular backfill
   Unit weight, $\gamma_b = 17$ kN/m^3
   Angle of internal friction, $\phi_b = 35°$
Allowable strength of geotextile, $\sigma_G = 20$ kN/m
Factor of safety against geotextile rupture = 1.5
Factor of safety against geotextile pullout = 1.5

Calculate the length of the geotextile layers, spacing of layers and lap lengths at depth $z = 2$ m, 4 m, and 8 m.

SOLUTION
From Equation (5.7), the Rankine earth pressure coefficient is

$$K_a = \tan^2(45° - \frac{35°}{2}) = 0.27.$$

At $z = 2$ m,
From Equation (5.9),

$$S_v = \frac{\sigma_G}{\sigma_a FS_{(R)}} = \frac{20}{17 \times 2 \times 0.27 \times 1.5} = 1.45 \text{ m}. \quad\quad \text{Answer}$$

From Equation (5.13),

$$l = \frac{S_v K_a [FS_{(P)}]}{2 \tan \phi_r} + \frac{H - z}{\tan(45° + \phi_b/2)} = \frac{1.45 \times 0.27 \times 1.5}{2 \times \tan(\frac{2}{3} \times 35°)} + \frac{8 - 2}{\tan(45° + 35°/2)}$$

$$= 0.68 \text{ m} + 3.12 \text{ m} = 3.80 \text{ m}. \quad\quad \text{Answer}$$

From Equation (5.14),

$$l_l = \frac{S_v K_a [FS_{(P)}]}{4 \tan \phi_r} = \frac{1.45 \times 0.27 \times 1.5}{4 \times \tan(\frac{2}{3} \times 35°)} = 0.34 \text{ m}. \quad\quad \text{Answer}$$

At $z = 4$ m,
From Equation (5.9),

$$S_v = \frac{\sigma_G}{\sigma_a FS_{(R)}} = \frac{20}{17 \times 4 \times 0.27 \times 1.5} = 0.73 \text{ m}. \quad\quad \text{Answer}$$

182 Analysis and design concepts

From Equation (5.13),

$$l = \frac{S_v K_a [FS_{(P)}]}{2 \tan \phi_r} + \frac{H-z}{\tan(45° + \phi_b/2)} = \frac{0.73 \times 0.27 \times 1.5}{2 \times \tan(\frac{2}{3} \times 35°)} + \frac{8-4}{\tan(45° + 35°/2)}$$

$$= 0.34 \text{ m} + 2.08 \text{ m} = 2.42 \text{ m}. \qquad \textbf{Answer}$$

From Equation (5.14),

$$l_1 = \frac{S_v K_a [FS_{(P)}]}{4 \tan \phi_r} = \frac{0.73 \times 0.27 \times 1.5}{4 \times \tan(\frac{2}{3} \times 35°)} = 0.17 \text{ m}. \qquad \textbf{Answer}$$

At $z = 8$ m,
From Equation (5.9),

$$S_v = \frac{\sigma_G}{\sigma_a FS_{(R)}} = \frac{20}{17 \times 8 \times 0.27 \times 1.5} = 0.36 \text{ m}. \qquad \textbf{Answer}$$

From Equation (5.13),

$$l = \frac{S_v K_a [FS_{(P)}]}{2 \tan \phi_r} + \frac{H-z}{\tan(45° + \phi_b/2)} = \frac{0.36 \times 0.27 \times 1.5}{2 \times \tan(\frac{2}{3} \times 35°)} + \frac{8-8}{\tan(45° + 35°/2)}$$

$$= 0.17 \text{ m} + 0 \text{ m} = 0.17 \text{ m}. \qquad \textbf{Answer}$$

From Equation (5.14),

$$l_1 = \frac{S_v K_a [FS_{(P)}]}{4 \tan \phi_r} = \frac{0.36 \times 0.27 \times 1.5}{4 \times \tan(\frac{2}{3} \times 35°)} = 0.08 \text{ m}. \qquad \textbf{Answer}$$

Note: Keeping field aspects and construction simplicity in view, one can use $S_v = 0.5$ m, $l = 5$ m, $l_1 = 1$ m for $z \leq 4$ m, and $S_v = 0.3$ m, $l = 2.5$ m, $l_1 = 1$ m for $z > 4$ m.

It is to be noted that the typical reinforcement spacing for geotextile-wrapped walls varies between 0.2 and 0.5 m. For spacings greater than 0.6 m, unless the wall has a rigid face, intermediate geotextile layers will be required to prevent excessive bulging of the wall face between the geotextile layers.

5.4 Embankments

The basic design approach for an embankment over the soft foundation soil with a basal geosynthetic layer is to design against the mode (or mechanism) of failure. The potential failure modes are as follows:

1. Overall slope stability failure (Fig. 5.5(a)).
2. Lateral spreading (Fig. 5.5(b)).

3   Embankment settlement (Fig. 5.5(c)).
4   Overall bearing failure (Fig. 5.5(d)).
5   Pullout failure (Fig. 5.5(e)).

These failure modes indicate the types of analysis that are required. In fact, each failure mode generates required or design value for the embankment geometry or the tensile strength of the geosynthetic. The conventional geotechnical design procedures, based on the total stress approach, are generally utilized with a modification for the presence of the geosynthetic layer.

1 Overall slope stability failure   This is the most commonly considered failure mechanism, where the failure mechanism is characterized by a well-defined failure surface cutting the embankment fill, the geosynthetic layer and the soft foundation soil (Fig. 5.5(a)). This mechanism can involve tensile failure of the geosynthetic layer or bond failure due to insufficient anchorage of the geosynthetic extremity beyond the failure surface. The analysis proceeds along the steps of conventional slope stability analysis with the geosynthetic providing an additional stabilizing force, T, at the point of intersection with the failure surface being the geosynthetic thus provides the additional resisting moment required for minimum required factor of safety. Figure 5.6 shows such a conventional stability model, usually preferred for preliminary routine analyses. Opinions differ on the calculation of the resisting (or stabilizing) moment, ΔM_g, due to the tension in the geosynthetic layer: $\Delta M_g = T R$ or $\Delta M_g = T y$ or any other value, where R is the radius of critical slip arc, and y is the moment arm of the geosynthetic layer. For circular failure arcs and horizontal geosynthetic layers, it is conservative to take $\Delta M_g = T y$ and to neglect any other possible effects on soil stresses. This approach

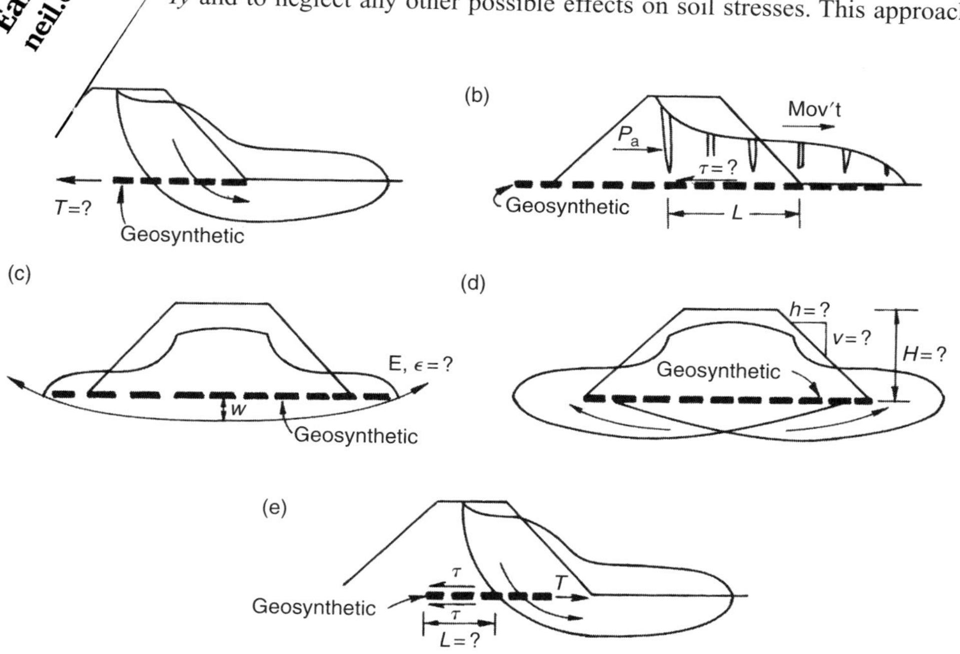

Figure 5.5 Design models for the embankments with a basal geosynthetic layer over soft foundation soils (after Fowler and Koerner, 1987).

184 Analysis and design concepts

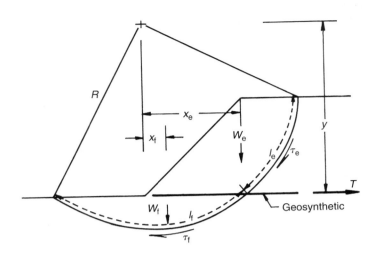

Figure 5.6 Overall slope stability analysis.

is conservative because it neglects any possible geosynthetic reinforcement along the alignment of the failure surface, as well as any confining effect of the geosynthetic. The factor of safety against the overall stability failure is then given as follows:

$$\text{FS} = \frac{\text{Resisting moment}}{\text{Driving moment}} = \frac{\tau_e l_e R + \tau_f l_f R + Ty}{W_e x_e + W_f x_f}, \quad (5.15)$$

where τ_e, τ_f are the shear strengths of embankment and foundation materials, respectively; l_e, l_f are the arc lengths within embankment and foundation soil, respectively; W_e, W_f are the weights of soil masses within the embankment and foundation soil, respectively; and x_e, x_f are the moment arms of W_e and W_f, respectively, to their centres of gravity.

Trials are made to find the critical failure arc for which the necessary force in the geosynthetic is maximum. Usually, a target value for the safety factor is established and the maximum necessary force is determined by searching the critical failure arc. It is to be noted that different methods of analysis or forms of definition of the safety factor may affect the result obtained. The design must consider the fact that a geosynthetic can resist creep if the working loads are kept well below the ultimate tensile strength of the geosynthetic. The recommended working load should not exceed 25% of the ultimate load for polyethylene (PE) geosynthetics, 40% of the ultimate load for polypropylene (PP) geosynthetics and 50% of the ultimate load for polyester (PET) geosynthetics.

2 *Lateral spreading* The presence of a tension crack through the embankment isolates a block of soil, which can slide outward on the geosynthetic layer (Fig. 5.5(b)). The horizontal earth pressures acting within the embankment mainly cause the lateral spreading. In fact, the horizontal earth pressures cause the horizontal shear stresses at the base of the embankment, which must be resisted by the foundation soil. If the foundation soil does not have adequate shear resistance, it can result in failure. The lateral spreading can therefore be prevented if the restraint provided by the frictional bond between the embankment and the

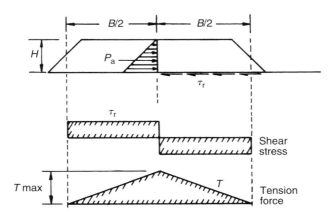

Figure 5.7 Block sliding analysis.

geosynthetic exceeds the driving force resulting from active soil pressures (and/or hydrostatic pressures in the case of water-filled cracks) within the embankment. For the conditions as sketched in Figure 5.7, the resultant active earth pressure P_a and the corresponding maximum tensile force T_{max} are calculated as follows:

$$P_a = \frac{1}{2}\gamma H^2 K_a \tag{5.16}$$

$$T_{max} = \frac{\tau_r B}{2} = \frac{(\gamma H \tan \phi_r) B}{2}, \tag{5.17}$$

where γ is the unit weight of the embankment material; H is the embankment height; B is the embankment base width as shown in the figure; K_a is the active earth pressure coefficient for the soil; τ_r is the resisting shear stress; and ϕ_r is the soil–geosynthetic interface shear resistance angle.

For no lateral spreading, one can get

$$\frac{T_{max}}{P_a} \geq 1 \tag{5.18}$$

or,

$$\tan \phi_r \geq \frac{H K_a}{B}. \tag{5.19}$$

It is general practice to consider a minimum safety factor of 1.5 with respect to strength and a geosynthetic strain limited to 10%. The required geosynthetic strength T_{req} and modulus E_{req} therefore are

$$T_{req} = 1.5 T_{max} \tag{5.20}$$

186 Analysis and design concepts

$$E_{req} = \frac{T_{max}}{\varepsilon_{max}} = 10 T_{max}. \tag{5.21}$$

The lateral spreading failure mechanism becomes important only for steep embankment slopes on reasonably strong subgrades and very smooth geosynthetic surfaces. Thus, it is not the most critical failure mechanism for soft foundation soils.

ILLUSTRATIVE EXAMPLE 5.2
A 4 m high and 10 m wide embankment is to be built on soft ground with a basal geotextile layer. Calculate the geotextile strength and modulus required in order to prevent block sliding on the geotextile. Assume that the embankment material has a unit weight of 18 kN/m³ and angle of shearing resistance of 35° and that the geotextile–soil interface angle of shearing resistance is two-thirds of that value.

SOLUTION
From Equation (5.17),

$$T_{max} = \frac{(\gamma H \tan \phi_r) B}{2} = \frac{18 \times 4 \times [\tan(\tfrac{2}{3} \times 35°)] \times 10}{2} = 155.29 \text{ kN/m}.$$

From Equation (5.20),

$$T_{req} = 1.5 T_{max} = 1.5 \times 155.29 = 232.94 \text{ kN/m}. \qquad \textbf{Answer}$$

From Equation (5.21),

$$E_{req} = 10 T_{max} = 10 \times 155.29 = 1552.9 \text{ kN/m}. \qquad \textbf{Answer}$$

3 *Embankment settlement* The embankment settlement takes place because of the consolidation of the foundation soil (Fig 5.5(c)). The settlement can also occur due to the expulsion of the foundation soil laterally. This mechanism may occur for heavily reinforced embankments on thin soft foundation soil layers (Fig. 5.8). The factor of safety against soil expulsion, F_e, can be estimated from (Palmeira, 2002):

$$F_e = \frac{P_p + R_B + R_T}{P_A}, \tag{5.22}$$

where P_p is the passive reaction force against block movement, R_T is the force at the top of the soil block, R_B is the force at the base of the soil block, and P_A is the active thrust on the soil block.

The active and passive forces can be evaluated by earth pressure theories, while the forces at the base and top of the soil block can be estimated as a function of the undrained strength

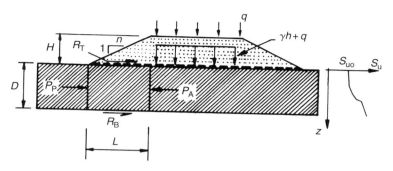

Figure 5.8 Embankment settlement due to lateral expulsion of foundation soil.

S_u at the bottom of the foundation soil and adherence between the reinforcement layer and the surface of the foundation soil, respectively.

The geosynthetic layer may reduce differential settlement of the embankment somewhat, but little reduction of the magnitude of its total final settlement can be expected, since the compressibility of the foundation soils is not altered by the geosynthetic, although the stress distribution may be somewhat different. The embankment settlement can result in excessive elongation of the geosynthetic. However, it is general practice to limit the total strain in the geosynthetic to 10% in order to minimize settlements within the embankment. Therefore, the modulus of the geosynthetic to be selected should be 10 T_{req}, where T_{req} is based on the overall stability calculation. For getting this benefit significantly from the geosynthetic layer, its edges must be folded back similar to 'wraparound' in retaining walls or anchored in trenches properly or weighted down by berms. Prestressing the geosynthetic in field, if possible, along with the edge anchorage can further reduce both the total and the differential settlements within the embankment (Shukla and Chandra, 1996a).

4 *Overall bearing failure* The bearing capacity of an embankment foundation soil is essentially unaffected by the presence of a geosynthetic layer within or just below the embankment (Fig. 5.5(d)). Therefore, if the foundation soil cannot support the weight of the embankment, then the embankment cannot be built. Overall bearing capacity can only be improved if a mattress like reinforced surface layer of larger extent than the base of the embankment will be provided. The overall bearing failure is usually analysed using classical soil mechanics bearing capacity methods. These analyses may not be appropriate if the soft foundation soil is of limited depth, that is, its depth is small compared to the width of the embankment. In such a situation, a lateral squeeze analysis should be performed. This analysis compares the shear forces developed under the embankment with the shear strength of the corresponding soil. The overall bearing failure check helps in knowing the height of the embankment as well as the side-slope angles that can be adopted on a given foundation soil. Construction of an embankment higher than the estimated value would require using staged construction that allows the underlying soft soils time to consolidate and gain strength.

5 *Pullout failure* Forces transferred to the geosynthetic layer to resist a deep-seated circular failure, that is, the overall stability failure must be transferred to the soil behind the

slip zone as shown in Figure 5.5(e). The pullout capacity of a geosynthetic is a function of its embedment length behind the slip zone. The minimum embedment length, L, can be calculated as follows:

$$L = \frac{T_a}{2(c_a + \sigma_v \tan \phi_r)} \quad (5.23)$$

where, T_a is the force mobilized in the geosynthetic per unit length; c_a is the adhesion of soil to geosynthetic; σ_v is the average vertical stress; and ϕ_r is the shear resistance angle of soil–geosynthetic interface.

If the high strength geosynthetic is used, then embedment length required is typically very large. However, in confined construction areas, this length can be reduced by folding back the edges of the geosynthetic similar to 'wraparound' in retaining walls or anchored in trenches properly or weighted down by berms.

The design procedure for embankment with basal geosynthetic layer(s) is given in the following steps:

Step 1: Define geometrical dimensions of the embankment (embankment height, H; width of crest, b; side slope, vertical to horizontal as 1:n)

Step 2: Define loading conditions (surcharge, traffic load, dynamic load). If there is possibility of frost action, swelling and shrinkage, and erosion and scour, then loading caused by these processes must be considered in the design.

Step 3: Determine the engineering properties of the foundation soil (shear strength parameters, consolidation parameters). Chemical and biological factors that may deteriorate the geosynthetic must be determined.

Step 4: Determine the engineering properties of embankment fill materials (compaction characteristics, shear strength parameters, biological and chemical factors that may deteriorate the geosynthetic). The first few lifts of fill material just above the geosynthetic layer should be free draining granular materials. This requirement provides the best frictional interaction between the geosynthetic and fill, as well as providing a drainage layer for excess pore water to dissipate from the underlying soils.

Step 5: Establish geosynthetic properties (strength and modulus, soil–geosynthetic friction). Also establish tolerable geosynthetic deformation requirements. The geosynthetic strain can be allowed up to 2–10%. The selection of geosynthetic should also consider drainage, constructability (survivability) and environmental requirements.

Step 6: Check against the modes of failure, as described earlier. If the factors of safety are sufficient, then the design is satisfactory, otherwise the steps should be repeated by making appropriate changes, wherever possible.

Since for reinforced surcharge/areal fills, used as parking lots, storage yards, and construction pads, applied loads are close to axisymmetric, the design strengths and strain considerations are generally the same in all directions. The analysis for geosynthetic requirements remains the same as those discussed above. However, special seaming techniques must often be considered to meet the required strength requirements. It is to be noted that since for embankments over soft soils the geosynthetic reinforcement is needed only during construction and foundation consolidation, a short-term constant rate of strain tensile test can be used for the design purpose.

5.5 Shallow foundations

The basic design approach for geosynthetic-reinforced foundation soils must consider their modes (or mechanisms) of failure. The possible potential failure modes are as follows:

1. *Bearing capacity failure of soil above the uppermost geosynthetic layer* (Fig. 5.9(a)): This type of failure appears likely to occur if the depth of the uppermost layer of reinforcement (u) is greater than about 2/3 of the width of footing (B), that is, $u/B > 0.67$, and if the reinforcement concentration in this layer is sufficiently large to form an effective lower boundary into which the shear zone will not penetrate. This class of bearing capacity problems corresponds to the bearing capacity of a footing on the shallow soil bed overlying a strong rigid boundary.
2. *Pullout of geosynthetic layer* (Fig. 5.9(b)): This type of failure is likely to occur for shallow and light reinforcement ($u/B < 0.67$ and number of reinforcement layers, $N < 3$).
3. *Breaking of geosynthetic layer* (Fig. 5.9(c)): This type of failure is likely to occur with long, shallow, and heavy reinforcement ($u/B < 0.67$, $N > 3$ or 4). The reinforcement layers always break approximately under the edge or towards the centre of the footing. The uppermost layer is most likely to break first, followed by the next deep layer and so forth.
4. *Creep failure of geosynthetic layer* (Fig. 5.9(d)): This failure may occur due to long-term settlement caused by sustained surface loads and subsequent geosynthetic stress relaxation.

The first three modes of failure were first reported, by Binquet and Lee (1975a) in the case of footing resting on sand reinforced by metallic reinforcement on the basis of observations made during laboratory model tests (Binquet and Lee, 1975b). The fourth mode of failure, that is, creep failure, was discussed by Shukla (2002c) and Koerner (2005).

A large number of studies have been carried to evaluate the beneficial effects of reinforcing the soils with geosynthetics as related to the load-carrying capacity and the settlement characteristics of shallow foundations (Shukla, 2002b). They all point to the conclusion that the geosynthetic reinforcement increases the load-carrying capacity of the

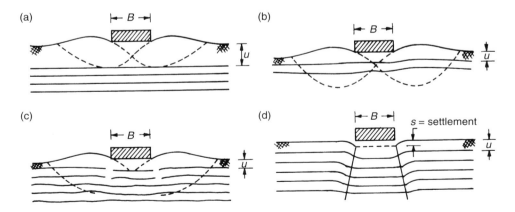

Figure 5.9 Possible modes of failure of geosynthetic-reinforced shallow foundations: (a) bearing capacity failure of soil above the uppermost geosynthetic layer; (b) pullout of the geosynthetic layer; (c) breaking of the geosynthetic layer; (d) creep failure of geosynthetic layer (after Binquet and Lee, 1975b, Shukla, 2002c, Koerner, 2005).

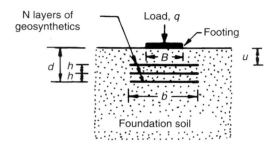

Figure 5.10 Geometrical parameters of the geosynthetic-reinforced foundation soil.

foundation soil and reduces the depth of the granular fill for the same settlement level. Through the laboratory model tests, Guido *et al.* (1985) and many other research workers have studied the parameters affecting the load-bearing capacity of a geosynthetic-reinforced foundation soil. All these parameters can be summarized as follows (Fig. 5.10):

- The width of footing, B
- Strength of foundation soil, τ_s
- The depth below footing of the first geosynthetic layer, u
- The number of geosynthetic layers, N
- The vertical spacing of the geosynthetic layers, h
- The width of geosynthetic layers, b
- The tensile strength of geosynthetic, σ_G.

The parameters, u, N, and h cannot be considered separately, as they are dependent on each other. It has been reported that more than three geosynthetic layers are not beneficial, and the optimum size of the geosynthetic layer is about three times the width of the footing, B. For beneficial effects, the geosynthetic layers should be laid within a depth equal to the width of footing. The optimum vertical spacing of the geosynthetic reinforcement layers is between $0.2B$ and $0.4B$. For a single layer reinforced soil, the optimum embedment depth is approximately 0.3 times the footing width.

For expressing the improvement conveniently, as well as for comparing the test data from studies, *bearing capacity ratio* (BCR), a term introduced by Binquet and Lee (1975a,b), is commonly used. This term is defined as follows:

$$\mathrm{BCR} = \frac{q_{(R)}}{q_u}, \tag{5.24}$$

where q_u is the ultimate load-bearing capacity of the unreinforced soil, and $q_{(R)}$ is the load-bearing capacity of the geosynthetic-reinforced soil at a settlement corresponding to the settlement s_u at the ultimate load-bearing capacity q_u for the unreinforced soil (Fig. 5.11).

Several workers carried out load-bearing capacity analysis considering limited roles of geosynthetics in improving load-bearing capacity and taking different sets of assumptions. For more details, the readers can refer to the book edited by Shukla (2002c).

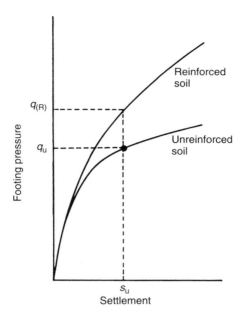

Figure 5.11 Typical load-settlement curves for a soil with and without reinforcement.

Among the reinforcement practices for buildings, roads and embankments constructed on soft ground; the use of a geocell foundation mattress is a unique method, in which the mattress is placed upon the soft foundation soil of insufficient bearing capacity so as to withstand the weight of the superstructure. The geocell foundation mattress is a honey-combed structure formed from a series of interlocking cells (Fig. 5.12). These cells are fabricated directly on the soft foundation soil using uniaxial-polymer geogrids in a vertical orientation connected to a biaxial base grid and then filled with granular material resulting in a structure usually 1 m deep. This arrangement forms not only a stiff platform, which provides a working area for the workers to push forward the construction of the geocell itself, and subsequent structural load, but also a drainage blanket to assist the consolidation of the underlying soft foundation soil. The incorporation of a geocell foundation mattress provides a relatively stiff foundation to the structure and this maximizes the bearing capacity of the underlying weak soil layer. The geocell mattress is self-contained and, unlike constructions with horizontal layers of geotextiles, it needs no external anchorage beyond the base of main structure. As a consequence of the flexible interaction with the supporting foundation soil underneath, even locally or unevenly applied vertical load propagates within the mattress and is transmitted widely to the supporting foundation soil.

Ochiai *et al*. (1994) described a conventional approach for the assessment of the improvement of the bearing capacity due to the placement of the geogrid-mattress foundation as described above. In this approach, a vertical load of intensity p and width B, applied on the mattress, is transmitted widely to the supporting foundation soil with the corresponding intensity p_m and width B_m (Fig. 5.13). The ultimate bearing capacity q_u without the use of

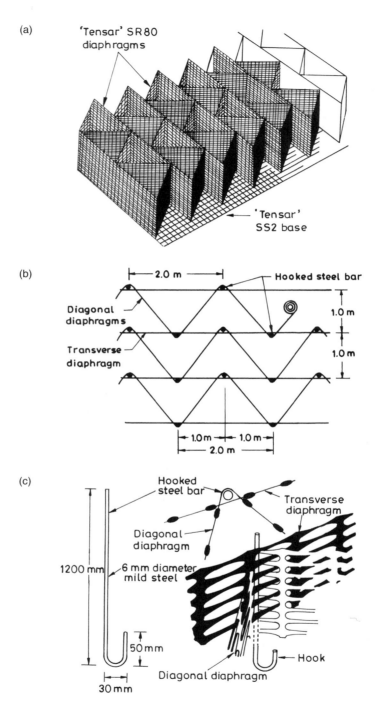

Figure 5.12 (a) Geocell mattress configuration; (b) plan view of geocell mattress; (c) connection details (after Bush et al., 1990).

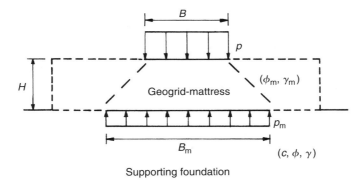

Figure 5.13 Load-bearing capacity analysis of geogrid mattress foundation (after Ochiai et al., 1994).

the mattress may be given by Terzaghi's equation as follows:

$$q_u = cN_c + \frac{1}{2}\gamma B N_\gamma, \tag{5.25}$$

where c is the cohesion, γ is the unit weight of the supporting foundation soil, and N_c and N_γ are bearing capacity factors. On the other hand, the ultimate bearing capacity q_m with the use of mattress may be given as follows; assuming that the placement of the geogrid mattress has a surcharge effect on the bearing capacity of the supporting foundation:

$$q_m = cN_c + \gamma_m H N_q + \frac{1}{2}\gamma B_m N_\gamma, \tag{5.26}$$

where γ_m is the unit weight of the mattress, H is the thickness of the mattress, and N_q is the bearing capacity factor. Therefore, the increase in the bearing capacity Δq due to the placement of the mattress can be given as follows:

$$\Delta q = \gamma_m H N_q + \frac{1}{2}\gamma(B_m - B)N_\gamma. \tag{5.27}$$

It is therefore found that the evaluation of the bearing capacity improvement requires the estimation of the width B_m. The experimental studies have revealed that the width of supporting foundation soil over which the vertical stress is distributed becomes larger as the thickness of geogrid-mattress becomes greater, and as the vertical stiffness of the supporting foundation soil becomes lower. It was suggested, from the design point of view, that the width of the geogrid-mattress should be at least large enough to accommodate the vertical stress distribution, which takes place under the mattress.

In addition to the load-bearing capacity analysis of geosynthetic-reinforced foundation soil, the designer must carry out the settlement analysis for the structural and functional safety of the structures resting on geosynthetic-reinforced foundation soils. One can make

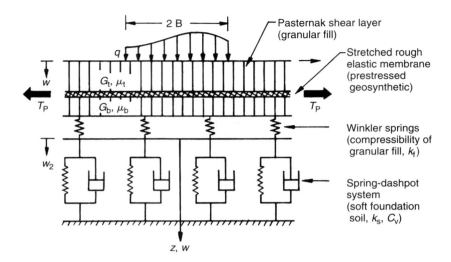

Figure 5.14 Mechanical foundation model (after Shukla and Chandra, 1994a).

such an analysis using the mechanical foundation model presented by Shukla and Chandra (1994a) (see Fig. 5.14). This model allows the study of time-dependent settlement behaviour of the geosynthetic-reinforced granular fill–soft soil system. In this model, the geosynthetic reinforcement and the granular fill are represented by the stretched rough elastic membrane and the Pasternak shear layer, respectively. The general assumptions are that the geosynthetic reinforcement is linearly elastic, rough enough to prevent slippage at the soil interface and has no shear resistance. A perfectly-rigid plastic friction model is adopted to represent the behaviour of the soil–geosynthetic interface in shear. The compressibility of the granular fill is represented by a layer of Winkler springs attached to the bottom of the Pasternak shear layer. The saturated soft foundation soil is idealized by the Terzaghi's spring-dashpot system. The spring represents the soil skeleton and the dashpot simulates the dissipation of the excess pore water pressure. The spring constant is assumed to have a constant value with depth of the foundation soil and also with time. Yin (1997a, b) further improved the mechanical foundation model by incorporating a nonlinear constitutive model for the granular fill and a nonlinear spring model for the soft soil. In the process of developing simple foundation models, Shukla and Yin (2003) suggested a model based on Timoshenko beam concept for time-dependent settlement analysis of a geosynthetic-reinforced granular fill–soft soil system when the granular fill is relatively dense.

The equations governing the response of the mechanical foundation model (Shukla, 1994a) are as follows:

$$q = \overline{X_1} \frac{k_f k_s w}{k_s + k_f U} - \{G_t H_t + \overline{X_2}(T_p + T)\cos\theta + \overline{X_1} G_b H_b\} \frac{\partial^2 w}{\partial x^2} \quad (5.28)$$

$$\frac{\partial T}{\partial x} = -\overline{X_3}\left(q + G_t H_t \frac{\partial^2 w}{\partial x^2}\right) - \overline{X_4}\left(\frac{k_f k_s w}{k_s + k_f U} - G_b H_b \frac{\partial^2 w}{\partial x^2}\right) \quad (5.29)$$

where

$$\overline{X_1} = \frac{1 + K_{0R}\tan^2\theta - (1 - K_{0R})\mu_b \tan\theta}{1 + K_{0R}\tan^2\theta + (1 - K_{0R})\mu_t \tan\theta} \tag{5.30a}$$

$$\overline{X_2} = \frac{1}{1 + K_{0R}\tan^2\theta + (1 - K_{0R})\mu_t \tan\theta} \tag{5.30b}$$

$$\overline{X_3} = \mu_t \cos\theta (1 + K_{0R}\tan^2\theta) - (1 - K_{0R}) \sin\theta \tag{5.30c}$$

$$\overline{X_4} = \mu_b \cos\theta (1 + K_{0R}\tan^2\theta) + (1 - K_{0R}) \sin\theta \tag{5.30d}$$

Note that q is the applied load intensity; $w(x, t)$ is the vertical surface displacement; $T(x, t)$ is the tensile force per unit length mobilized in the membrane; T_p is the pretension per unit length applied to the membrane; G_t and H_t are the shear modulus and thickness of the upper shear layer respectively; G_b and H_b are the shear modulus and thickness of the lower shear layer, respectively; μ_t and μ_b are the interface friction coefficients at the top and bottom faces of the membrane; k_f is the modulus of subgrade reaction of the granular fill; k_s is the modulus of subgrade reaction of the soft foundation soil; K_{0R} is the coefficient of lateral stress at rest at an overconsolidation ratio (R), which is defined here as the ratio of the maximum stress, to which the granular fill is subjected through compaction, to the existing stress under the working load; θ is the slope of the membrane; U is the degree of consolidation of soft foundation soil; C_v is the coefficient of consolidation; x is the distance measured from the centre of the loaded region along the x-axis; B is the half width of loading; and t is any particular instant of time measured from the instant of loading. It should be noted that Equations (5.28), (5.29) and (5.30) governing the model response are applicable for plane strain loading conditions. For axiymmetric loading conditions, the readers can refer to the work of Shukla and Chandra (1998).

The parameters of mechanical foundation models can be determined as per the guidelines suggested by Selvadurai (1979), and Shukla and Chandra (1996b). The parametric studies carried by Shukla and Chandra (1994a) show the effects of various parameters on the settlement response of geosynthetic-reinforced granular fill–soft soil system. Figure 5.15(a) shows the settlement profiles for a typical set of parameters at various stages of consolidation of soft foundation soil. The trend of results obtained using the above generalized model is in good agreement with other reported works.

It is now well established that geosynthetics, particularly geotextiles, show their beneficial effects only after relatively large settlements, which may not be a desirable feature for many structures resting on geosynthetic-reinforced foundation soils. Hence there is a need for a technique, which can make geosynthetics more beneficial without the occurrence of large settlements. Prestressing the geosynthetics can be one of the techniques to achieve this goal. The study, carried out by Shukla and Chandra (1994b), has shown that an improvement in the settlement response increases with an increase in prestress in the geosynthetic reinforcement within the loaded footing and is most significant at the centre of the loaded footing with a reduction in differential settlement (Fig. 5.15(b)).

It should be noted that the compaction level of the granular fill also affects the settlement behaviour of the geosynthetic-reinforced soil. For reduced settlements, a higher degree of compaction is always desirable; however, beneficial effects of the geosynthetic layer decrease with increase in the degree of compaction of the granular fill (Shukla and Chandra, 1994c; Shukla and Chandra, 1997).

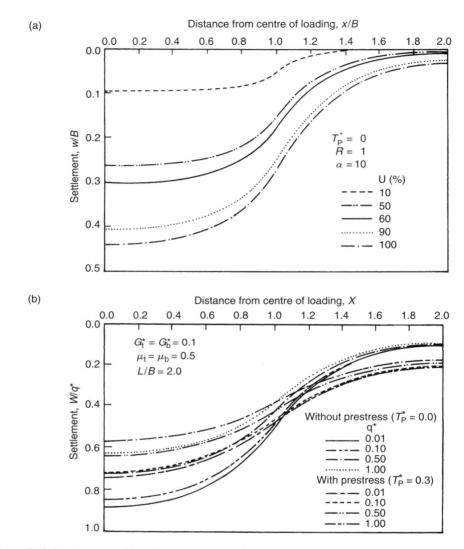

Figure 5.15 Settlement profiles of geosynthetic-reinforced granular fill–soft soil system: (a) for various degrees of consolidation of soft saturated foundation soil; (b) effect of prestressing for various load intensities (after Shukla and Chandra, 1994a,b).

Note
[$X = x/B$; $W = w/B$; $G_t^* = G_t H_t/k_s B^2$; $G_b^* = G_b H_b/k_s B^2$; $q^* = q/k_s B$; $T_p^* = T_p/k_s B^2$; $\alpha = k_f/k_s$].

5.6 Roads

5.6.1 Unpaved roads

A few design methods are available for unpaved road constructions with geosynthetics. Research work is still continuing for the development of new design methods and for the improvement of these in the existing ones. Some of the manufacturers have developed their own unpaved road design charts for use with their particular geosynthetics. All these design charts recommend greater savings of granular material, required in construction, as the soil subgrade becomes softer, showing logical results. A design method based on the specific, well-defined geosynthetic property, such as geosynthetic modulus, is generally acceptable by all. Such a design method is described as a *reinforcement function design method*.

Reinforcement function design method (RFDM)

Giroud and Noiray (1981) presented a design method for geotextile-reinforced unpaved roads by combining the quasi-static analysis and the empirical formula. This method evaluates the risk of failure of the foundation soil and of the geotextile. The geotextile is considered to function as only reinforcement. The failure of the granular layer is not considered; thus it is assumed that

1. The friction coefficient of the granular layer is large enough to ensure the mechanical stability of the layer.
2. The friction angle of the geotextile in contact with the granular layer under the wheels is large enough to prevent the sliding of the granular layer on the geotextile.

It is also assumed that

1. Thickness of the granular layer is not significantly affected by the subgrade soil deflection.
2. The granular layer provides a pyramidal distribution with depth of the *equivalent tyre contact pressure*, p_{ec}, applied on its surface (Fig. 5.16(a)).

Therefore,

$$p_{ec}LB = (B + 2h_0 \tan \alpha_0)(L + 2h_0 \tan \alpha_0)(p_0 - \gamma h_0) \tag{5.31}$$

in the absence of geotextile, and

$$p_{ec}LB = (B + 2h \tan \alpha)(L + 2h \tan \alpha)(p - \gamma h) \tag{5.32}$$

in the presence of geotextile.

In Equations (5.31) and (5.32), L and B are the length dimensions of the equivalent rectangular tyre contact area; h_0 is the thickness of granular layer in the absence of geotextile; h is the thickness of granular layer in the presence of geotextile; α_0 is the load diffusion angle in the absence of geotextile; α is the load diffusion angle in the presence of geotextile; p_0 is the pressure at the base of the granular layer in the absence of geotextile; p is the pressure at the base of the granular layer in the presence of geotextile; and γ is the unit weight of the granular fill material.

The equivalent tire contact pressure is given as

$$p_{ec} = \frac{P}{2LB} \tag{5.33}$$

where, P is the axle load.

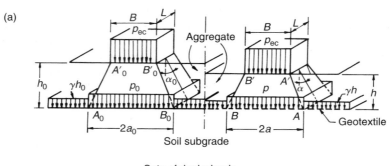
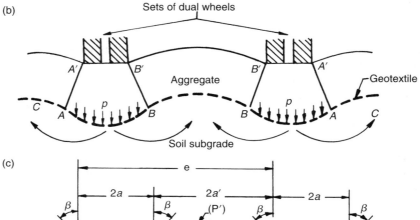

Figure 5.16 (a) Load diffusion model; (b) kinematics of subgrade deformation; (c) shape of the deformed geotextile (after Giroud and Noiray, 1981).

From Equations (5.31), (5.32), and (5.33), the following equations are obtained:

$$p_0 = \frac{P}{2(B + 2h_0 \tan \alpha_0)(L + 2h_0 \tan \alpha_0)} + \gamma h_0 \tag{5.34}$$

in the absence of geotextile, and

$$p = \frac{P}{2(B + 2h \tan \alpha)(L + 2h \tan \alpha)} + \gamma h \tag{5.35}$$

in the presence of geotextile.

Load diffusion angles, α and α_0, may vary in their values, however they are assumed both equal to $\tan^{-1}(0.6)$ in the present design method. This assumption implies that the presence of the geotextile layer does not modify significantly the load transmission mechanism through the granular layer.

On the application of the wheel load, the geotextile exhibits a wavy shape; consequently, it is stretched. This happens if the soil subgrade, having a low permeability, is saturated, and behaves in an undrained manner under traffic loading. This incompressible nature of the soil subgrade results in settlement under the wheels and heave between and beyond the wheels (Fig. 5.16(b)). Under such a situation, the volume of soil subgrade displaced downwards by settlement must be equal to the volume of soil displaced upwards by heave, which may be called *volume conservation* of the undrained soil subgrade. In the stretched position of the geotextile, the pressure against its concave face is higher than the pressure against its convex face. This reinforcing mechanism is known as the *membrane effect* of the geotextile, which provides the following two beneficial effects:

1. confinement of the soil subgrade between and beyond the wheels;
2. reduction of the pressure applied by the wheels on the soil subgrade.

The pressure applied on the subgrade soil by the portion AB of the geotextile is

$$p^* = p - p_g, \tag{5.36}$$

where p_g is the reduction of pressure resulting from the use of a geotextile. The pressure reduction, p_g, is a function of the mobilized tension in the geotextile, which depends on its elongation; thus its deflected shape is significant.

Since the soil subgrade confinement provided by the geotextile helps in keeping the deflection small for all applied pressures less than the ultimate load-bearing capacity, q_u, of the soil subgrade, as given by Equation (5.37) below, the pressure p^* can be as large as q_u.

$$q_u = (\pi + 2)c_u + \gamma h, \tag{5.37}$$

where c_u is the undrained cohesion or shear strength of the soil subgrade.

$$p^* = q_u = (\pi + 2)c_u + \gamma h \tag{5.38}$$

From Equations (5.36) and (5.38), one gets

$$p - p_g = (\pi + 2)c_u + \gamma h \tag{5.39}$$

In the absence of the geotextile, an equation similar to Equation (5.39) can be obtained by equating p_0 to the elastic bearing capacity of the soil subgrade given as

$$q_e = \pi c_u + \gamma h \tag{5.40}$$

in order to avoid large deflection under the wheel. Thus,

$$p_0 = \pi c_u + \gamma h \tag{5.41}$$

in the absence of the geotextile.

200 Analysis and design concepts

Equations (5.34) and (5.41) lead to

$$c_u = \frac{P}{2\pi(B + 2h_0 \tan \alpha_0)(L + 2h_0 \tan \alpha_0)}, \quad (5.42)$$

which is applicable in the absence of geotextile.

The shape of the deformed geotextile is assumed to consist of portions of parabolas connected at A and B, points located on the initial plane of the geotextile (Fig. 5.16(c)). The reduction of pressure, p_g, is due to the tension of the geotextile in parabola (P). In fact, p_g is a uniform pressure applied on AB and is equivalent to the vertical projection of the tension T of the geotextile at points A and B:

$$ap_g = T \cos \beta \quad (5.43)$$

According to the property of parabolas:

$$\tan \beta = \frac{a}{2s} \quad (5.44)$$

From the definition of secant modulus, E (in N/m), of the geotextile, one gets

$$T = E\varepsilon \quad (5.45)$$

where, ε is the per cent elongation.

Combining Equations (5.43), (5.44), and (5.45), one gets

$$p_g = \frac{E\varepsilon}{a\sqrt{1 + \left(\frac{a}{2s}\right)^2}} \quad (5.46)$$

Equations (5.35), (5.39), and (5.46) lead to

$$(\pi + 2)c_u = \frac{P}{2(B + 2h \tan \alpha)(L + 2h \tan \alpha)} + \frac{E\varepsilon}{a\sqrt{1 + \left(\frac{a}{2s}\right)^2}}, \quad (5.47)$$

which is applicable in the presence of geotextile.

In Equations (5.42) and (5.47), the following expressions can be used for L and B:

$$L = \frac{B}{\sqrt{2}}, \text{ and} \quad (5.48a)$$

$$B = \sqrt{\frac{P}{p_c}} \quad (5.48b)$$

for on-highway trucks

$$L = \frac{B}{2}, \text{ and} \quad (5.49a)$$

$$B = \sqrt{\frac{P\sqrt{2}}{p_c}} \qquad (5.49b)$$

for off-highway trucks where, p_c is the tyre inflation pressure.

Solving Equation (5.42) for h_0, and Equation (5.47) for h allows us to determine the reduction of granular layer thickness, Δh, due to reinforcement function of geotextile as per quasi-static analyses. Thus,

$$\Delta h = h_0 - h \qquad (5.50)$$

A further assumption is that the value of Δh remains unchanged under repeated traffic loading, thus allowing it to uncouple the reinforcement effect and its analysis from the cyclic nature of loading. Therefore,

$$h' = h'_0 - \Delta h \qquad (5.51)$$

where, h' is the required granular layer thickness of the unpaved road in the presence of the geotextile and under traffic loading, and h_0' is the required granular layer thickness of the unpaved road in the absence of the geotextile and under traffic loading.

Under traffic loading, the required granular layer thickness, h_0', of the unpaved road without the geotextile is determined using an empirical method originally developed by Webster and Alford (1978) for a rut depth of $r = 0.075$ m and simplified by Giroud and Noiray under the form

$$h_0' = \frac{0.19 \log_{10} N_s}{(CBR)^{0.63}}, \qquad (5.52)$$

where N_s is the number of passages of standard axle with a load $P_s = 80$ kN; and CBR is the California Bearing Ratio of soil subgrade.

Giroud and Noiray extended Equation (5.52) to other values of axle load and rut depth using the following relationships:

$$\frac{N_s}{N_p} = \left(\frac{P}{P_s}\right)^{3.95} \qquad (5.53)$$

$$\log_{10} N_s \rightarrow [\log_{10} N_s - 2.34 (r - 0.075)], \qquad (5.54)$$

where $\rightarrow$ indicates 'replaced by'.

They also introduced the undrained cohesion of the subgrade using the following empirical formula:

$$c_u \text{ (in Pa)} = 30{,}000 \times CBR \qquad (5.55)$$

Combining Equations (5.52), (5.53), and (5.55), and replacing $\log_{10} N_s$ as per Equation (5.54), the following expression is obtained:

$$h'_0 = \frac{119.24 \log_{10} N + 470.98 \log_{10} P - 279.01r - 2283.34}{c_u^{0.63}} \tag{5.56}$$

This formula is based on extrapolation and therefore, it should not be used when the number of passages exceeds 10,000.

A design chart for a particular set of parameters, based on the analysis presented above, is shown in Figure 5.17. The following two features of this chart are noteworthy:

1. Δh can never be higher than h_0.
2. No granular layer is needed on top of the geotextile when Δh versus c_u curve is above the h_0' versus c_u curve.

The design chart provides values of Δh and h_0'. The subtraction of Δh from h_0', according to Equation (5.51), results in the value of granular layer thickness, h'. A set of curves, giving the geotextile elongation, ε, versus subgrade soil cohesion, c_u, in the design chart allows the user of the chart to check that, in the considered case, the geotextile is not subjected to excessive elongation.

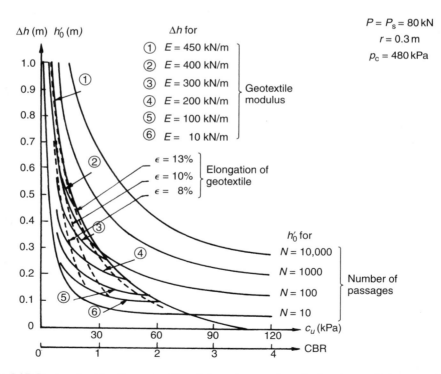

Figure 5.17 Design chart for the geotextile-reinforced unpaved road related to on-highway truck with standard axle load (after Giroud and Noiray, 1981).

The reduction in granular layer thickness can be of the order of 20–60%, as in some typical cases considered by Giroud and Noiray. To be safe, it is recommended not to use the design chart in Figure 5.17 for numbers of passages larger than 10,000.

ILLUSTRATIVE EXAMPLE 5.3
Consider:

> Number of passages, $N = N_s = 340$
> Single axle load, $P = P_s = 80$ kN
> Tyre inflation pressure, $p_c = 480$ kPa
> Subgrade soil CBR $= 1.0$
> Modulus of geotextile, $E = 90$ kN/m
> Allowable rut depth, $r = 0.3$ m

What is the required thickness of the granular layer for the unpaved road in the presence of geotextile?

SOLUTION
The design chart, presented in Figure 5.17, provides:

$h_0' = 0.35$ for CBR $= 1.0$ and $N = 340$
$\Delta h = 0.15$ for CBR $= 1.0$ and $E = 90$ kN/m

The required thickness of the granular layer for the unpaved road in the presence of the geotextile is calculated using Equation (5.51) as:

$h' = h_0' - \Delta h = 0.35 - 0.15 = 0.20$ m **Answer**

From the design chart,
elongation of the geotextile $\varepsilon \approx 10\%$
It should be checked that the elongation at failure of the geotextile, as obtained from practical test, is larger than this value.

Note: This example was explained by Giroud and Noiray (1981).

It may be noted that among the assumptions made in RFDM, the adoption of different limit bearing pressures for the soil subgrade in unreinforced and reinforced cases leads to the results that may seem theoretically inconsistent. According to RFDM, the computed performance of an unreinforced road should be similar to that of the same road reinforced with a zero-modulus geotextile, which is not a fact. It has been recognized in practice that even very low modulus geotextiles are beneficial in reducing the granular layer thickness because of their *separation function*. RFDM does not consider this reinforcing mechanism in analysis for granular layer thickness. In addition to the determination of the granular layer thickness by RFDM, computations should be completed with verifications of the tensile resistance and lateral anchorage of the geotextile (Bourdeau and Ashmawy, 2002). However, mainly because of simplicity, RFDM is widely used for designing the geosynthetic-reinforced unpaved roads for a common range of parameters.

Based on the field observations on unpaved roads with geosynthetics, Fannin and Sigurdsson (1996) reported that Giroud and Noiray's RFDM is found to be appropriate for unpaved roads that do not experience compaction of the granular base layer during traffic loading. Compaction will lead to an overprediction of performance at small ruts. It was also reported that separation appears to be very important on the thinnest granular base course, where geotextiles outperform the geogrid. The geogrid outperforms the geotextiles on the thicker granular base layers, for which reinforcement rather than separation dominates in a system that is less deformed by vehicle loading.

Since for roads, the geosynthetic reinforcement needs to support repeated loads, it is the response of the geosynthetic to rapid and cyclic loads that should be considered for design purposes.

Separation function design method (SFDM)

Steward et al. (1977) presented a design method for the geosynthetic-reinforced unpaved roads, considering mainly the separation function of the geosynthetic, which is more important for thin roadway sections with relatively small live loads where ruts, approximating less than 75 mm, are anticipated. This design method is based on theoretical analysis and empirical (laboratory and full-scale field) tests, and it allows the designer to consider vehicle passes, equivalent axle loads, axle configurations, tyre pressures, soil subgrade strengths and rut depths, along with the following limitations:

1. The granular layer must be (a) cohesionless (non-plastic), and (b) compacted to CBR 80.
2. Vehicle passes less than 10,000.
3. Geotextile survivability criteria must be considered.
4. Soil subgrade undrained shear strength less than about 90 kPa (CBR < 3).

Steward et al. presented design charts to determine the required thickness of the granular layer (Fig. 5.18). The main concept involved in developing these design charts is the presentation of stress level acting on the soil subgrade in terms of bearing capacity factor, similar to those commonly used for the design of shallow foundations (continuous footings) on cohesive soils using the following expression for ultimate bearing capacity, q_u:

$$q_u = c_u N_c + \gamma D \tag{5.57}$$

where, c_u is the undrained cohesion of the soil subgrade; N_c is the bearing capacity factor; γ is the unit weight of the granular material above the geosynthetic layer; and D is the depth of the granular layer.

The bearing capacity factor is adjusted when a geosynthetic, especially geotextile, is introduced between the soft soil subgrade and the granular base course, as per the values given in Table 5.3.

ILLUSTRATIVE EXAMPLE 5.4
Consider:

> Number of passes, $N = 6000$
> Single axle load, $P = 90$ kN
> Tyre inflation pressure, $p_c = 550$ kPa
> Cohesive subgrade soil CBR $= 1.0$
> Allowable rut depth, $r = 40$ mm

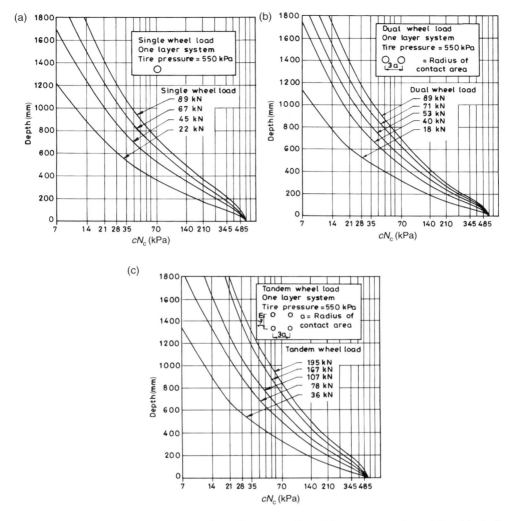

Figure 5.18 US Forest Service design charts for geotextile-reinforced unpaved road for: (a) single wheel load; (b) dual wheel load; (c) tandem wheel load (after Steward et al., 1977).

Table 5.3 Bearing capacity factors for different ruts and traffic conditions both with and without geotextile separators (after Steward et al., 1977)

Field site situation	Ruts (mm)	Traffic (passes of 80 kN axle equivalents)	Bearing capacity factor, N_c
Without geotextile	Less than 50	Greater than 1000	2.8
	Greater than 100	Less than 100	3.3
With geotextile	Less than 50	Greater than 1000	5.0
	Greater than 100	Less than 100	6.0

What is the required thickness of the granular layer for the unpaved road without geotextile, and with geotextile?

SOLUTION
Single wheel load = (90 kN)/2 = 45 kN
From Table 5.3, for 6000 passes and 40 mm rut,

N_c = 2.8 without a geotextile layer, and
N_c = 5.0 with a geotextile layer.

Using Equation (5.55), for CBR = 1.0, $c_u = c$ = 30 kPa
Without a geotextile layer, cN_c = 30 × 2.8 = 84 kPa
With a geotextile layer, cN_c = 30 × 5.0 = 150 kPa
The design chart, presented in Figure 5.18 (a), provides:

without a geotextile layer, thickness of granular layer, $h_0 \approx$ 500 mm   **Answer**
and
with a geotextile, thickness of granular layer, $h \approx$ 350 mm   **Answer**

It should be noted that about 150 mm granular layer thickness can be saved by placing a geotextile layer as a separator at the interface of soil subgrade and the granular base layer in unpaved roads.

It is to be noted that a design method like SFDM, which assumes no reinforcing effect, is generally conservative.

It is important to note that the geotextile, recommended for use in unpaved roads, should meet the minimum hydraulic requirements in addition to minimum installation survivability requirements, as discussed in Sec. 4.11.

Richardson (1997a) presented a simple separation function design method (SFDM) for geosynthetic-reinforced unpaved roads, described in the following steps:

Step 1: Use a granular layer thickness that produces a subgrade pressure $p = 4c_u$. This results in sufficient granular material being placed initially to fill the ruts that will develop as the geotextile/granular layer deforms.
Step 2: Determine the geotextiles's minimum survivability requirements.
Step 3: Determine the geotextile's minimum hydraulic requirements.
Step 4: Select a suitable geotextile that meets the criteria in steps 2 and 3. Almost any woven and nonwoven geotextile can be used if it meets the requirements in steps 2 and 3.

In the first step of design, one can use a simple 60° angle for estimating the distribution of the applied surface load through the granular layer.

The design method, suggested by Richardson, is based on the field observations made by Fannin and Sigurdsson (1996) on the stabilization of unpaved roads with geosynthetics. This method can be used in routine applications; however, it is suggested that the design values be compared with those obtained from other methods especially for a few initial problems as this would act as a confidence building measure.

Modified CBR design method

This method uses a multiplier to the in situ CBR of the soil subgrade, in order to get an equivalent CBR when using a geosynthetic layer at the interface of soft soil subgrade and the granular fill layer. The multiplier is assumed to be equal to the *reinforcement ratio*, which is the ratio of loads at the specified deflection, as determined from load versus deflection test in the CBR mould both with and without geosynthetic at the interface of soft soil subgrade and granular fill layer. It has been observed that the reinforcement ratio increases as both the deflection and the water content in soil subgrade increase. The thickness of the granular fill layer is calculated using the following equation for both the cases, without and with geosynthetic, taking in situ CBR, and modified CBR, respectively at the specified deflection (U.S. Army Corps of Engineers Modified CBR Design Method, WES TR 3-692 as reported by Koerner, 1994).

$$h = (0.1275 \log_{10} C + 0.087) \left(\frac{P}{8.1 \times \text{CBR}} - \frac{A}{\pi} \right)^{1/2} \tag{5.58}$$

where, h is the design thickness of the granular layer in inches; C is the anticipated number of vehicle passes; P is the single or equivalent single wheel loads in pounds; and A is the tyre contact area in square inches.

ILLUSTRATIVE EXAMPLE 5.5
Consider:

Number of passes, $C = 10,000$
Single wheel load, $P = 20,000$ lb
Tyre contact area = 12 in. × 18 in.
Subgrade soil CBR = 1.0
Reinforcement ratio, as determined from modified CBR test, $R = 2.0$

What is the required thickness of the granular layer for the unpaved road without geotextile and with geotextile?

SOLUTION
Using Equation (5.58),
In the absence of the geotextile, the required thickness is

$$h'_0 = (0.1275 \log_{10} 10,000 + 0.087) \left(\frac{20,000}{8.1 \times 1.0} - \frac{12 \times 18}{\pi} \right)^{1/2} \approx 29.2 \text{ in.} \quad \text{Answer}$$

Modified CBR = (subgrade soil CBR) × R = 1.0 × 2.0 = 2.0
In the presence of the geotextile, the required thickness is

$$h' = (0.1275 \log_{10} 10,000 + 0.087) \left(\frac{20,000}{8.1 \times 2.0} - \frac{12 \times 18}{\pi} \right)^{1/2} \approx 20.4 \text{ in.} \quad \text{Answer}$$

It is noted that the savings in the granular layer thickness is 8.8 in. ($\approx 30\%$) in the presence of the geotextile layer at the interface of soil subgrade and the granular base layer in the unpaved road.

5.6.2 Paved roads

Geosynthetic layer at the subgrade level

The ruts with depth in excess of approximately 25 mm are generally not acceptable in paved roadways, which are utilized as permanent roadways for safe, efficient and economical transport of passengers and goods. If the geosynthetic layer is used only for the construction lift (or stabilization lift), then the thickness of the granular subbase/base layer required to adequately carry the design traffic loads for the design life of the paved roadway is generally not reduced. The paved roadways with geosynthetic layers are usually designed for structural support using normal pavement design methods, as described by various agencies (AASHTO, 1993; IRC: 37-2001; IRC: 58-2002), without providing any allowance for the geosynthetic layers.

If the soil subgrade is susceptible to pumping and granular base course intrusion, an additional granular layer thickness above that required for structural support is needed. In the presence of a geosynthetic layer, especially a nonwoven geotextile, at the interface of granular subbase/base layer and the soil subgrade, the required additional granular layer thickness can be reduced by approximately 50% keeping the project cost effective (Holtz et al., 1997). Savings of granular material can also be made by placing a geosynthetic layer in the granular stabilizer lift that can tolerate even 75 mm of rutting under construction equipments. The stabilizer lift with a geosynthetic layer is generally designed, taking it to be a geosynthetic-reinforced unpaved road and this has been described in Sec. 5.6.1 in detail.

As a final design step, the recommended geosynthetic should be checked to meet both the minimum hydraulic requirements and the minimum survivability requirements, as discussed in Sec. 4.11.

Geosynthetic layer at the asphalt overlay base level

The fluid barrier function of the geosynthetic should be achieved in field application, keeping in view the fact that the water (coming from rain, surface drainage or irrigation near pavements), if allowed to infiltrate into the base and subgrade, can cause pavement deterioration by one or more of the following processes:

- softening the soil subgrade
- mobilizing the soil subgrade into the road base stone, especially if a separation/filtration geosynthetic is not used at the road base and subgrade interface
- hydraulically breaking down the base structures, including stripping bitumen-treated bases and breaking down chemically stabilized bases
- freeze/thaw cycles.

The selected paving grade geosynthetic should meet the physical requirements described in Sec. 4.11. Prior to laying the paving fabric, the tack coat should be applied uniformly to the prepared dry pavement surface at the rate governed by the following equation (IRC: SP: 59-2002):

$$Q_d = 0.36 + Q_s + Q_c, \tag{5.59}$$

where Q_d is the design tack coat quantity (kg/m^2); Q_s is the saturation content of the geotextile being used (kg/m^2) to be provided by the manufacturer; and Q_c is the correction based on tack coat demand of the existing pavement surface (kg/m^2).

The quantity of tack coat is critical to the final membrane system. Too much tack coat will leave an excess between the fabric and the new overlay resulting in a potential sliding failure surface and potential bleeding problems, while too little will fail to complete the bond and create the impermeable membrane. In fact, the misapplication of the tack coat can make the difference between paving fabric installation success and failure. The asphalt tack coat forms a low-permeability layer in the fabric and bonds the system to the existing pavement and overlay. The fabric allows slight movement of the system, while holding the tack coat layer in place and maintaining its integrity.

The actual quantity of tack coat will depend on the relative porosity of the old pavement and the amount of bitumen sealant required to saturate the paving fabric being used. The quantity of sealant required by the existing pavement is a critical consideration. The saturation content of the fabric depends primarily on its thickness and porosity; that is, its mass per unit area. It is to be noted that the more the mass per unit area of the geotextile, the greater tack coat required to saturate the fabric. For typical paving fabrics in the 120–135 g/m^2 mass per unit area range, most manufacturers recommend fabric–bitumen absorption of about 900 g/m^2, or application rates of about 1125 g/m^2. For the full waterproofing and stress-relieving benefits, the paving fabric must absorb at least 725 g/m^2 of bitumen. The remaining part of the applied bitumen helps in bonding the system with the existing pavement and the overlay. Additional tack coat may be required between the overlap to satisfy the saturation requirements of the fabric.

A review of projects with unsatisfactory paving fabric system performance shows the importance of the tack coat to the whole system. From a study of the records of 65 projects, which were completed over a 16-year period, it is clear that the tack coat applications was too light (less than 725 g/m^2) in an overwhelmingly high percentage of failure cases. This is shown graphically in Figure 5.19. In the laboratory tests it has been observed that the waterproofing benefit of a paving fabric is negligible until the fabric absorbs at least 725 g/m^2 of tack coat (Fig. 5.20). Inadequate tack coat may result in rutting, shoving, or, occasionally, complete delamination of the overlay. It has been found that the structural problems such as

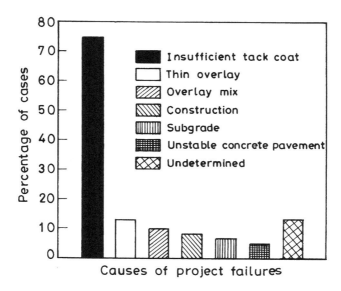

Figure 5.19 Causes in 65 project failures investigated in the United States between 1982–1997 (after Baker, 1998).

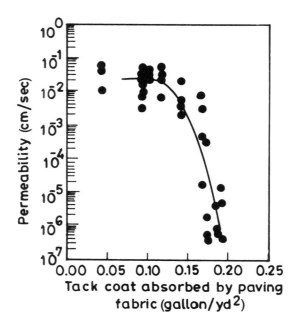

Figure 5.20 Laboratory-prepared paving fabric tests, demonstrating the permeability's sensitivity to the amount of tack coat on the paving fabric (after Marienfeld and Baker, 1998).

overlay slippage and delamination begins to occur where the tack coat quantity absorbed by the fabric is less than about 450 g/m^2.

In addition to low application amounts of tack coat, there can be another set of conditions that may result in a low tack amount in the paving fabric. Inadequate rolling or, low overlay temperatures may create conditions in which the tack may not be taken up by the fabric. In fact, overlays less than 40 mm thick are seldom recommended with paving fabric, in part, because of their rapid heat loss.

The controlled studies have shown that an overlay thickness designed to retard reflective cracking can be reduced by up to 30 mm for equal performance, with the additional waterproofing advantage if a paving fabric interlayer is included in the system (Marienfeld and Smiley, 1994). Equations are available that enable the designer of a geosynthetic-reinforced overlay to design an appropriate overlay thickness and corresponding geosynthetic. The major drawback to currently available design techniques is that they allow us to address the potential failure modes (traffic load induced, thermally induced and surface initiated) separately but not together. In reality, all three modes occur together – a condition that can only be evaluated using sophisticated finite element modeling (Sprague and Carver, 2000). For routine applications, the thickness of the overlay may not be reduced with the use of a geosynthetic interlayer, and one can design the overlay as per the guidelines in design standards on overlays without the geosynthetic interlayer. This is mainly because the major purpose of introducing a paving fabric is to enhance the performance of the pavement not to reduce the thickness of the overlay.

5.7 Railway tracks

Geosynthetics in railway tracks are designed to perform several functions: separation, filtration, drainage and reinforcement/confinement. Keeping these functions in view, the following design procedure can be followed for the design of the geotextile layer (Tan, 2002).

- Design geotextile as a separator – this function is always required. Burst strength, grab strength, puncture resistance and impact resistance should be considered.
- Design geotextile as a filter – this function is also usually required. The general requirements of adequate permeability, soil retention and long-term soil-to-geotextile flow equilibrium are needed as in all filtration designs. Note however, that railway loads are dynamic; thus pore pressures must be rapidly dissipated. For this reason high permittivity is required.
- Consider geotextile flexibility if the cross-section is raised above the adjacent subgrade. Here a very flexible geotextile is an advantage in laterally confining the ballast stone in its proper location. Quantification of this type of lateral confinement is, however, very subjective.
- Consider the depth of the geotextile beneath the bottom of the tie. The very high dynamic load of railway acting on the ballast imparts accelerations to the stone that gradually diminish with depth. If the geotextile location is not deep enough, it will suffer from abrasion at the points of contact with the ballast. The studies conducted at Canadian rail sites indicate that the major damage occurs within 250 mm of the tie, and at depth greater than 350 mm, damage is not noticeable (see Fig. 5.21). From this data, it can be safely concluded that the minimum depth for geotextile placement is 350 mm for abrasion protection. If this depth is excessive, a highly abrasion resistant geotextile must be used. An example of abrasion damage to geotextile due to inadequate ballast thickness is shown in an exhumed geotextile in Figure 5.22.
- The last step is to consider the survivability of the geotextile during installation. To compact ballast under ties, the railroad industry uses a series of vibrating steel prongs

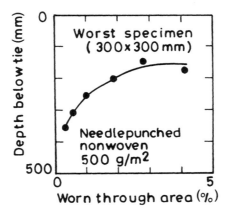

Figure 5.21 Depth below tie/sleeper base of exhumed geotextile versus damage assessment in terms of complete worn through area from Canadian Rail sites (after Raymond, 1999).

Figure 5.22 Abrasion failures of geotextiles placed too close to the track structures (after Raymond, 1982).

forced into the ballast. Considering both the forces exerted and the vibratory action, high geotextile puncture resistance is required. Hence it is necessary to keep the geotextile deep or to use special high puncture resistant geotextile.

It is important to underline that the selected geotextile must meet the following four durability criteria:

- It must be tough enough to withstand the stresses of the installation process. Properties required are: tensile strength, burst strength, grab strength, tear strength and resistance to UV light degradation for two weeks, exposure with negligible strength loss.
- It must be strong enough to withstand static and dynamic loads, high pore pressures, and severe abrasive action to which it is subjected during its useful life. Properties required are: puncture resistance, abrasion resistance and elongation at failure.
- It must be resistant to excessive clogging or blinding, allowing water to pass freely across and within the plane of the geotextile, while at the same time filtering out and retaining fines in the subgrade. Properties required are: cross-plane permeability (permittivity), in-plane permeability (transmissivity) and apparent Opening Size (AOS).
- It must be resistant to rot, and attacks by insects and rodents. It must be resistant to chemicals such as acids and alkalis and spillage of diesel fuel.

A standard specification for the use of geotextiles in railway track stabilization has been developed and published by the American Railway Engineering Association (AREA, 1985). The specification recommends minimum physical property values for three categories of nonwoven geotextiles; Regular, Heavy, and Extra Heavy. Selections of one

Table 5.4 Properties of geotextiles recommended by AREA (American Railway Engineering Association)

Test Methods	Regular	Heavy	Extra heavy
Puncture resistance, N ASTM D4833-88	500	675	900
Abrasion resistance, N ASTM D3884 (Taber test at 1000 rev; 1 kg load/wheel)	675	810	1080
Grab strength, N ASTM D4632	900	1080	1440
Elongation, % ASTM D4632	50	50	50
Trapezoidal tear strength, N ASTM D4533	450	540	720
Cross-plane permeability, cm/sec ASTM D4491	0.2	0.2	0.2
Permittivity, 1/sec ASTM D4491	0.5	0.4	0.3
In-plane transmissivity, sq m/min $\times 10^{-4}$ ASTM D4716	2	4	6
AOS (Apparent Opening Size), US standard sieve, microns, ASTM D4751	70	70	70

of these, while based on subgrade conditions, are somewhat subjective. Therefore, many designers recommend the Heavy and Extra Heavy geotextiles, as the cost of geotextiles is small compared to overall cost of track rehabilitation work being done at the time of installation.

Table 5.4 shows the index properties recommended by AREA for average roll values that should be considered when specifying geotextiles for railway tracks.

Raymond (1982, 1983a,b, 1986a), and Raymond and Bathurst (1990) evaluated a number of exhumed geotextiles from beneath Canadian and US railroads and found that many were pockmarked with abrasion holes. Their studies examined soil fouling content, change in permeability ratio, change in geotextile strength and other geotextile properties. The primary functions/properties of railway rehabilitation geotextile have been established as separation, drainage, filtration, abrasion resistance and elongation. Based on extensive laboratory tests on both unused and exhumed geotextiles from railway track installations in Canada and the USA, the following manufacturing specifications for geotextiles were recommended for railway rehabilitation works (Raymond, 1999):

- *Mass* – 1050 g/m² or greater for track rehabilitation without the use of capping sand;
- *Type* – Needle-punched nonwoven with 80 penetrations per square centimetre (80 p/cm²) or greater;
- *Fibre size* – 0.7 tex or less;
- *Fibre strength* – 0.4 newton per tex (N/tex) or greater;
- *Fibre polymer* – Polyester;
- *Yarn length* – 100 mm or greater;
- *Filtration opening size* – 75 microns or less;
- *In-plane coefficient of permeability* – 0.005 cm/s or greater;

- *Fibre bonding by resin treatment or similar* – not less than 5% and nor more than 20% by weight of low modulus acrylic resin or other suitable non-water soluble resins that leaves the geotextile pliable;
- *Elongation* – 60% or more to ASTM D4632;
- *Colour* – Must not cause 'Snow blindness' during installation;
- *Abrasion resistance* – 1050 g/m² geotextile must withstand 200 kPa on 102 mm burst sample after 5000 revolutions of H-18 stones, each loaded with 1000 grams of rotary platform double head abraser (ASTM D3884);
- *Width and length without seaming* – To be specified by client;
- *Seams* – No longitudinal seams permitted;
- *Wrapping* – 0.2 mm thick black polyethylene or similar;
- *Packaging* – Must be weatherproofed and clearly identified at both ends stating manufacturer, width, length, type of geotextile and date of manufacture.

It is important to underline that the surficial inspection of Canadian tracks in 1980/81 showed that all geotextiles installed in the previous five years with a mass per unit area less than 500 g/m² had failed. Nonwoven geotextiles with a mass per unit area greater than 500 g/m² showed considerably less distress. These assessments were confirmed by excavations of failed geotextiles at several locations. Finally, a mass per unit area of 1050 g/m² was found to be desirable for a rehabilitation geotextile, placed on the undercut surface, to remain durable so as to continue to function as a separation and a filtration layer. Geotextiles installed without a capping sand and meeting the specifications given above are still showing excellent durability after 18 years of service in the very physically harsh environment of the North American track (Raymond, 1999).

Jay (2002) has reported that the use of geotextiles directly on clay and silt soils beneath the ballast may slow the pumping process, but will not prevent it altogether. In this situation, it has been recommended to use a blanket of well graded, fine-to-medium sand subballast overlain by a geotextile (Fig. 5.23). The fine sand filters the silty clay subgrade, and the geotextile filters the sand, preventing intermixing with the ballast. A sand blanket layer of 15 mm will form an effective filter, but for practical construction reasons a thickness of not less than 50 mm is usually specified. This solution is safe even over wet ground, over a high water table, or even over artesian ground water conditions which can occur, for example in railway cuttings. The benefit of the geotextile is to reduce the need for a deep layer of blanketing sand, thus reducing the cost.

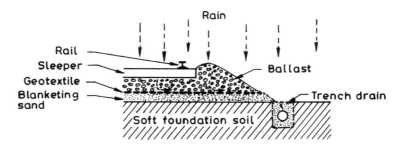

Figure 5.23 A typical cross-section of railway track showing the use of a geotextile and sand blanket below the ballast.

5.8 Filters and drains

Today the application of geotextiles as filters is the most common use of geosynthetics in civil engineering constructions. The ability of a geotextile to allow sufficient water flow without migration of soil particles is a critical design requirement for filtration and drainage applications. Design approaches for geotextile filters are based largely on experience and are wholly empirical in nature. Proper geotextile performance is required for long-term serviceability of the structure. Various elements of the filtration system (soil/waste, filter, drain, water/leachate) must be considered along with external conditions such as unidirectional or bidirectional flow, construction equipment and survivability, static and/or dynamic loading, and long-term durability. To achieve satisfactory filter performance by geosynthetics, especially geotextiles, the following functions must be fulfilled during the design life of the application under consideration:

1. Maintain adequate permeability (or hydraulic conductivity)/permittivity to allow flow of water from the soil layer without significant flow impedance so as not to build up excess hydrostatic pore water pressure behind the geosynthetic (*permeability/permittivity criterion*).
2. Prevent significant washout of soil particles, that is, soil piping (*retention or soil-tightness or piping resistance criterion*).
3. Avoid accumulation of soil particles within the geosynthetic structure, called clogging, resulting in complete shut off of water flow (*anti-clogging criterion*).
4. Survive the installation stresses and any other long-term mechanical, biological or chemical degradation impacts for the lifetime of the structure to perform effectively (*survivability and durability criterion*).

It may be noted that the permeability criterion places a lower limit on the pore size of the geotextile, whereas the retention criterion places an upper limit on the pore size of the geotextile. In other words, the permeability criterion requires a large pore size because the permeability of a geosynthetic filter increases with its increasing pore size; on the other hand the retention criterion requires a reduction of the pore size to restrict the migration of soil particles. These two criteria are, in principle to some extent, contradictory if they have to be fulfilled simultaneously. However, in the majority of cases, it is possible to find a filter that meets both the permeability criterion and the retention criterion. Several different geosynthetic filter criteria have been developed (Giroud, 1982, 1996; Lawson, 1982, 1986; Hoare, 1982; Wang, 1994) largely based on the conventional granular filter criteria, which were first formulated by Terzaghi and Peck (1948). All of these criteria are applicable for specific filter applications. These criteria use soil permeability and compare it with the geotextile permeability for establishing the permeability criterion, whereas they compare soil particle size distribution with geotextile pore size distribution for establishing the retention criterion.

While establishing geosynthetic filter criteria for drainage applications, the following basic filtration concepts are kept in mind:

1. If the largest pore size in the geotextile filter is smaller than the larger particles of soil, then the filter will retain the soil.
2. If the smaller openings in the geotextile are sufficiently large enough to allow smaller particles of soil to pass through the filter, then the geotextile will not *blind* or *clog*.

3   A large number of openings should be present in the geotextile so that proper flow can be maintained even if some of the openings later become clogged.

It must be noted that the filter criteria and the design method for field application of filters should be developed on the basis of data obtained from detailed soil–geotextile performance testing in the field or the laboratory. However, in the absence of such real data, the criteria discussed in the current section can be considered.

It is a general misconception that the pore sizes of a filter should be smaller than the smallest particle size of the soil to be protected, because it would lead to using quasi-impermeable filters (which, of course, would not meet the permeability criterion). In some cases, the filter openings can be larger than the largest soil particles and the filter will still retain the soil (Giroud, 1984). It should be noted that soil retention does not require that the migration of all soil particles be prevented. Soil retention simply requires that the soil behind the filter remain stable; in other words, some small particles may migrate into and/or through the filter provided this migration does not affect the soil structure, that is, does not cause any further movement of the soil mass. At the same time, the filter and the drainage medium located downstream of the filter should be such that they can accommodate the migrating particles without clogging.

The filtration mechanism as explained in Sec. 4.7 shows that the geotextile filter acts essentially as the catalyst, which induces the formation of the natural filter in the soil. It is basically the soil filter zone, which most significantly controls water flows. The sooner the natural filter is established, the smaller the number of particles that will migrate. For the ideal geotextile–filter performance, the permeability (or the hydraulic conductivity) of the particles network at the soil/filter interface, as well as of the geotextile filter itself, should always be equal to or greater than the permeability of the parent soil. It is important that after an initial period of instability during the formation of the soil filter, the permeability of the soil filter system should remain relatively constant over the time.

The permeability criteria of geotextile filters, commonly suggested, are in the following form:

$$k_n \geq A k_s, \tag{5.60}$$

where k_n is the coefficient of the cross-plane permeability of the geotextile; k_s is the coefficient of permeability of the protected soil; and A is a dimensionless factor varying over a wide range, say 0.1 to 100.

The permeability criterion, $k_n \geq k_s$, has long been advocated by many researchers on the assumption that the geotextile needs to be no more permeable than the protected soil. Christopher and Holtz (1985) recommend the criterion, $k_n \geq 10 k_s$, for critical soil and hydraulic conditions in which clogging has been shown to cause roughly an order of magnitude decrease in the geotextile permeability. The criterion, $k_n \geq 0.1 k_s$, was proposed by Giroud (1982) on the premise that a geotextile with only 10% of the permeability of the soil would still have a much greater flow capacity than the soil because the length of the flow path is directly related to the flow rate through a porous media.

The presence of a filter, even when very permeable, disturbs the flow in the soil located immediately upstream. The selected filter should have permeability such that the disturbance to the flow – for example, the pore water pressure and the flow rate – is small and acceptable. For geotextile filters, typical permeability criteria for some specific applications

are as follows (Giroud, 1996):
For a standard drainage trench:

$$k_n > 10k_s \tag{5.61a}$$

For a typical dam-toe drain:

$$k_n > 20k_s \tag{5.61b}$$

For dam clay cores:

$$k_n > 100\, k_s \tag{5.61c}$$

It must be noted that the critical applications may require the design of even higher k_n/k_s ratio values, due to the high gradient that can occur in the filter vicinities.

Federal Highway Administration (FHWA) also established the following permittivity requirements for subsurface drainage applications (Holtz et al., 1997):
For < 15% passing 75 μm:

$$\psi \geq 0.5\ \text{s}^{-1} \tag{5.62a}$$

For 15–50% passing 75 μm:

$$\psi \geq 0.2\ \text{s}^{-1} \tag{5.62b}$$

For > 50% passing 75 μm:

$$\psi \geq 0.1\ \text{s}^{-1} \tag{5.62c}$$

Retention criteria govern the upper (piping) filtration limit for filters and ensure that the soil to be protected is not continually piped through the geotextile filter and into the drainage medium. Failure to adopt appropriate retention criteria for filter design can have costly and potential catastrophic consequences. The retention criteria of geotextile filter commonly suggested are in the following form:

$$O_f \leq BD_s \tag{5.63}$$

where O_f is a certain characteristic opening size of geotextile filter; D_s is a certain characteristic particle diameter of the soil to be protected, it indicates particle diameter, such that s%, by weight, of the soil particles are smaller than D_s; and B is a dimensionless factor varying over a certain range.

The magnitude of B depends on a number of factors, including soil types, hydraulic gradient, allowable amount of soil to be initially piped, the test method to determine O_f and D_s, and state of loading (confined and unconfined) (Faure and Mlynarek, 1998; Lawson, 1998). It is commonly determined by permeameter testing, which has the advantage of allowing near-field conditions to be modelled. Figure 5.24(a) shows the procedure used to determine retention criteria for a specific soil type based on such testing. The relationship

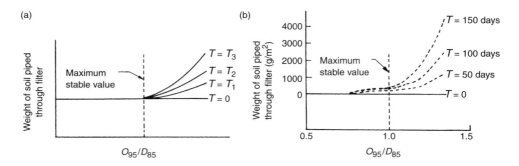

Figure 5.24 (a) General procedure for determining the piping limit; (b) determination of piping limit for Hong Kong Completely Decomposed Granite (CDG) soils (after Lawson, 1998).

between the weight of soil passing through the geotextile filter is plotted against O_f/D_s, say O_{95}/D_{85} ratios tested for various times. The piping limit conventionally is established as the maximum stable ratio of O_{95}/D_{85} below which soil is not continually piped through the geotextile filter. Having derived the appropriate value of B from the permeameter evaluation, an appropriate retention criterion can be presented for this individual soil type in terms of the format shown in Equation (5.63). Figure 5.24(b) shows a series of test results for a specific tropical residual soil of essentially granular structure.

To overcome the time element associated with permeameter testing of individual soil types, standard retention criteria have been developed over a wide range of groups in the past. In general, for filtering granular soils, values of B range from 0.5 to 1.0. For filtering fine-grained soils with a plasticity index less than 10%, values of B range from 2 to 3. To filter cohesive soils with a plasticity index greater than 10%, the required geotextile AOS is normally independent of soil particle size (cohesive soils do not behave as individual particles) and consequently, $O_{95} \leq 0.2$ mm normally would suffice (Lawson, 1998).

All the existing retention criteria for geotextile filters reported in the literature are functions of various opening sizes of the geotextile such as O_{95}, O_{90}, O_{50}, and O_{15}, and diameter of soil particles such as D_{90}, D_{85}, D_{50}, and D_{15} depending mostly on the uniformity coefficient of the soil, C_u ($= D_{60}/D_{10}$). Most of the criteria are given in the form of the O_f/D_s ratio (called *soil tightness number*) not exceeding a certain value or a range. Typical ranges of variations of O_{95}/D_{50}, O_{95}/D_{85}, and O_{90}/D_{90} are, respectively, 1–6, 1–3 and 1–2.

For geotextile filters, retention criteria as per FHWA guidelines developed by Christopher and Holtz (1985) are as follows:

Steady-state flow conditions

$$O_{95} \leq BD_{85} \tag{5.64a}$$

where, for a conservative design, $B = 1$, or for a less conservative design, where $D_{50} > 75\,\mu\text{m}$:

$$B = 1 \quad \text{for } C_u \leq 2 \text{ or } \geq 8 \tag{5.64b}$$
$$B = 0.5\, C_u \quad \text{for } 2 \leq C_u \leq 4 \tag{5.64c}$$
$$B = 8/C_u \quad \text{for } 4 \leq C_u \leq 8 \tag{5.64d}$$

and, for $D_{50} \leq 75$ μm:

$B = 1$ for wovens (5.64e)
$B = 1.8$ for nonwovens (5.64f)

For cohesive soils (Plasticity Index, PI > 7):

$$O_{95} \leq 0.3 \text{ mm} \tag{5.65}$$

Dynamic, pulsating and cyclic flow (if geotextile can move)

$$O_{95} \leq 0.5 \, D_{85} \tag{5.66}$$

Lawson (1998) has pointed out that if geotextile filters that fall outside the boundaries indicated by appropriate retention criteria are used, immediate catastrophic failures do not occur. Over time, with continual soil piping, a loss of serviceability in the immediate filter area may arise. This may be evidenced by undue deformations, structural cracking, etc. It is only if these serviceability problems are allowed to persist without maintenance that subsequent collapse can occur.

A more comprehensive approach to soil retention criteria has been suggested by Luettich *et al.* (1992). It has been proposed by Lafleur *et al.* (1993) that the filter opening size must fall within a narrow range. If it is too large, erosion will take place; if it is too small, blocking or clogging can occur near the interface, resulting in decreased system discharge capacity. For all applications where the geotextile can move, and when it is used as sandbags, it is recommended that samples of the site soils should be washed through the geotextile to determine its particle-retention capabilities (Holtz *et al.*, 1997).

Long-term flow capability of geosynthetics (generally geotextiles) with respect to the hydraulic load coming from the upstream soil is of significant practical interest. Filtration tests, such as the gradient ratio test for cohesionless soils or the hydraulic conductivity ratio test for cohesive soils, as stated in Sec. 3.5, must be performed to recommend anti-clogging criterion, especially for critical/severe applications. In these tests, GR ≤ 3.0 or HCR ≥ 0.3 should generally be satisfied as anti-clogging criterion in order to ensure satisfactory filter performance in the field. In the absence of such real data, particularly for less critical applications, the selected geotextile filter should satisfy the following anti-clogging criteria (Christopher and Holtz, 1985):

$$O_{95} \geq 3 \, D_{15}, \text{ for soil with } C_u > 3 \tag{5.67}$$

For soil with $C_u \leq 3$, select geotextile with maximum opening size possible from the retention criteria.

Other qualifiers: For soils with percentage passing 75 μm > 5%

Porosity, $n \geq 50\%$ (5.68a)

for nonwoven geotextile filters, and

POA $\geq 4\%$ (5.68b)

for woven geotextile filters.

For soils with percentage passing 75 μm < 5%:

Porosity, $n \geq 70\%$ (5.68c)

for nonwoven geotextile filters, and

$POA \geq 10\%$ (5.68d)

for woven geotextile filters.

In order to avoid the possibility of clogging of the geotextile filter, its opening size or per cent open area cannot be too small. If the base soil, that is the soil to be protected is internally stable, then there is less possibility of occurrence of clogging. A soil is said to be internally stable (or self-filtering) if its own fine particles do not move through the interconnected pores of its coarser fraction. Internal stability has been found to depend on the shape of the gradation curve for cohesionless soils and on the dispersive ability of cohesive soils (Kenney and Lau, 1985; Mlynarek and Fannin, 1998). Typically, plastic soils, uniformly graded granular soils (coefficient of uniformity, C_u less than approximately 3) and well-graded soils ($C_u > 4$, and Coefficient of curvature, $C_c > 1$) behave as stable soils (Fig. 5.25 (a)). The unstable soils cannot perform self-filtration, that is, they have the potential to pipe internally. Such soils may include gap-graded soils, non-plastic broadly graded soils, dispersive soils (sugar sands and rock flour), wind-blown silt deposits (i.e. loess-type soils) with interbedded sand seams and other highly erodible soils. Gap-graded soils have a coarse and fine fraction, but very little medium fraction is present (Fig. 5.25(b)). If there is an insufficient quantity of soil particles in the medium fraction, fine soil particles pipe through the coarser fraction and a soil filter bridge behind the geotextile. In broadly graded soils (with concave upward grain-size distributions and having uniformity coefficient $C_u > 20$), the gradation is distributed over a very wide range of particle sizes, such that fine soil tends to pipe through coarser particles. Dispersive soils are fine-grained natural soils that deflocculate in the presence of water and, therefore, are highly susceptible to erosion and piping (Sherard et al., 1972).

Situations such as those involving internally unstable, high hydraulic gradients and very high alkalinity groundwater have been identified to create severe clogging problems. Iron, carbonate, and some organic deposits can chemically clog the geotextiles. Under such situations, one should avoid the use of geotextile filters and should use a granular filter, or should open up the geotextile to the point where some soil loss will occur, if the downstream conditions permit such soil loss. In fact, unstable soils require a more rigorous geotextile evaluation, if one wants to use it as a filter.

Certain filtration and drainage applications, such as in landfills, may expose the geotextile to chemical or biological activity that can drastically influence its durability as well as its filtration and drainage properties. If biological clogging is a concern, a higher porosity geotextile may be used. It is also better to have an inspection and maintenance programme to flush the drainage system in the drain design and operation. Thus, it is important to note that the properties of the soil and the geosynthetic as well as the characteristics of the fluid passing through the filter influence the hydraulic characteristics of the soil–geosynthetic systems.

For filtration and drainage applications, the geotextile should also meet certain minimum standards of strength and endurance to survive the installation stresses as well as the

Analysis and design concepts 221

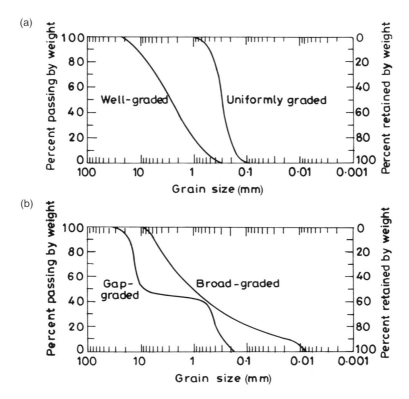

Figure 5.25 Typical grain-size distribution curves: (a) for stable well-graded and uniformly graded granular soils; (b) for potentially unstable soils (after Richardson and Christopher, 1997).

long-term degradation impacts. The limits must be established on the basis of site-specific evaluation, testing and design. However, for routine projects, the users can have some specific guidelines in selecting geotextiles from the available guidelines in standards/codes of practice/manuals, such as AASHTO M288-00 geotextile specifications, as described in Sec. 4.11. These survivability requirements are the default criteria that can be used in the absence of site-specific information. It should be noted that data on static puncture are necessary for the filtration function. When the site loading conditions are such that there is a potential risk of static puncture of the filter, data on static puncture should be obtained on priority.

For designing a geotextile filter, one should identify the conditions under which it will be required to perform. These can include the project conditions (critical nature of the project) and the physical conditions (soil, hydraulic and stress conditions). The critical nature of the project will help determine the level of design effort, and the physical conditions will establish the geotextile requirements (Christopher, 1998). For critical projects, design should be based on a thorough analysis, including performance test results.

Flow directions and hydraulic gradients can significantly influence the behaviour and stability of an engineered filter. Flow can be classified as either steady state or dynamic.

Steady-state flow is usually present in trench drains used to lower the groundwater level beneath roads and parking lots, behind retaining walls, under foundations and below recreation fields. Steady-state flow suggests that water movement occurs in one principal direction; this is the simplest application of a geotextile filter. Dynamic pulsating flow is usually encountered in edge drains used to remove surface infiltration water from roads; dynamic cyclic flow is encountered in permanent erosion control applications for shorelines, stream banks, and canals. In contrast to steady-state flow, dynamic flow may occur in more than one direction. If the geotextile is not properly weighted down and in intimate contact with the soil to be protected, or if dynamic, cyclic, or pulsating loading conditions produce high localized hydraulic gradients, then soil particles can continuously move without formation of any natural soil filter behind the geotextile, severely taxing its ability to perform (Christopher, 1998). While recommending geotextile filters for wave attack applications, or any other situation in which turbulent or two-dimensional flow conditions can occur – for example erosion control systems, one should be very careful.

While designing with geotextiles in filtration applications, the basic concepts are essentially the same as when designing with granular filters. The geotextile must allow the free passage of water and prevent the erosion and migration of soil particles into the drainage system or into the armour of the revetment depending on the type of application throughout the design life of the structure. The simplified design procedures for a geotextile filter for stable soils in subsurface drainage systems can be summarized in the following steps:

Step 1: Evaluate the soil to be filtered (the retained soil). As a minimum, this should include:

- visual classification
- consistency limits
- particle size distribution analysis.

Step 2: Determine the minimum survivability requirements.
Step 3: Determine the minimum permeability using the permeability criterion.
Step 4: Determine the maximum opening size using the retention criterion as well as clogging criterion.
Step 5: Select the geotextile in accordance with Steps 2, 3, and 4.
Step 6: Perform a filtration test to meet the requirements of retention and anti-clogging criterion, if the application is critical.

For unstable soils, one should consult the subject expert and plan on performing soil-specific laboratory testing.

For designing the drainage system, the maximum seepage flow into the system must be estimated. In the case of in-plane drainage with thick geotextile or geocomposite, the flow rate per unit width of the geosynthetic should be compared with the flow rate per unit width requirement of the drainage system. Since the in-plane flow capacity for geosynthetic drains reduces significantly under compression as well as with time due to creep, the final design must be based on the performance test under the anticipated design loading conditions with a safety factor for the design life of the project. It must be noted that the objective of design is to ensure stability throughout the design life by a reduction in pore pressure or depth of water table. The steps required in a design can be summarized as follows:

Step 1: Keeping the critical nature and site conditions of the application in view, define the stability requirement and design life.
Step 2: Determine the particle size distribution curve and the coefficient of permeabilty of soil samples from the site.
Step 3: Select drainage aggregate, if it has to be used along with geotextile filter.
Step 4: Assess the reduction in pore water pressure or reduction in water table depending on the requirement in a particular application and estimate the water flow into and through the drainage system based on the hydraulic gradient and the permeability of the soil.
Step 5: Determine the type and dimensions of the drainage system.
Step 6: Determine the geosynthetic requirements considering the permeability criterion, retention criterion, anti-clogging criterion and survivability criterion, as discussed above, and then select the proper geosynthetic accordingly.
Step 7: Monitor installation during construction and observe the drainage system during and after storm periods.

If geocomposite drains are being used for drainage applications, then their design must satisfy the following criteria (Corbet, 1992):

1. The core must resist the applied loads (normal and shear) without collapsing.
2. Under sustained load the core must not reduce significantly in thickness (compression creep).
3. The core must allow the expected water flow to reach the discharge point without the buildup of water pressure in the core.
4. The core must support the geotextile filter.

Geosynthetic drains in the form of band drains, as described in Sec. 1.5, are nowadays frequently installed within the saturated soil mass to provide vertical drainage, which can be obtained in the conventional method by constructing sand drains of appropriate diameter. In such specific applications, the complete drainage design of band drains as per the radial consolidation theory requires estimation of equivalent sand drain diameter, D_e, which can be calculated using the following expression:

$$D_e = \frac{2(B + \Delta x)}{\pi}, \qquad (5.69)$$

where B is the width and Δx is the thickness of band drain. The above expression is derived based on the fact that the effectiveness of a drain depends, to a great extent, upon the circumference of its cross-section but very little upon its cross-sectional area (Kjellman, 1948).

The vertical compression test for geocomposite pavement panel drains may be conducted to simulate vertical, horizontal, and eccentric loading resulting from an applied vertical load under various backfill conditions. The results of the test may be used to evaluate the vertical strain of the panels and the core area change for a given load.

The design, specification and construction of any drainage or filter system should recognize that backfill conditions and materials must be selected, placed and compacted so that

224 Analysis and design concepts

the geosynthetic product and soil act in concert to carry the applied loads without excessive strains, either vertical, horizontal, or at any load angle. Construction forces and in-service static and dynamic load-induced compression must be considered properly. Appropriate filter gradation criteria must be followed in the selection of granular backfill material to minimize migration of soil fines into the voids of backfill material in the presence of hydraulic gradients. Backfill material selection and placement method should be based primarily on achieving adequate compaction without damaging the drainage and filter materials, while also achieving intimate contact with the soil face. Permeability of the backfill material must also be considered in its selection to promote higher ground water flow to the drainage system. To enhance placement, especially around geocomposite drains and to prevent damage to these structures, the aggregate size should not generally exceed 19 mm.

ILLUSTRATIVE EXAMPLE 5.6
A geotextile-wrapped trench drain is to be constructed to drain a soil mass. Determine the appropriate hydraulic properties of the geotextile to function as a filter in a critical application with the following soil properties:

$D_{10} = 0.14$ mm
$D_{15} = 0.18$ mm
$D_{60} = 0.65$ mm
$D_{85} = 1.1$ mm
$k_s = 2 \times 10^{-4}$ m/s
percentage passing 75 μm < 5%

SOLUTION
From Equation (5.61), the cross-plane permeability (k_n) of the geotextile should meet the following requirement:

$$k_n > 10 k_s$$

$$\Rightarrow \quad k_n > 10 \times 2 \times 10^{-4} \text{ m/s} = 2 \times 10^{-3} \text{ m/s}$$

Since the soil has less than 15% passing 75 μm, therefore from Equation (5.62a),

$$\psi \geq 0.5 \text{ s}^{-1}$$

Coefficient of uniformity,

$$C_u = \frac{D_{60}}{D_{10}} = \frac{0.65 \text{ mm}}{0.14 \text{ mm}} = 4.6$$

Since C_u lies between 4 and 8, therefore the value of factor B for its use in Equation (5.64a) can be calculated from Equation (5.64d) as:

$$B = \frac{8}{C_u} = \frac{8}{4.6} = 1.7$$

Now, from Equation (5.64a),

$$O_{95} \leq B D_{85}$$

$$\Rightarrow \quad O_{95} \leq 1.7 \times 1.1 \text{ mm} = 1.87 \text{ mm}$$

Since $C_u > 3$, the geotextile should meet the following anti-clogging criterion (Eq. (5.67)).

$O_{95} \geq 3 D_{15}$
$\Rightarrow \quad O_{95} \geq 3 \times 0.18 \text{ mm} = 0.54 \text{ mm}.$

Also, from Equations (5.68c) and (5.68d),

Porosity, $n \geq 70\%$

for nonwoven geotextile filter, and

POA $\geq 10\%$

for woven geotextile filter.
Thus, the geotextile filter should have the following hydraulic properties:

$k_n > 2 \times 10^{-3}$ m/s
$\psi \geq 0.5 \text{ s}^{-1}$
$0.54 \text{ mm} \leq \text{AOS} \leq 1.87 \text{ mm}$
$n \geq 70\%$ for nonwoven geotextile filter
POA $\geq 10\%$ for woven geotextile filter. **Answer**

ILLUSTRATIVE EXAMPLE 5.7
A geosynthetic has to be selected to provide drainage behind a 10-m high retaining wall with a vertical backface as shown in following figure. The coefficient of permeability of the soil backfill is 1×10^{-5} m/s. Determine the required transmissivity of the geosynthetic to function as a drain. Would an ordinary single layer of nonwoven geotextile be adequate?

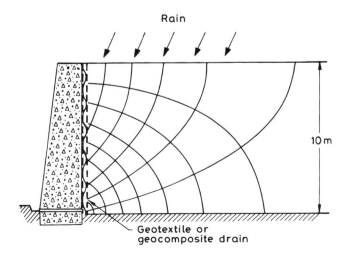

Solution

The flow net can be used to estimate the rate of flow per unit length of the wall, q, as

$$q = k_s H \frac{N_f}{N_d},$$

where k_s is the coefficient of permeability of soil backfill, H is the total head loss, N_f is the number of flow channels, and N_d is the number of potential drops.
In the present problem, $k_s = 1 \times 10^{-5}$ m/s, $H = 10$ m, $N_f = 6$, and $N_d = 8$. Therefore,

$$q = 1 \times 10^{-5} \times 10 \times \frac{6}{8} \, \text{m}^2/\text{s} = 7.5 \times 10^{-5} \, \text{m}^2/\text{s}.$$

Hydraulic gradient, $i = 10 \text{ m}/10 \text{ m} = 1.0$
Transmissivity, θ, can be calculated using the following expression (Eq. (3.8)):

$$Q_p = \theta \, i \, B$$
$$\Rightarrow Q_p/B = \theta \, i$$
$$\Rightarrow q = \theta \, i$$
$$\Rightarrow \theta = q/i = 7.5 \times 10^{-5} \, \text{m}^2/\text{s}. \qquad \textbf{Answer}$$

Typical values of transmissivity for the most common nonwoven geotextiles fall into the range of 10^{-4}–10^{-6} m²/s depending greatly on normal stress acting on the geotextile. We can compare the required transmissivity to the actual value obtaining a factor of safety as follows:

$$FS = \frac{\theta_{allowable}}{\theta_{required}} = \frac{10^{-4} - 10^{-6}}{7.5 \times 10^{-5}} = 1.33\text{--}0.013 \text{ which is not adequate.}$$

Therefore, a single layer of geotextile is not suited for drainage application behind the retaining wall. In fact, a geocomposite having much greater in-plane flow capacity should be used. **Answer**

5.9 Slopes

5.9.1 Erosion control

Revetment systems are very effective in erosion control of slopes including coastal shorelines, stream banks, canal banks, hill slopes and embankment slopes. In conventional systems graded granular layers are used as filters beneath ripraps and other revetment systems. In the past four decades, geotextile layers have been used as a replacement of graded granular filters in riprap erosion control systems. To evaluate the stability of the revetment (cover layer and sublayers), information is required about the hydraulic conditions, the structural properties and the possible failure mechanisms. When designing revetments the designer should note that the geotextile filter is only one of the structural components involved, and there are a few more components, as shown in Figure 5.26(a), to be designed.

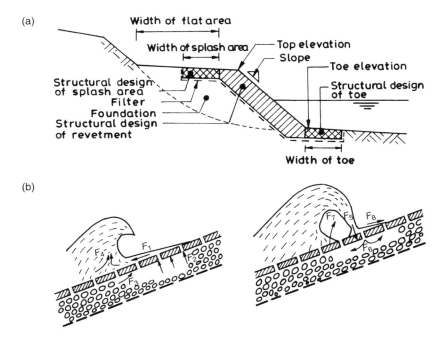

Figure 5.26 (a) Design components of a typical revetment structure (after Pilarczyk, 2000); (b) various forces due to water waves that may act on the revetment system.

Notes
F_1 – forces due to down-rush; F_2 – uplift forces due to water in filter; F_3 – uplift forces due to approaching wave front; F_4 – forces due to change in velocity field; F_5 – wave impact; F_6 – uplift forces due to mass of water falling on slope; F_7 – force caused by low pressures on slope due to air entrainment; F_8 – forces due to up-rush.

Failure of any one component may cause the failure of the entire revetment structure. To achieve a perfect erosion control system for a slope, the following aspects must be taken into account in the design process:

1 stability of cover layer, sublayers, subsoil considering the whole system as well as the individual element;
2 flexibility, that is, ability to follow settlement;
3 durability of cover layer and geotextile filter;
4 possibility of inspection of failure;
5 easy placement and repair;
6 low construction and maintenance cost;
7 overall performance.

The design of revetments like other hydraulic structures must be based on an integral approach of the interaction between the structure and the subsoil. The main geotechnical limits that should be evaluated in the design of the revetments on sloping ground are:

1 overall stability of slopes;
2 settlements and horizontal deformations due to weight of the structure;

3   seepage of groundwater;
4   piping or internal erosion due to seepage flow;
5   liquefaction caused by cyclic loading of water due to wave actions or earthquakes.

A complete analytical approach to the design of revetments incorporating geotextiles does not currently exist. While certain aspects, particularly in the hydraulic field, can be relatively accurately predicted, the effect of various forces (Fig. 5.26b) on the revetment cannot be represented with confidence in a mathematical form for all possible configurations and systems. Therefore, the designer must make use of empirical rules or past experiences. Using this approach, it is likely that the design will be conservative. Since there is a great variety of possible composition of erosion control systems, it is not possible to describe the complete designs of all these systems in this section. However, the design of geotextile filter applicable to all the systems is being discussed. The readers can find more details, in the works of Fuller (1992), on the design, particularly of typical articulating block system (ABS) revetments with a geotextile filter in the coastal conditions.

Since filtration is the primary function of geotextiles, the design steps for geotextile layer remains essentially the same as the design for geotextile filters in subsurface drainage systems discussed in Sec. 5.8. However, while designing the geotextile filter for erosion control systems, the following special considerations should be given:

1   Since the riprap stones or concrete blocks may cover some portions of the geotextile filter, it is essential to evaluate the flow rate required through the open area of the system, and select a geotextile that meets those flow requirements.
2   For < 15% passing 75 μm, $\psi \geq 0.7$ s^{-1}.
3   The largest opening in the geotextile should be small enough to retain even the smaller particles of the base soil. It means that the value of B in retention criterion should be reduced to 0.5 or less. Usually, no transport of soil particles should be allowed and thus the washout of soil particles should be completely prevented, independent of the level of hydraulic loading, because settlement and loss of stability it could result in (see Fig. 5.27). In this situation, the geosynthetic filter is called a *geometrically tight filter*. However, a very limited washout is sometimes acceptable; in that case the filter is called a *geometrically open filter* which has the openings larger than the size of certain soil particles.

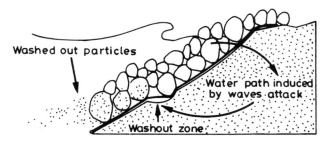

Figure 5.27 Development of filter failure resulting from washout of fines (after Mlynarek and Fannin, 1998).

4   Where the geotextile can move, an intermediate layer of gravel-sized particles may be placed over the geotextile and the riprap of sufficient weight should be placed to prevent dynamic flow action from moving either riprap stone or geotextile.
5   Keeping in view the severe hydraulic conditions caused by continual or even reversing dynamic flows, soil-geotextile filtration tests (in accordance with ASTM D5101-01 for cohesionless soils, and ASTM 5567-94, reapproved 2001 for cohesive soils) should always be performed with site soil samples for appropriate selection of geotextiles.
6   Since the placement of riprap is generally more severe than the placement of drainage aggregate, Class 2 classification for monofilament and Class 1 for all others should be considered to meet the survivability requirements, as discussed in Sec. 4.11, in the absence of any site-specific evaluation, testing and design.

A number of geosynthetic manufacturers have developed their own design manuals. However, a proper basic knowledge in the background of the design methodology must be used to verify the real value of the design procedure of a particular product before its use in field applications. Full-scale prototype testing is a good method of verifying designs but costs may limit application to only major projects. On smaller projects, a physical modelling of the cover layer under hydraulic attack can be carried out to verify a design or to refine the results of mathematical modelling.

ILLUSTRATIVE EXAMPLE   5.8
Consider a revetment system as shown in the following figure with the following parameters:

α = angle of slope in degrees
γ_w = unit weight of water in kN/m^3
W'_c = submerged weight per unit area of protective covering material in kN/m^2
Δh = head loss across the geotextile in m

Under what condition will there be no geotextile uplift?

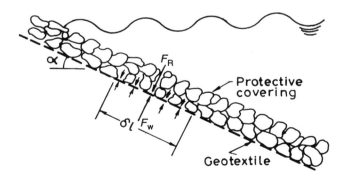

SOLUTION
There will be no geotextile uplift if the force, F_R, from the riprap perpendicular to the slope exceeds the force, F_w, from the water pressure beneath the geotextile.

Mathematically, for no geotextile uplift

$F_R > F_w$

$\Rightarrow (W'_c \cos\alpha) \Delta l > (\gamma_w \Delta h) \Delta l$, where Δl is length of any part of the slope.

$\Rightarrow \Delta h < (W'_c \cos\alpha)/\gamma_w$ **Answer**

For routine field applications, the stability of the covering material in waterway revetments incorporating geotextiles can be assessed using an analytical approach based on a stability number, S_N, defined as below (Pilarczyk, 1984a):

$$S_N = \frac{H}{\gamma'_R D}, \qquad (5.70)$$

where H is the wave height in metre, D is the depth of protective covering material in metre, and γ'_R is the submerged relative unit weight of the covering material (dimensionless) as defined below:

$$\gamma'_R = \frac{\gamma_c - \gamma_w}{\gamma_w}, \qquad (5.71)$$

where γ_c is the unit weight of protective covering material (kN/m^3) and γ_w is the unit weight of water. Usually γ'_R varies from 1.24 to 1.38.

The minimum depth of the protective covering material required to withstand the wave action can be determined from the table of required stability numbers (Pilarczyk, 1984a,b; Tutuarima and Wijk, 1984) listed in Table 5.5. It should be noted that in each case, it is assumed that the permeability of the geotextile exceeds that of the soil. If the geotextile permeability is only equal to that of the soil, then the above required stability numbers should be reduced by 40% or to 2.0, whichever gives the higher value (Tutuarima and Wijk, 1984).

It should be noted that the design principle for waterway revetments based on stability number approach as described above can also be applied to coastal erosion control. Since waves of larger wave heights are usually encountered in coastal environment, heavier revetments will be required to control the coastal erosion. It has been reported that for very high waves, the stability number approach seriously underestimates the stability of riprap armour

Table 5.5 Required stability numbers for waterway revetment systems

Protective covering	Required stability number
Unbonded riprap	<2
Free blocks	<2
Asphalt grouted open aggregate	<4.3
Sand-filled mattresses	<5
Articulated blocks	<5.7
Grouted articulated blocks	<8

stones. In such a situation, the following formula should be used to determine the appropriate stone weight, W (Hudson, 1959):

$$W = \frac{\gamma_s H^3 \tan \alpha}{\lambda_D (G_s - 1)^3},\tag{5.72}$$

where H is the wave height, γ_s is the unit weight of solid stones, G_s is the specific gravity of the stones, λ_D is the damage coefficient, and α is the slope angle. For no damage and no overtopping of the revetment, $\lambda_D = 3.2$.

The stone weight found from Equation (5.72) can be converted into an average stone diameter, D_{50}, using the following expression:

$$D_{50} = \sqrt[3]{0.699W} \tag{5.73}$$

where, W is in tonne and D_{50} is in metre.

The size of the stone obtained from Equation (5.73) is likely to be too large for direct placement on a geotextile-filter sheet. In such a situation the intermediate layer or layers of smaller stones of suitable grading that will not cause any damage to the geotextile should be provided between the large stone armour having a minimum thickness of $2D_{50}$ and the geotextile.

ILLUSTRATIVE EXAMPLE 5.9
Consider a waterway revetment system with the following parameters:

Wave height, $H = 1.2$ m
Unit weight of protective covering material (unbonded riprap), $\gamma_c = 23$ kN/m^3

Assume that the permeability of the geotextile is greater than the permeability of soil to be protected. Determine the minimum depth of the protective covering. Take unit weight of water, $\gamma_w = 10$ kN/m^3

SOLUTION
From Equation (5.71), the relative submerged unit weight of the covering material

$$\gamma'_R = \frac{23 - 10}{10} = 1.3$$

From Equation (5.70), the stability number

$$S_N = \frac{1.2}{1.3\,D}$$

From Table 5.5, for unbonded riprap

$$S_N < 2$$

$\Rightarrow \quad \dfrac{1.2}{1.3\,D} < 2$

$\Rightarrow \quad D > 0.46$ m

Thus, the minimum depth of protective covering is 0.46 m. **Answer**

232 Analysis and design concepts

ILLUSTRATIVE EXAMPLE 5.10
Determine the average size of the armour stone in a revetment system with geotextile filter required to protect a 25° slope from waves up to 2.5 m high assuming that no overtopping of the revetment occurs. Take specific gravity of stone, $G_s = 2.70$.

SOLUTION
From Equation (5.72),

$$W = \frac{\gamma_s H^3 \tan \alpha}{\lambda_D (G_s - 1)^3} = \frac{2.70 \times (2.5)^3 \times \tan 25°}{3.2 \times (2.70 - 1)^3} = 1.25 \text{ t}$$

From Equation (5.73),

$$D_{50} = \sqrt[3]{0.699 W} = \sqrt[3]{0.699 \times 1.25} = 0.96 \text{ m} \qquad \textbf{Answer}$$

If the erosion control systems are required using bags, tubes and mats, the geotextile must be permeable but soil-tight, usually O_{90} about 0.1 to 0.2 mm. Large bags (> 1 m³) are fabricated usually of strong polyester geotextile with mass per unit area larger than 500 g/m² and tensile strength >10 kN/m. These bags must be filled in place using a pumped sand slurry or concrete. Their large size makes them more resistant to movement under water wave attacks. A schematic overview of the failure mechanisms of a stacked sand-filled geotube structure is presented in Figure 5.28. The designer for each particular project should carefully examine all these failure modes. The material durability and the long-term behaviour of geosystems need special attention. Systematic monitoring of realized projects, including failure cases, and evaluation of the prototype and laboratory data may provide useful information for verification purposes, as well as for design.

Rolled erosion control products (RECPs) are initially selected on the basis of a combination of somewhat arbitrary factors. By understanding predictive methods for soil loss, one can predict the soil loss rate and understand the role of these rolled erosion control products and field techniques in limiting the process. For this purpose the Universal Soil Loss

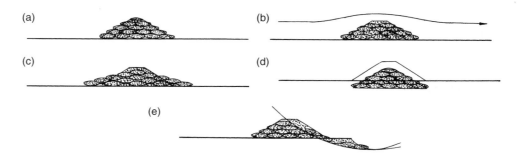

Figure 5.28 Possible failure modes of geosystem structures: (a) theoretical cross section; (b) hydraulic stability; (c) internal stack stability; (d) stability against squeezing/settlement; (e) geotechnical stability (after ACZ, 1990).

Equation (USLE), developed by the US Soil Conservation Service in the 1930s, can be used. This equation can be stated as follows (Ingold, 2002):

$$X = R \times K \times S \times L \times C \times P, \tag{5.74}$$

where X is the annual soil loss per unit area (metric tons per ha); R is a rainfall erosion index that reflects the erosion potential from regional precipitation; K is the soil-erodibility factor; S is the slope-steepness factor; L is the slope length factor; C is the cover and management factor (protected and unprotected conditions); and P is an erosion control practice factor that reflects the maintenance activities of the facility. For more details about all these factors, readers can refer to the contribution by Ingold (2002).

Well-established vegetation is an effective and attractive form of protection for slopes exposed to mild and moderate surface erosion. When designing grassed slopes and waterways, it is required to take into account immediate and long-term flow resistance based upon longevity of the erosion control products being used. Two design concepts are used to evaluate and define a channel configuration that will perform within acceptable limits of stability. These methods are defined as the *permissible velocity approach* and the *permissible tractive force (boundary shear stress) approach*. Under the permissible velocity approach, the channel is assumed stable if the adopted velocity is lower than the maximum permissible velocity. The tractive force approach focuses on stresses developed at the interface between flowing water and the materials forming the channel boundary (Chen and Cotton, 1988; Theisen, 1992).

The permissible velocity approach uses Manning's equation, in which, given the depth of flow, D, the mean velocity, V, in the cross-section may be calculated as:

$$V = 1.49 \, R^{2/3} S^{1/2}/n \tag{5.75}$$

where, n is the Manning's roughness coefficient; R is the hydraulic radius equal to the cross-sectional area, A, divided by the wetted perimeter, P; and S is the friction slope of the channel, approximated by the average bed slope for uniform flow conditions.

The tractive force approach uses a simplified shear stress analysis which is as follows:

$$\tau = \gamma DS, \tag{5.76}$$

where τ is the tractive force; γ is the unit weight of water; D is the maximum depth of flow; and S is the average bed slope or energy slope.

Design based on flow velocity may be limited because maximum velocities vary widely with channel length (L), shape R and roughness coefficients (n). In reality, the force developed by the flow, not the flow velocity itself, challenges the performance of erosion control systems. The maximum shear stress criterion is thus necessary to ensure properly engineered design of channel lining erosion control systems. It should be noted that velocity and tractive force are not directly proportional. Under certain conditions, a decrease in velocity may increase depth of flow, thereby increasing shear stress. Flow duration is another significant parameter affecting the design of erosion control systems, particularly the grassed systems. As the duration of flow progresses, the resistance of grassed surface reduces. Short-term performance of fully vegetated surfaces is impressive at nearly 4 m/s (Theisen, 1992).

As discussed in Sec. 4.8.1, jute geotextile is capable of reducing the erosive effects of rain drops and controlling migration of soil particles of the exposed surface. On biodegradation, jute geotextile forms mulch and fosters quick vegetative growth. The choice of the right type of jute geotextile and plant species is critical for effective results. The species of vegetation needs to be selected carefully considering the local soil and climatic conditions. The Indian standard, IS 14986(2001), suggests the names of plants useful for stabilization of bunds, terrace fences and steep slopes and gullies. The choice of jute geotextile basically depends on the type of soil to be protected. It must be ensured, primarily, that the slope to be protected from rain water erosion is geotechnically stable. The selection of jute geotextile is also required to be done in consideration of the extreme rainfall in a limited time-span at that location as the intensity of rainfall is more important than the average annual rainfall at a place for assessing the erosion index (R) and deciding on the choice of a particular type of jute geotextile. IS 14986 (2001) provides broad guidelines to the choice of the jute geotextile type by the users.

5.9.2 Stabilization

The higher strength of the reinforced soil structure allows for the construction of steep slopes. Compared with other alternatives, geosynthetic-reinforced soil slope structures are cost-effective option for slope stabilization.

From stability considerations, a given or proposed slope should meet the safety requirements, viz. soil mass under given loads should have an adequate safety factor with respect to shear failure, and the deformation of the soil mass under the given loads should not exceed required tolerable limits. The analyses are generally made for the worst conditions, which seldom occur at the time of investigation. Methods, originally developed for analysing unreinforced slopes, have been extended to analyse reinforced slopes taking care of the presence of reinforcements. There are basically four methods for analysing geosynthetic-reinforced soil slopes (Shukla, 2002d):

1. limit equilibrium method,
2. limit analysis method,
3. slip line method,
4. finite element method.

Limit equilibrium method is most widely used to design geosynthetic-reinforced soil slopes. Various limit equilibrium methods have been used in different studies (Ingold, 1982a; Murray, 1982; Leshchinsky and Volk, 1985, 1986; Schmertmann *et al.*, 1987; Jewell, 1990; Wright and Duncan, 1991). In these methods of analysis, it is considered that failure occurs along an assumed or a known failure surface. At the moment of failure, the shear strength is fully mobilized all the way along the failure surface, and the overall slope and each part of it are in static equilibrium. The shear strength required to maintain a condition of limiting equilibrium is compared with the available shear strength, giving the average factor of safety along the failure surface as below:

$$FS = \frac{\text{Shear strength available}}{\text{Shear strength required for stability}} \quad (5.77)$$

Analysis and design concepts   235

The shear strength of the soil is normally estimated by using Mohr-Coulomb strength criterion. Allowable tensile strength of geotextile layers is taken into account while calculating available shear strength. Several slip surfaces are considered and the most critical one is identified; the corresponding (smallest) factor of safety is then taken to be the factor of safety of the slope. It should generally be greater than 1.3. The problem is generally considered in two dimensions, that is, conditions of plane strain are used. A two-dimensional analysis is found to give a conservative result compared to a three-dimensional analysis (dish-shaped surface).

For an assumed circular arc failure plane within the shallow slope (inclination $\beta \leq 45°$) reinforced with horizontal geosynthetic layers (Fig. 5.29), the factor of safety, in terms of shear strength parameters of soil and allowable tensile strength of geosynthetic, can be obtained as below, following the method of slices, commonly used for slope stability analysis of unreinforced soil slopes.

$$FS = \frac{\text{Moment of shear strength of soil and allowable tensile strength of geosynthetic along failure arc}}{\text{Moment of weight of failure mass}}$$

$$= \frac{\sum_{i=1}^{n}(N_i \tan\phi + c\Delta l_i)R + \sum_{j=1}^{m} T_j y_j}{\sum_{i=1}^{n}(w_i \sin\theta_i)R}, \quad (5.78)$$

where w_i is the weight of ith slice; θ_i is the angle made by the tangent to the failure arc at the centre of the ith slice with horizontal; $N_i = w_i \cos\theta_i$; Δl_i is the arc length of ith slice; R is the radius of circular failure arc; c and ϕ are shear strength parameters, cohesion and angle of shearing resistance (total or effective depending upon field situations), respectively; T_j is the allowable geosynthetic tensile strength for the jth layer; y_j is the moment arm for jth geosynthetic layer; n is the number of slices; and m is the number of geosynthetic layers.

The stability of steep reinforced slopes (inclination $\beta > 45°$) can be analysed by the tieback wedge analysis approach used for vertical retaining walls, as described in Sec. 5.3.

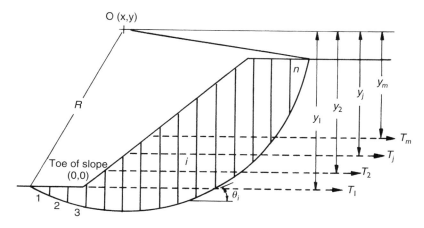

Figure 5.29 Details of method of slices for circular slip analysis.

Limit equilibrium methods do not furnish any information on soil deformations. Nevertheless, these methods have been very useful in solving slope stability problems and need less computational efforts. By means of suitable factors of safety, whose choice is largely governed by experience, the amount of deformation can be limited. It is required to consider separate factors of safety for the soil and the geosynthetic because their deformational characteristics are different.

Limit analysis is a universal method for correct and accurate solution of the slope stability problems (Sawicki and Lesniewska, 1989; Michalowski and Zhao, 1993, 1994, 1995; Zhao, 1996; Jiang and Magnan, 1997; Porbaha and Lesniewska, 1999; Porbaha et al., 2000). It is based on plasticity theory. This method can be applied to slopes (and also other structures) of arbitrary geometry, complicated loading conditions and homogeneous as well as heterogeneous plastic materials. Using the limit theorems, it is possible to bracket the collapse load even if it cannot be determined exactly. In the lower bound approach, we determine whether there exists an equilibrium stress field, which is in equilibrium with the applied load and with which the plastic yield condition is nowhere violated in the slope. If such a stress field exists, it can be ascertained that the applied load is less than the limit load and no plastic failure will occur in the slope. In the upper bound approach, we search for a kinematically admissible velocity field; we then calculate the corresponding internal and external plastic power dissipations. If the external power dissipation is higher than the internal one, the load can be said to be greater than the limit load. In this way, the limit load can be defined as the load under which there exists a statically admissible stress field; yet a free plastic flow can occur. An efficient and accurate numerical technique like then finite element method is vital to make limit analysis applicable to complicated problems of slope stability.

Slip line method is based on the derived failure criterion describing the failure of a homogenized geosynthetic-reinforced soil composite and the application of the method of stress characteristics (Anthoine, 1989; de Buhan et al., 1989). The derivation of the failure criterion for a geosynthetic-reinforced soil composite was presented by Michalowski and Zhao (1995). The limit loads on geosynthetic-reinforced soil slopes can be calculated using the slip line method described by Zhao (1996). This approach is expected to have a wider application in the analysis of slopes with less conventional reinforcements such as continuous filament or for fibre-reinforced soil slopes.

Finite element method of analysis is generally based on a quasi-elastic continuum mechanics approach in which stresses and strains are calculated. Since geosynthetic-reinforced soil slopes exhibit large deformations during the stage construction process, it is appropriate to adopt a nonlinear soil model for the stress–strain analysis with a suitable failure criterion (e.g. Mohr-Coulomb Criterion). Such models of varying degrees of complexity have been developed. They require additional parameters, but these can usually be furnished by the standard triaxial test if shear and volumetric strain measurements can be carried out with sufficient accuracy. The geosynthetics are also required to be modelled by an appropriate constitutive model. More details on this method can be found in the works of Rowe and Soderman (1985), Almeida et al. (1986), Ali and Tee (1990), and Porbaha and Kobayashi (1997).

Among the available methods of analysing the stability of geosynthetic-reinforced slopes, limit equilibrium methods are most popular. Essentially, in each method, a failure mechanism is assumed and some of the limit equilibrium requirements are satisfied. Most of the limit equilibrium methods, with their inappropriately oriented slip surfaces, are not correct from the viewpoint of the mathematical theory of plasticity, and they do not furnish any

Analysis and design concepts 237

information on soil deformations. Ideally, other methods of slope stability analysis described in this chapter are attractive, but they are really only suited to research studies

Stabilization of slopes is one of the most challenging tasks for the geotechnical engineers. Standardization is not possible due to a variety of cases observed under field conditions. Use of geosynthetics allows a reduction of earthwork by changing the geometry as also allowing the use of soils with average mechanical properties.

Geosynthetic-reinforced slopes are designed to provide the following three basic modes of stability (Simac, 1992):

1 internal stability (Fig. 5.30(a))
2 global stability (Fig. 5.30(b))
3 surficial stability (Fig. 5.30(c)).

The factor of safety must be adequate for both the short-term and long-term conditions and for all possible modes of failure, similar to those for unreinforced slopes.

Internal stability controls the quantity, strength, length and vertical spacing of the primary reinforcement elements. As the system moves towards failure, the soil deforms, creating tension in the reinforcement across the failure plane. A limit equilibrium analysis can be carried out to determine the amount of reinforcement tension necessary to maintain equilibrium with a reasonable factor of safety (minimum FS = 1.3).

Conventional slope stability analysis for potential failures completely around the reinforced soil mass is designated as *global stability*. The global stability of the entire reinforced soil mass is usually controlled by the foundation upon which it rests. The influence of foundation soil strength, groundwater conditions, soil stratigraphy, imposed surcharge loadings and slope geometry must be analysed to ensure satisfactory performance. A compound failure mode should also be analysed, where the potential failure passes partially through the reinforced mass and the soil that is being retained by the reinforced slope.

Surficial stability determines the secondary reinforcement requirements of the reinforced slope to preclude surficial sloughing of the slope face during and after construction. Depending on the magnitude of erosive forces imposed on the slope face, erosion control measures can range from temporary to permanent armoured systems. The available experience

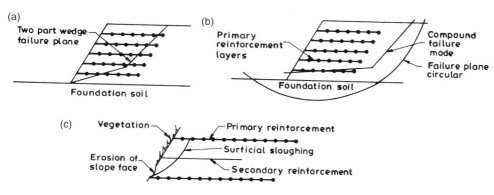

Figure 5.30 Modes of failure: (a) internal stability; (b) global stability; (c) surficial stability (after Simac, 1992).

suggests a maximum 60 cm vertical spacing between the secondary reinforcement layers, recommended to be a minimum 120 cm long. The secondary reinforcement can also be recommended at each compaction lift, about 22–30 cm intervals, in case of some cohesionless soils.

Field and numerical model test results indicate that the limit equilibrium approach to reinforced slope design provides a suitable though conservative design approach (Christopher et al., 1990). A step-by-step design procedure incorporating the circular arc approach may be found in Christopher and Leshchinsky (1991). Simplified charts for the design of geosynthetic-reinforced slopes have been proposed by many research workers (Jewell and Woods, 1984; Jewell et al., 1984; Christopher and Holtz, 1985). These charts can be used to evaluate the preliminary stability of geosynthetic-reinforced slopes before more thorough design procedures are performed. Critical reinforced slopes as well as permanent slopes (having design life greater than 1 to 3 years) should be designed using comprehensive slope stability analyses. The factor of safety against slope stability should be taken from the critical surface requiring the maximum reinforcement.

The major steps for the design of a reinforced slope can be given as follows:

Step 1: Define geometrical dimensions of the slope (slope height, H; slope angle, β).
Step 2: Define loading conditions (surcharge load, temporary live load and dynamic load).
Step 3: Determine the engineering properties (permeability, shear strength and consolidation parameters) of the foundation soils and the slope soils.
Step 4: Locate the groundwater table. For slope and landslide repair projects, identify the cause of instability and locate the previous failure surface.
Step 5: Determine the properties (gradation and plasticity index, compaction characteristics, shear strength parameters and chemical composition that may affect the durability of geosynthetic reinforcement) of available reinforced fill.
Step 6: Establish geosynthetic properties (strength and modulus, soil-geosynthetic interface friction). Also establish tolerable geosynthetic deformation requirements. The geosynthetic strain can be allowed up to 2–10%. The selection of geosynthetic should also consider drainage, constructability (survivability) and environmental requirements.
Step 7: Determine the factor of safety of the unreinforced slope and determine the geosynthetic reinforcement requirement (vertical spacing, and length) based on the internal stability analysis.
Step 8: Check the factors of safety against external stability including sliding, load-bearing failure, foundation settlement, deep-seated slope failure and dynamic stability as carried out for conventional retaining wall designs assuming that the geosynthetic-reinforced soil mass acts as a rigid body in spite of the fact that it is really quite flexible. The minimum values of factor of safety against sliding, load-bearing failure, deep-seated failure and dynamic loading are generally taken to be 1.5, 2, 1.3 and 1.1, respectively.
Step 9: Check the requirements for surface and subsurface water control, where surface water runoff and drainage are critical for maintaining slope stability.

The design concepts for some popular stabilization methods are described in Sec. 6.3.9 along with the application guidelines. It should be noted that reinforced slopes essentially are mechanically stabilized (MSE) structures with similar behaviour properties and design criteria as those of vertical-faced, gravity MSE walls such as reinforced earth or a concrete segmental unit combined with geogrid reinforcement. Therefore, for convenience in designs, one can also consider reinforced slopes as the gravity earth retaining structures with a sloped face.

Reinforced slope design can ideally be carried out using a conventional slope stability computer program modified to account for the stabilizing effect of reinforcement. However, to facilitate the complete design of geosynthetic-reinforced slopes, many types of software are commercially available, though some are limited to specific soil and reinforcement conditions. Leshchinsky (1997) has mentioned such a software, called 'Reslope'. For a given problem, including the ultimate strength of the reinforcement layers, the software yields the optimal length and spacing of the geosynthetic layers. This layout satisfies the various specified factors of safety input by the user.

5.10 Containment facilities

5.10.1 Landfills

Landfill design and construction technology has advanced rapidly in recent years. The most important requirement of a landfill is that it should not pollute or degrade its environment, that is, it should always be environmentally friendly. So, the primary engineering assignment in designing, constructing and operating landfills is to provide efficient barriers against contamination. This requirement is achieved by both careful siting and by adopting proper design/construction method. The site of the landfill must be geologically, hydrologically, and environmentally suitable. A detailed design of several landfill elements is necessary. The proper functioning of each of these elements is essential to construct and maintain a landfill in an environmentally sound manner.

Since water is the most important transporting agent for pollutants, the infiltration of water into and the extraction of water out of the waste body must be controlled by reliable technical means, such as leachate collection and removal systems. Drainage facilities must maintain minimum gradients to facilitate gravitational flow. The physical and biological properties of the waste, as well as the availability of construction material, are important parameters for the design of liners and covers. Aiming at a justifiable degree of safety with respect to the environment, nationally or regionally responsible authorities issue minimum requirements and some basic rules for the design of landfill liners and covers. These rules differ from one country to another and sometimes even within one country. The readers can find the details of the German practice on liners and landfill covers in the contribution by Zanzinger and Gartung (2002a).

Special attention is paid to the properties and the placement of the waste material. The waste body is considered a barrier by itself. The refuse should be in such a condition that the stability of the landfill is granted, and there is little or no tendency for harmful material to be dissolved and transported with the seeping water, and the deformation due to settlements should be predictable and small. So the integrity of the cover would not be impaired in the long-term. In summary, the landfill structure forms a multibarrier system. Each of the barriers has to meet certain minimum technical requirements, independent of the performance of the other barriers.

A municipal solid waste (MSW) landfill must be designed and constructed to accept highly variable waste system. It must be able to prevent groundwater pollution, collect leachate, permit gas venting and provide for groundwater and gas monitoring. Most, if not all, of the design and construction principles for MSW landfills apply equally to hazardous waste landfills.

The landfill project must consider the various regulations, formed by responsible agencies such as the United States Environmental Protection Agency (USEPA), governing landfill siting, design, construction, operation, groundwater and gas monitoring, landscaping plan, closure monitoring and maintenance for the design life that may be 30 years.

The properties, advantages/limitations, and design requirements for compacted clay liner, geomembrane liner and geosynthetic clay liner are summarized in the following paragraphs.

Compacted clay liner (CCL)

The compacted clay liner for a landfill must have a low permeability to prevent or minimize leachate leakage, adequate shear strength for stability and minimal shrinkage potential to prevent desiccation cracking. Since the compaction on wet-side of optimum moisture content minimizes hydraulic conductivity (Lambe, 1958; Mitchell et al., 1965; Boynton and Daniel, 1985), the clay soil liners must be designed for wet-side. The range typically varies from 0–4% points wet of standard or modified Proctor optimum. Typically, clay liners must have a hydraulic conductivity, $k \leq 1.0 \times 10^{-9}$ m/s. The water content and the dry unit weight must be established in a range so that the compacted clay liner will also have adequate strength to withstand high overburden pressures and shear stresses depending on the height of the landfill that may be up to 75 m high. Note that the practice of wet-side compaction may cause problems in arid regions where near-surface clays may desiccate during periods of drought. It is a difficult task to find a way to compact clay soil with both low hydraulic conductivity and low shrinkage potential. However, the solutions such as using soils rich in sand, placing the soil at the lowest practical water content or avoiding the use of highly plastic soils can be adopted to meet the design requirements appropriately. In designing the compacted clay liner, the causes of failure must be considered: subsidence, desiccation cracking and freeze–thaw cycling. The required minimum thickness of the compacted clay liner is normally 0.5 m. For hazardous waste landfills, the thickness may be increased up to 1.5 m placed in lifts of 0.25 m each.

Geomembrane liner

The most widely used geomembrane in the landfill engineering is high density polyethylene (HDPE), because this offers excellent performance for landfill liners and covers. If greater flexibility than HDPE is required, then linear low density polyethylene (LLDPE) geomembranes can be used. The geomembranes with textured surfaces on one side or both sides can be used for improving stability on slopes. The introduction of textured geomembranes significantly helps, allowing for 3(H):1(V) side slopes and even steeper ones in some cases. The soil–geomembrane and geomembrane–geotextile interface friction values should be correctly estimated because these values are critical for the proper design of geomembrane-lined side slopes of landfills. Cover soils can slide over the surface of the geomembrane. Alternatively, the geomembrane can also fail by pulling out of the anchor trench and slipping downhill.

For an intact geomembrane, the transfer of moisture or gas transmission across the membrane occurs by diffusion and the rates are very low. For example, the water vapour transmission rate through 30 mil (0.75 mm) thick HDPE geomembrane is 0.02 g/m²/day, whereas the methane gas transmission rate through 24 mil (0.6 mm) thick HDPE geomembrane is 1.3 ml/m²-day-atm (USEPA, 1988). In order to warrant a sufficient robustness of the geomembrane in handling, the specified minimum thickness of approved geomembranes is 2.5 mm. This thickness also happens to be very satisfactory with respect to the sealing function. However, HDPE geomembranes of 2.5 mm are not very flexible. The minimum width of the geomembrane is 5 m in order to minimize the amount of field seaming needed to create large waterproof sheets (Zanzinger and Gartung, 2002a).

The tensile stresses, developed due to unbalanced friction forces (Fig. 5.31(a)) and/or due to localized subsidence (Fig. 5.31(b)), must be properly analysed. The former situation arises when a material with high interface friction (like sand or gravel) is placed above the geomembrane and a material with low interface friction (like high moisture content clay) is placed beneath the geomembrane. The geomembrane goes into a state of pure shear and carries a tensile force. The factor of safety, FS_T, for geomembrane against tensile failure is expressed as

$$FS_T = \frac{T_a}{T_r}, \quad (5.79)$$

where T_a is the allowable tensile force per unit width in the geomembrane, and T_r is the required tensile force per unit width in the geomembrane

The second situation arises whenever localized subsidence occurs beneath a geomembrane that is supporting a cover soil, often happening in landfill closure situation where the underlying waste has been poorly and nonuniformly compacted. The out-of-plane forces from the overburden cause some induced tensile stresses in the geomembrane, depending upon the dimensions of the subsidence zone and the cover soil properties (Koerner and Hwu, 1991). The calculated value of the required tensile strength of geomembrane for the specific site situation must be compared with an appropriate laboratory simulation test (three-dimensional axisymmetric tension test) for allowable tensile force in the geomembrane.

Geotextiles can be recommended with geomembranes to function as cushion in order to enhance its puncture resistance during installation and in-service in containment systems including landfills. The design and selection of the geotextile for the specific geomembrane types and thickness consist of evaluating the local stress conditions. Determining the

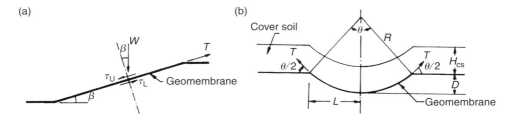

Figure 5.31 (a) Shear and tensile stresses acting on a geomembrane due to unbalanced forces; (b) tensile stresses in a geomembrane mobilized by cover soil and caused by subsidence.

puncture force, P, as per site conditions, the design puncture point load, P_L, is calculated as follows:

$$P_L = FS \times P, \tag{5.80}$$

where FS is factor of safety for long-term loading conditions, generally greater that 10. Based on the value of P_L, mass per unit area required for each type of geotextile can be considered. Figure 5.32 provides an example design chart for continuous filament polypropylene needle-punched nonwoven geotextiles on HDPE geomembranes. A typical example of a protective layer is a needle-punched nonwoven HDPE of 1200 g/m² plus 100–150 mm of sand or crushed stone of maximum 8 mm grain size.

The landfill cover should be designed with the same degree of attention and care that is applied to the soil liner. Note that unlike a liner, a soil cover acts as a hydraulic barrier against infiltrating water from outside only; it is not required to act as a barrier against leakage of leachate solutes under combined *advection* and *diffusion*. Note that the advection refers to the process by which solutes are transported simultaneously, along with the flowing fluid/solvent in a porous medium under a hydraulic gradient, whereas the diffusion refers to the movement of solutes/dissolved substances under a chemical and concentration gradient. The solutes can diffuse in the same direction as the advective movement, or they can diffuse in an opposite or counter direction. Figure 5.33 shows a schematic diagram of conventional landfill design showing advective and diffusive flows acting in the same direction. The relative importance of diffusion as a leakage pathway increases as the hydraulic conductivity of the barrier decreases (Qian et al., 2002).

Geomembrane-lined soil slopes (Fig. 5.34) require proper stability checks. The stability of the overlying materials (soil/drainage geosynthetic) as well as the tensile stresses that may be induced in the underlying geomembranes should always be performed. The interface friction values between the geomembrane and the overlying materials, generally evaluated from simulated direct shear tests, play a great role in the stability analysis. Both the stability of the overlying soil materials and the reduction of tensile stresses in the geomembrane can be

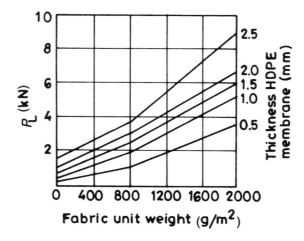

Figure 5.32 Example design chart for continuous filament polypropylene needle-punched nonwoven geotextiles on HDPE geomembranes (after Werner et al., 1990).

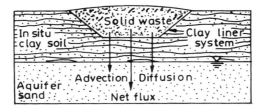

Figure 5.33 Schematic diagram of conventional landfill design showing advective and diffusive flow acting in the same direction (after Qian et al., 2002).

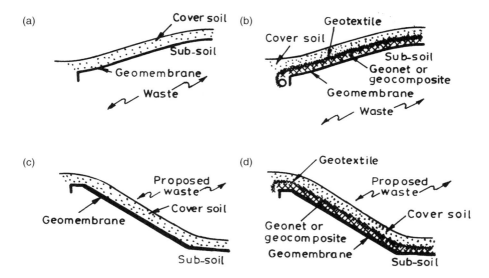

Figure 5.34 Geomembrane-lined slopes: (a) landfill cover with soil above geomembrane; (b) landfill cover with drainage geosynthetic above geomembrane; (c) landfill liner with soil above geomembrane; (d) landfill liner with drainage geosynthetic above geomembrane (after Koerner and Hwu, 1991).

accommodated by reinforcing the cover soil with either the geogrids or geotextiles (Koerner and Hwu, 1991).

Recommendations must be made for the use of geomembrane without any hole or opening. If there is a hole in a geomembrane liner, the leachate will move easily through the hole and seepage will takes place through the soil subgrade (Fig. 5.35(a)). With a clay soil liner alone, seepage takes place over the entire area of liner (Fig. 5.35(b)). With a composite liner, only a limited amount of leachate will pass through any hole in the geomembrane, but it will then encounter low-permeability clay soil, which will impede further migration of the limited amount of leachate passing through the hole (Fig. 5.35(c)). Thus, a composite liner (i.e. geomembrane on low-permeability soil) is more effective in reducing the rate of leakage through the liner than either a geomembrane alone or a soil liner alone (Giroud and Bonaparte, 1989). The designer must keep in mind that the geomembrane should not be

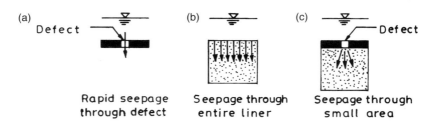

Figure 5.35 Seepage patterns through: (a) geomembrane liner; (b) clay soil liner; (c) composite liner (after Qian et al., 2002).

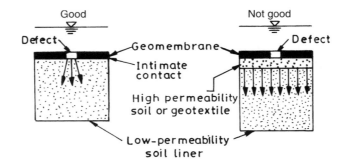

Figure 5.36 Proper design of composite liner for intimate hydraulic contact between geomembrane and compacted soil (Daniel, 1993).

separated from the clay liner with permeable materials, such as a bed of sand or a geotextile, because this would jeopardize the intimate contact (Fig. 5.36).

For designing the leak detection monitoring system, the leakage assessment should be based on analytical approaches supported by empirical data from other existing operational facilities of similar design. It must be noted that the leakage is significantly affected by the performance of the geomembrane liner controlled by the defects or penetrations of the liner, including imperfect seams, punctures or pinholes caused by construction defects.

Geosynthetic clay liner (GCL)

Geosynthetic clay liners are used as a substitute for compacted clay liners in cover systems and composite bottom liners. They are installed on horizontal surfaces as well as on slopes by unrolling and overlapping the edges and ends of the panels. Overlaps self-seal when the bentonite comes in contact with water, that is, hydrates. When a geosynthetic clay liner hydrates, the bentonite swells in the pores, thereby forming a watertight sheet that also offers a protection to the overlying geomembrane liner. It has many advantages over a CCL including the following (USEPA, 1993; Snow et al., 1994):

- easily shipped to any site and thus can be made easily available at any site;
- simple and rapid installation;
- no requirement of heavy equipments for installation;
- less requirement of vehicular traffic and less energy use for installation;

- lower thickness, approximately 0.5 in. (13 mm), and hence conservation of landfill space;
- material quality (consistency and uniformity) maintained in a controlled environment;
- lower consumption of construction water, dust generation, and vehicular traffic during installation;
- lower susceptibility to desiccation cracking;
- self-healing capabilities if punctured;
- better resistance to freeze/thaw and wet/dry cycles;
- availability of tensile strength developed by the geotextiles or geomembranes;
- can tolerate significantly more differential settlement;
- relatively simple, straightforward, common-sense procedures for quality assurance, thus making economical assurance system.

Note that compacted clay liners have also some advantages over the geosynthetic clay liners, such as the large thickness (approximately 2–3 ft (600–900 mm)) that makes them virtually puncture proof, and greatly increases breakthrough time by diffusion, and at the same time there is a long history of use of compacted clay liners. In fact, the substitution of a GCL for a CCL should be decided based on the evaluation of contaminant transport equivalency between them. This evaluation should be based on comparing not only advective mass fluxes through the liner, but also diffusive mass flux during the lifetime of the landfill. The diffusive mass flux decreases and the advective mass flux increases with time for both the GCL and the CCL. Thus, the contaminant transport by diffusion is important during the early stages. However, the contaminant mass transported by diffusion is relatively small and the effect of diffusion can be ignored during the later stages (Qian et al., 2002).

There are several factors that affect the hydraulic conductivity of geosynthetic clay liners, such as type of permeants and confining stress. The effects of wet–dry cycling and freeze–thaw cycling also must be considered when selecting geosynthetic clay liners in the bottom liner and final cover systems for landfills. With increase in confining stresses, the hydraulic conductivity of a geosynthetic clay liner generally decreases significantly, mainly because of lower void ratio of bentonite resulting from higher confining stresses. Alternate wetting and drying may occur in a geosynthetic clay liner in final cover systems of landfills and site remediation projects. When exposed to freeze–thaw cycling, the hydraulic conductivity of a geosynthetic clay liner does not get any significant changes, although the compacted clay liners generally undergo large increases in hydraulic conductivity when exposed to freeze–thaw cycling. The designers must note that in general, GCL falls between compacted clay liners and geomembranes in terms of ability to maintain their hydraulic integrity during distortion such as that induced by differential settlement in landfill final covers. Hydraulic conductivity of typical GCL are generally equal to or less than 1.0×10^{-11} m/s to 5.0×10^{-11} m/s.

The stability of geosynthetic clay liners is an important design consideration because of the low shear strength of the bentonite after hydration. For higher shear strength applications, reinforced geosynthetic clay liners (e.g. geotextile-encased, stitch-bonded or geotextile-encased, needle-punched) should be recommended. The design should consider the possibility of a shearing failure involving a geosynthetic clay liner at the following three locations (Daniel et al., 1998):

1   the external interface between the top of the geosynthetic clay liner and the overlying material (soil or geosynthetic);

2   internally within the geosynthetic clay liner;
3   the external interface between the bottom of the geosynthetic clay liner and the underlying material (soil or geosynthetic).

Design values of internal shear strength of geosynthetic clay liners should be measured on a product-specific basis from laboratory direct shear tests under conditions closely simulating those expected in the field (Fox *et al.*, 1998). The reduction of internal or interface shear strength from peak to residual is dependent on the reinforcement type of the GCL or contact materials at the interface. As a general guideline, unreinforced GCLs are not recommended for slopes steeper than 10(H):1(V) (Frobel, 1996; Richardson, 1997b). In fact, one should not design slopes that exceed the safe slope angle for the geosynthetic clay liners or their respective interfaces within the systems. Stitch-bonded and needle-punched GCLs probably are suited equally for applications involving a low normal stress (e.g. pond and lagoon liners and cover systems), whereas needle-punched geosynthetic clay liners are probably the better choice for applications where a high normal stress is applied (e.g. landfill bottom liners).

All landfills have at their base a leachate drainage layer consisting of a natural soil (sands and gravels) and/or a geosynthetic drainage material (i.e. geocomposite, such as geotextile bonded to one or both surfaces of a geonet). Landfills with a double composite liner system have both primary and secondary leachate drainage layers, called leachate collection and leak detection layers, respectively. The most essential requirement for a landfill leachate drainage layer is that it should have adequate drainage capacity to handle the maximum leachate flow produced during landfill operations. The leachate head buildup in the drainage layer should generally be less than 12 in. (0.3 m) (Qian *et al.*, 2002). Leachate pipes are generally installed in trenches that are filled with gravel. The trenches are lined with geotextile to minimize entry of fines from the liner into the trench and eventually into the leachate collection pipe. Typical trench details are shown in Figure 5.37. Usually the design shown in Figure 5.37(a) is used in landfills in which liner material is clay and the design shown in Figure 5.37(b) is used in landfills in which the primary liner material is geomembrane. It is essential to have a deeper excavation below the collection trench so that the liner has the same minimum design thickness even below the trench. The geotextile, which acts as a filter, should be folded over the gravel. Alternatively, a graded sand filter may be designed to minimize the infiltration of fines into the trench from waste. The design of geotextile filters and drains has already been discussed in Sec. 5.8. A leachate pipe may fail due to clogging crushing or faulty design. The design and maintenance of leachate pipes for each of these situations must be considered properly.

Starting at the bottom, a typical double composite liner system consists of a minimum 2-ft (0.6 m) thick compacted clay liner (or an alternative liner that can be equivalent to a 2 ft (0.6 m) thick compacted clay liner), followed by a secondary geomembrane liner, secondary leachate collection (or leak detection) layer, a minimum 2 ft (0.6 m) thick primary compacted clay liner (or geosynthetic clay liner that can be equivalent to a 2 ft (0.6 m) thick compacted clay liner), a primary geomembrane and a primary leachate collection system. A 2-ft (0.6 m) thick protective sand blanket tops off this sequence. The leachate collection system consists of a layer of geonet and geotextile. The former provides good in-plane drainage conveyance and the latter good cross-plane drainage together with the ability to exclude (filter out) fines. The geomembranes must be at least 1.5 mm (60 mils) thick if HDPE, or 0.75 (30 mils) thick if made from other polymers. The permeability of the subbase and CCL must not exceed 1.0×10^{-9} m/s.

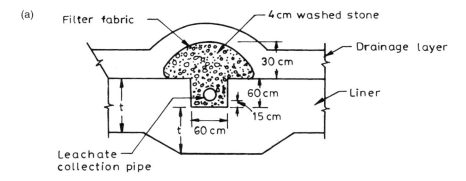
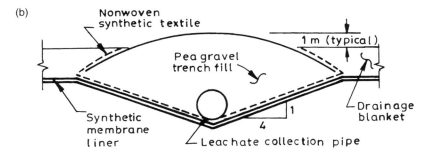

Figure 5.37 Typical leachate collection trench details: (a) for clay liner; (b) for geomembrane liner (after Bagchi, 1994).

Proper closure is essential to complete a filled waste landfill, particularly of a hazardous type. The cover system must be protected from burrowing animals, wind and water erosion, wet–dry cycles and freeze–thaw cycles. In fact, it should be devised at the time the site is selected and the plan and design of the landfill containment structure is chosen. The location, the availability of low permeability of soil, the stockpiling of good topsoil, the availability and use of geosynthetics to improve performance of the cover system, the height restrictions to provide stable slopes and the use of the site after the post-closure care period are typical considerations. The design goals of the cover system are that further maintenance is minimized and that human health and the environment are protected. For hazardous waste facilities, a final cover with minimum requirements (Fig. 5.38(a)) consists of, from bottom to top the following:

1. a 60 cm (24 in.) layer of compacted natural or amended soil with a hydraulic conductivity of 1×10^{-9} m/s in intimate contact with a minimum 0.5 mm (20 mil) geomembrane liner;
2. *a drainage layer*: a minimum 30 cm soil layer having a minimum hydraulic conductivity of 1×10^{-4} m/s, or a layer of geosynthetic materials with the same performance characteristics;
3. *a top, vegetation/soil layer*: a top layer with vegetation (or an armoured top surface) and a minimum of 60 cm of soil graded at a slope between 3% and 5% to prevent erosion and to promote drainage from the area.

Where the type of waste may create gases, vent structures (either soil or geosynthetic) must be included in the cover (Fig. 5.38(b)). Plant roots or burrowing animals (collectively called bio-intruders) may disrupt the drainage and the low-permeability layers to interfere with the drainage capability of the layers. A 90-cm (3 ft) biotic barrier of cobbles directly beneath the top vegetation layer (Fig. 5.38(b)) may stop the penetration of some deep-rooted plants and may stop the invasion of burrowing animals. Settlement and subsidence should be evaluated for all covers and designed into the final cover plans. The cover design process should consider the stability of all the waste layers and their intermediate soil covers, the soil and foundation materials beneath the landfill site, all the liner and leachate collection systems and all the final cover components. When a significant amount of settlement and subsidence is expected within a few years (2–5 years) of closure, an interim cover might be proposed – one that protects human health and environment. When settlement/subsidence is essentially complete, the interim cover should be replaced or incorporated into a final cover.

The complete design of a landfill requires a variety of calculations during the design process in order to demonstrate regulatory compliance and ensure proper design. For this purpose available computer software can be used by the designers. However, they should always keep the limitations of the software being used in mind.

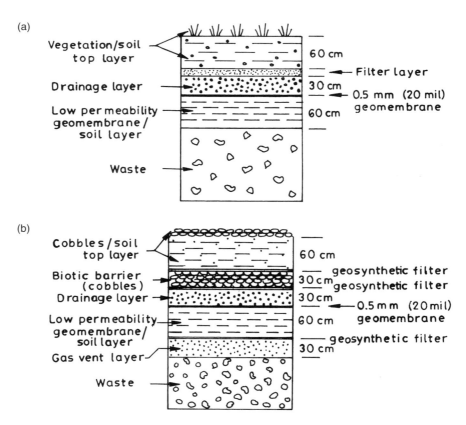

Figure 5.38 USEPA-recommended landfill cover designs (after Landreth and Carson, 1991).

5.10.2 Ponds, reservoirs, and canals

Ponds, reservoirs and canals require lining systems for their effective performance. The primary design consideration for any lining project is the loss of contained liquid throughout the intended service life. The intrinsic permeability of less than 1.0×10^{-14} m/s for HDPE geomembranes far exceeds the requirements for any containment and conveyance project. The main controlling factor is always the loss of water through seams and punctures resulting from damage during or after installation. In fact, leakage, and not permeability, is the primary concern when designing geosynthetic containment structures. Leakage can occur through poor seams, pinholes from manufacture, and puncture holes from handling, placement, or in-service loads. Leakage of geosynthetic liner/barrier systems is minimized by attention to design, specification, testing, quality control and quality assurance.

The geosynthetic barrier should be designed as per its role as a primary or secondary liner keeping in view in-service conditions, installation damages and durability. Note that installation of geomembranes or geosynthetic clay liners is a primary design consideration. Placement, handling, and soil covering operations can also affect geosynthetic design. Liquid depths also govern the design of the liner system. If a designer assumes that the surface liner system will not even leak, then the following provisions must be made (Richardson, 2002):

1 The use of such details as battens and conventional pipe penetration details that cannot be leak tested must be avoided. All components of the containment system must be pressure or vacuum tested.
2 The liner must be protected from harm during its surface life. Thus, if one can see the geomembrane, he must assume that he will get a defect and resultant leakage during the liner's service life. It is more reasonable to assume that the surface impoundment liner has a very minor rate of leakage and design to accommodate that leakage as follows:

- If the contained liquid may harm the environment, then a secondary liner/collection system should be used to monitor the performance of the primary liner.
- If the contained liquid will not harm the environment, then the ability of leakage to drain away from the bottom of the liner must be ensured. This may require a designed underdrain where natural subgrade soil has a low permeability.

The following points are some general design indications concerning the use of geosynthetics in ponds, reservoirs and canals (Duquennoi, 2002):

1 The bottom of the structure should form a slight slope, between 1% and 2% lengthways, and between 2% and 3% sideways.
2 The embankment slopes should be designed according to the state-of-the-art soil mechanics; it has to be underlined so that geomembrane lining systems cannot be used to reinforce slopes. For many applications a 1V:2H (1 vertical by 2 horizontal) slope should be advised, and 2V:3H has to be considered as a maximum.
3 The embankment top should be wide enough to enable geosynthetics anchoring; minimum anchoring length is generally 2 m for ponds and reservoirs and 1 m for canals, but specific designs are to be taken into account. It is generally not recommended to lay a geomembrane directly on the subgrade, except in particular cases when risks of geomembrane puncturing and underliner pore water or gas pressure have been catered

for. A better way to prevent the above-mentioned risks is to specifically design underliner systems. Underliner water drainage can be performed either by gravel layers, gravel-filled drainage trenches, or geosynthetic draining strips. Depending on the volume of water to be drained, perforated geopipes may supplement gravel-based drainage systems. Drain pipes are always connected to a main collecting pipe or manhole and then to pumped or gravity outlet.

4. A protective layer may be interposed between the geomembrane and the subgrade when the latter is not smooth enough to guarantee geomembrane safety, especially below high water head. Geotextiles are now generally preferred because of their possible combined functions of gas drainage and mechanical protection of underliner.

5. The core of a lining system is, of course, the impermeable material, that is, either a geomembrane or a GCL. The design and choice of a lining system should be decided considering economic, hydraulic, mechanical and durability aspects in addition to ease of installation and seam performance aspects as per site-specific requirements.

6. Beside single geomembrane or geosynthetic clay lining systems, it is possible to install double lining systems using two geomembranes with a drainage layer in between them. This solution is still rare in liquid containment and conveyance applications and is only applied where the risk of leaks must be greatly reduced.

7. One of the best ways to prevent anticipated ageing of geosynthetics in general and geomembranes in particular is to limit their exposition to weather action by covering them. The purpose of overliner layers is also to prevent liner damage by floating or transported solids (e.g. ice and wood), by operating vehicles or machines (e.g. mobile pumping equipment), by burrowing animals and plant roots and by vandalism or accidental human intervention. One usual design is to protect the geomembrane by a geotextile and then to cover it by a layer of granular material. Granular layers may be composed of several sublayers differing in granulometry, from the finest-grained (e.g. fine sand) directly over the geotextile up to the coarsest-grained material (e.g. riprap) on top of the granular layer. Other usual designs may consist in concrete covers, using precast blocks or slabs, in situ poured reinforced concrete layers or even shotcrete. Another purpose of overliner covers is to prevent geomembrane lining system uplift under wind action. Some installation procedures may include temporary ballast over the geomembrane in order to prevent uplift during installation, before installing permanent overliner protection layers.

8. Preventing hydraulic actions such as fluvial erosion in the case of canals requires the use of specifically designed systems, which are generally geosynthetic systems: geocells, geomattresses, geoarmours or geomats. These systems may also be used alone to prevent bank erosion, without covering any geosynthetic lining system.

9. All geosynthetic systems are to be anchored on top of the embankment slopes or on the slope itself, depending on the overall design. The most common anchoring design is the anchor trench, which is generally a square section trench in which the geomembrane is laid on one side and at the bottom; the trench is then backfilled with non-puncturing soil. It is generally recommended that anchor trenches should be deeper and wider than 0.5 m; they should be situated at least 0.7 m from the edge of the slope. For canals, especially, excess geomembrane width related to anchoring design may generate excess cost; anchoring characteristics must then be precisely derived from calculation or alternative techniques such as tying the geomembrane to stakes may be applied. In some applications (e.g. deep reservoirs) intermediary anchoring may be required alongside the slope (Fig. 5.39).

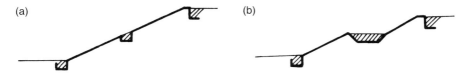

Figure 5.39 Typical examples of anchor trench on slopes.

10 Access roads and tracks are sometimes required, especially in large containment ponds where vehicles have to access the bottom of the pond for maintenance or exploitation purposes. Special attention has to be given to the protection of geomembranes under the road and to the stability of the road over geosynthetics. The subgrade has to be shaped to take the access road into account; extra protection of the geosynthetic lining system should be designed.
11 Connection to concrete structures usually poses the problem of waterproofing continuity. A lot of technical solutions are available, depending mainly on the geomembrane type. Metallic fixations are generally used in association with metallic and elastomeric plates and/or geomembrane overlaps.

Figure 5.40 shows design details of some typical lining systems for ponds, reservoirs and canals based on the case studies presented by Duquennoi (2002). It is important to underline that all the above-mentioned points are closely interrelated in terms of design. For example, it is impossible to select a geomembrane without taking the characteristics of the overliner protection layer into account, and conversely to design a geomembrane protection layer without considering the type of geomembrane. A geosynthetic lining system has thus to be designed as a whole, including subgrade preparation, underliner and overliner layers, and specific features such as the ones described above. Moreover, different geosynthetic lining systems may be equivalent in terms of hydraulic, chemical, or mechanical criteria and the difference may be finally only related to installation needs, economic criteria, or availability. As we can see, the basics of geosynthetic systems for liquid conveyance and containment are fairly simple; however applying them to specific works may be complex and requires more information and experience than what has been briefly presented here.

5.10.3 Earth dams

Safety is the main concern with dams. Although the design of an earth dam is a complex art, with each situation different from the other, the basic steps involved in the design, as mentioned below, are quite easy to follow.

1 A thorough exploration of the foundation and abutments, and an evaluation of the quantities and characteristics of all construction materials available within a reasonable distance of the site.
2 Selection of possible trial design.
3 An analysis of safety of the trial design.
4 The modification of the design in order to meet stability requirements.

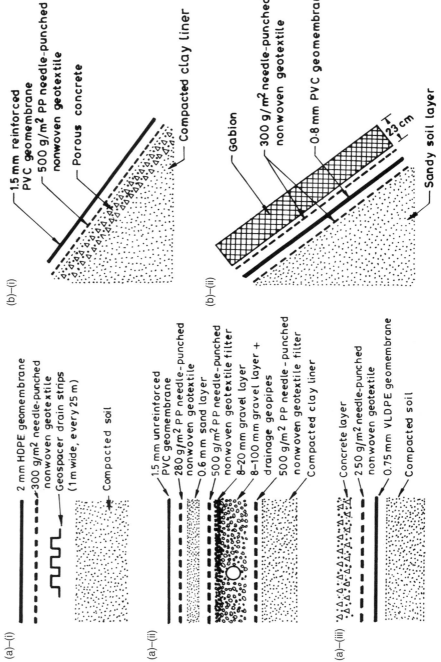

Figure 5.40 Design details of some typical lining systems for ponds, reservoirs, and canal: (a) bottom lining system; (b) slope lining system (after Duquennoi, 2002).

5   The preparation of the detailed cost estimation.
6   The final selection of the design which seems to offer the best combination of economy, safety, and convenience in construction.

Although a conventional design incorporates these steps to a great extent, some recent developments in embankment and dam construction have imposed several challenges in order to achieve perfection and an economical cross-section both in terms of time and money. It is observed that in recent times the use of geosynthetics, in conjunction with the conventional earth dam construction materials, is gaining ground. This imposes a challenging task to the civil engineering practices. Further, use of geosynthetics in earth dams affects their construction procedure and stability. In fact, efficient use of geosynthetics requires special attention. The properties of geosynthetics must be evaluated (see Chapter 3) based on specific criterion and functional requirements, such as acting as a water barrier, filter, drainage medium, protective layer or reinforcement.

In general, the design procedure is guided by the International Commission on Large Dams (ICOLD). A desktop analysis may be undertaken based on the available guidelines. Once a desktop analysis has been completed and suitable geosynthetics are identified, these are subjected to soil-geosynthetic compatibility testing, before making a final selection, which includes consideration of minimum strength and deformation requirements of the geosynthetics. These parameters need to be taken into account of both the short-term loading expected during installation and construction as well as post-construction loads and deformations. While overall embankment settlement may be low, local stresses and strains may be high due to differential settlements or shrinkage of the soil. It is for this reason that a geosynthetic needs to maintain its restraining characteristics, even after local concentration of stresses and strains takes place. There may be a substantial change in the pore size of the geotextiles due to elongation (Legge, 1986). However, the main concern is the extent to which woven tape and staple fibre products' pores elongate when the fabric is placed in tension. It is to be noted that all dams shall be designed on the understanding that there is a significant risk that the core will crack and that the possibility of internal erosion of the core has to be allowed for in the filter design (McKenna, 1989).

Geotextile filters used in dams play a critical role and, therefore, must be carefully selected. The designers of dams must not use the simplistic filter criteria that are sometimes used for non-critical applications. For example, if the soil in contact with the filter has a high coefficient of uniformity – a common situation for earth dams – some simplistic filter criteria may lead to the selection of a geotextile filter that does not prevent soil piping – a typical cause of dam failure. In cases where some particular soil characteristics make it difficult for geotextile filters to strictly meet the filter criteria, filtration tests simulating the conditions expected in the field can be conducted to evaluate the candidate geotextile filters.

When a transmissive geosynthetic is used to provide drainage in a dam, it is important that the transmissivity of the geosynthetic be measured in a laboratory test that simulates the conditions in which the geosynthetic will be used in the field. The design engineer must perform calculations to determine the maximum stresses that are expected on the geosynthetic in the field, and the laboratory team must plan a test where the field boundary conditions, including the maximum applied stresses, are accurately reproduced.

The durability of geosynthetics should always be an essential consideration when they are used in dams. It is also a key requirement in waste disposal applications. Based on knowledge accumulated in designing and constructing lining systems for waste containment, geosynthetics available today clearly have adequate durability for safe use in dams.

5.11 Tunnels

Geosynthetic design in tunnel applications requires that the geosynthetic system must provide watertight integrity for the life of the tunnel. It must withstand different kinds of stress and strain both during installation and after construction. It also must withstand variable and aggressive chemical environments. As both installation and service conditions are severe, it is considered essential that the geomembrane should be exceptionally resistant to tearing, puncturing and abrasion. The geotextile should fulfil the following criteria (Posch and Werner, 1993):

- *Mechanical resistance*: The geotextile must have certain minimum values for mechanical strength and elasticity. These are needed to absorb the stresses resulting from the installation and concreting pressures, the deformation of the inner lining of the tunnel due to load shifting and temperature variations and the joint water pressure increasing locally over time.
- *Chemical resistance*: The geotextile should resist all kinds of rock water, calcium hydroxide, and other constituents such as the binding agents in concrete and grout.
- *Water permeability in the plane*: Residual water, consisting of seepage and leakage, must be reliably drained in the geotextile plane to the bottom drain.

The typical geosynthetic properties required in tunnel applications are given in Table 5.6.

It should be noted that many software programs are available for the analysis, design and specification of geosynthetic-related applications. One can have internet sources for information about these commercially available software programs. The programs are mostly included under the categories of geosynthetics, reinforced slopes and walls, slope stability or ground improvement. Before using any software program, it is important to have a check for demonstrated validation to a standard procedure of analysis and design.

Table 5.6 Geosynthetic properties for tunnel applications (after Gnilsen and Rhodes, 1986)

Property	Minimum specifications	ASTM
A Geotextile (nonwoven polypropylene)		
Thickness (mm)	4.0	D1777
Mass per unit area (g/m^2)	500	—
Grab strength (N)	1150	D1682
Elongation (%)	80	D1682
Trapezoid tear strength (N)	440	D2263
Burst strength (kPa)	2760	D751
Chemical resistance (pH value)	2–13	—
Flammability	Self extinguishing	D568
B Geomembrane (PVC-soft)		
Thickness (mm)	1.5	D374
Ultimate tensile strength (kPa)	7600	D638
Ultimate elongation (%)	300	D638
Brittleness temperature	±7°C	D1790
Flammability	Self extinguishing	D568
Dimensional stability 6 hr at 80°C (%)	2	D1204

Self-evaluation questions

(Select the most appropriate answers to the multiple-choice questions from 1 to 16)

1. The design of a structure incorporating geosynthetics aims to ensure its

    (a) Strength.
    (b) Stability.
    (c) Serviceability.
    (d) All of the above.

2. Which one of the following is the most preferred design approach for geosynthetics?

    (a) Design-by-experience.
    (b) Design-by-cost-and-availability.
    (c) Design-by-specification.
    (d) Design-by-function.

3. The principal partial factors of safety generally employed in limit state design for reinforced soil structures can be

    (a) Less than 1.
    (b) Equal to 1.
    (c) Greater than 1.
    (d) Both (b) and (c).

4. The typical reinforcement spacing for geotextile-wrapped walls varies between

    (a) 0.1 and 0.5 m.
    (b) 0.5 and 1.0 m.
    (c) 1.0 and 2.0 m.
    (d) None of the above.

5. Which one of the following is not the most critical failure mechanism for embankments on soft foundation soils?

    (a) Overall slope stability failure.
    (b) Lateral spreading.
    (c) Settlement.
    (d) Overall bearing failure.

6. If B be the footing width, then for a single layer geosynthetic-reinforced soil, the optimum embedment depth of the geosynthetic layer is approximately

    (a) $0.1B$.
    (b) $0.3B$.
    (c) $0.5B$.
    (d) B.

7. Which one of the following is the incorrect assumption of RFDM for unpaved roads, suggested by Giroud and Noiray (1981)?

    (a) The friction coefficient of the granular layer is large enough to ensure the mechanical stability of the granular layer.

(b) The friction angle of the geotextile in contact with the granular layer under the wheels is large enough to prevent the sliding of the granular layer on the geotextile.
(c) Thickness of the granular layer is significantly affected by the subgrade soil deflection.
(d) The granular layer provides a pyramidal distribution with depth of the equivalent tyre contact pressure on its surface.

8. When a geotextile, within the unpaved road, gets deformed in a curved shape under loading, the pressure against its concave face is

   (a) Equal to the pressure against its convex face.
   (b) Lower than the pressure against its convex face.
   (c) Higher than the pressure against its convex face.
   (d) Equal to or higher than the pressure against its convex face.

9. The minimum thickness of bituminous overlays recommended with paving fabrics is

   (a) 20 mm.
   (b) 40 mm.
   (c) 75 mm.
   (d) None of the above.

10. For abrasion protection, the minimum depth of ballast below the tie for geotextile placement in a railway track is

    (a) 250 mm.
    (b) 300 mm.
    (c) 500 mm.
    (d) None of the above.

11. If k_n be the coefficient of cross-plane permeability of geotextile and k_s the coefficient of permeability of the protected soil, then for a dam clay core the geotextile-filter criterion can be

    (a) $k_n > k_s$.
    (b) $k_n > 10k_s$.
    (c) $k_n > 20k_s$.
    (d) $k_n > 100k_s$.

12. If the band drain's cross section is 100 mm × 5 mm, then the equivalent drain diameter will be approximately

    (a) 33 mm.
    (b) 67 mm.
    (c) 100 mm.
    (d) None of the above.

13. The geosynthetic strain in slope stabilization applications can be allowed up to

    (a) 2% to 10%.
    (b) 5% to 15%.
    (c) 10% to 20%.
    (d) None of the above.

14. In order to warrant a sufficient robustness of the geomembrane in handling, the specified minimum thickness of approved geomembranes in landfill liner systems is generally

    (a) 2.5 mm.
    (b) 5.0 mm.
    (c) 7.5 mm.
    (d) 10 mm

15. Which one of the following is the primary concern when designing geosynthetic containment structures?

    (a) Permeability.
    (b) Leakage.
    (c) Strength.
    (d) None of the above.

16. Geosynthetic design in tunnel applications requires that the geosynthetic system must

    (a) Provide watertight integrity for the life of tunnels.
    (b) Withstand different kinds of stress and strain both during installation and after construction.
    (c) Withstand variable and aggressive chemical environments.
    (d) All of the above.

17. What should be the role of a designer for geosynthetic applications?
18. What is the reason due to which the test property values of a geosynthetic are not directly considered as the design property values?
19. Describe the general approach of design-by-function method for geosynthetics. What are the limitations of this method?
20. What is the fundamental principle of limit state design for geosynthetics?
21. What are the advantages of limit state design approach over the traditional working stress design approach for geosynthetics?
22. If several geosynthetics are found to meet the required factor of safety for an application, then how will you select one of them for use in that particular application?
23. For geosynthetic-reinforced soil structures, the design factors of safety are generally kept lower than normally used for rigid structures. Why?
24. Can you recommend the fine-grained soils as a backfill material? Justify your answer.
25. What are the various failure modes of geosynthetic-reinforced soil retaining walls? Explain them briefly.
26. List the factors governing the lap length in wraparound facing of the geotextile-reinforced retaining wall.
27. Design a 10-m high geotextile-wrapped retaining wall with the following data:

    For the granular backfill
    Unit weight, $\gamma_b = 18$ kN/m³
    Angle of internal friction, $\phi_b = 32°$

    For geotextile
    Allowable tensile strength, $\sigma_G = 35$ kN/m

For the foundation soil
Cohesion, $c = 30$ kN/m^2
Unit weight, $\gamma = 16.8$ kN/m^3
Angle of internal friction, $\phi = 15°$

Foundation soil – geotextile interface shear parameters
Friction angle, $\phi_r = 0.95\phi$
Adhesion, $c_a = 0.90c$

Factor of safety against geotextile rupture $= 1.5$
Factor of safety against geotextile pullout $= 1.5$

28. What are the potential failure modes in an embankment on the soft foundation soil? Describe briefly.
29. A 3.5 m high and 7 m wide embankment is to be built on soft ground with a basal geotextile layer. Calculate the geotextile strength and modulus required in order to prevent block sliding on the geotextile. Assume that the embankment material has a unit weight of 17 kN/m^3 and angle of shearing resistance of 32° and that the geotextile–soil interface angle of shearing resistance is two-thirds of that value.
30. Why are the improvements in load-bearing capacity of geosynthetic-reinforced foundation soil not very significant at low deformations and considerably better at high deformations? Can you suggest a method to improve the low-deformation behaviour?
31. List the parameters affecting the load-bearing capacity of a geosynthetic-reinforced foundation soil. Describe the effects of the most significant parameters.
32. What is BCR? Can it be lower than one?
33. Compare a typical load-settlement curve for a geosynthetic-reinforced soil with that for an unreinforced soil.
34. What is the effect of prestressing the geosynthetic reinforcement on the settlement behaviour of a geosynthetic-reinforced soil?
35. Give a brief description of a geocell mattress foundation. How will you assess its load-bearing capacity.
36. Does the presence of a geosynthetic layer inside the granular fill modify the load transmission mechanism through it?
37. Does the reduction of granular fill thickness in the unpaved road, due to the presence of geosynthetic layer at the interface of granular fill and the soft subgrade soil, depend on the traffic loading?
38. Describe the basic principles of the following design methods for geosynthetic-reinforced unpaved roads:

    (a) Reinforcement function design method.
    (b) Separation function design method.
    (c) CBR design method.

39. Consider:

    Number of passes, $N = N_s = 1000$
    Single axle load, $P = P_s = 80$ kN
    Tyre inflation pressure, $p_c = 480$ kPa
    Subgrade soil CBR $= 1.0$

Modulus of geotextile, $E = 300$ kN/m
Allowable rut depth, $r = 0.3$ m

What is the required thickness of the granular layer for the unpaved road in the presence of geotextile? (Hint: Use RFDM)

40. Consider:

Number of passes, $N = 2000$
Single axle load, $P = 45$ kN
Tyre inflation pressure, $p_c = 550$ kPa
Cohesive subgrade soil CBR $= 1.0$
Allowable rut depth, $r = 75$ mm

What is the required thickness of the granular layer for the unpaved road without geotextile, and with geotextile? (Hint: Use SFDM)

41. Consider:

Number of passes, $N = 10,000$
Single wheel load, $P = 20,000$ lb
Tyre contact area $= 12$ in. $\times 18$ in.
Subgrade soil CBR $= 2.0$
Reinforcement ratio, as determined from modified CBR test, $R = 1.8$

What is the required thickness of the granular layer for the unpaved road without geotextile, and with geotextile?

42. Regarding the proper quantity of asphalt sealant for geotextile in paved road applications, why is sealant in excess of saturation or less than saturation a problem?

43. What are the major causes of paving fabric-related project failures as observed in the past?

44. What are the durability criteria to be satisfied by a selected geotextile for use in railway tracks?

45. Describe the major manufacturing specifications for geotextiles for use in railway track rehabilitation works.

46. What is the minimum depth at which a geotextile should be placed beneath the bottom of a railway track tie to prevent its abrasion? How can you get this value?

47. What are the essential features that must be considered in the design of filters?

48. The permeability and retention criteria for filters are, in principle, contradictory if they have to be fulfilled simultaneously. Do you agree with this statement? Give a proper justification in support of your answer.

49. What is the soil-tightness number? Explain its significance.

50. What are the main conditions known where soils are likely to cause excessive clogging of geotextiles? In such instances, what is the logical recommendation?

51. Describe the mechanism of filter failures resulting from washout of fines.

52. Determine the properties of the nonwoven geotextile (elongation, $\varepsilon > 50\%$) to be used as a subsurface filter adjacent to soil with 70% passing the 0.075 mm sieve under typical installation survivability conditions.

53. Determine the properties of the woven geotextile (elongation $= 60\%$) to be used as a separator between the firm soil subgrade (CBR $= 8$) and the granular subbase under harsh installation survivability conditions.

54. A geotextile-wrapped trench drain is to be constructed to drain a soil mass. Determine the appropriate hydraulic properties of the geotextile to function as a filter in a non-critical application with the following soil properties:

    $C_u = 2.5$
    $D_{15} = 0.15$ mm
    $D_{85} = 0.7$ mm
    $k_s = 1 \times 10^{-5}$ m/s
    Percentage passing 75 µm = 7%

55. A geosynthetic has to be selected to provide drainage behind a retaining wall with a vertical backface. The estimated vertical flow into the drain is 0.0018 m³/s. Determine the required transmissivity of the geosynthetic. Would an ordinary single layer of non-woven geotextile be adequate?

56. What are the special considerations for designing the geotextile filter to be used in erosion control systems?

57. What do you mean by geometrically tight filter?

58. Consider a waterway revetment system with the following parameters:

    Wave height, $H = 1.0$ m
    Unit weight of protective covering material (free concrete blocks), $\gamma_c = 24$ kN/m³
    Assume that the permeability of the geotextile is greater than the permeability of soil to be protected. Determine the minimum depth of the protective covering required to protect the soil. Take unit weight of water, $\gamma_w = 10$ kN/m³.

59. Determine the average size of the armour stone in a revetment system with geotextile filter required to protect a 22° slope from waves up to 3.0 m high assuming that no overtopping of the revetment occurs. Take specific gravity of stone, $G_s = 2.71$.

60. Suggest a suitable reinforcement layout to prevent the shallow slips in the shoulder of a 7.5 m high embankment to be built from London Clay with a 1:2 side slope.

61. List the limitations of limit equilibrium approach for designing the geosynthetic-reinforced soil slopes.

62. Given a 12 m high embankment at a slope angle of $\beta = 42°$, the soil strength parameters are $c = 20$ kPa, and $\phi = 20°$ in both the embankment and foundation sections. The unit weight of soil, $\gamma = 17$ kN/m³. For a failure circle located at coordinates $(+2, +15)$ with respect to the toe at $(0, 0)$ (see Fig. 5.29), determine the factor of safety assuming a radius of 20 m. How many layers of geotextiles spaced 250 mm apart and having an allowable tensile strength of 70 kN/m placed at the interface of the foundation and the embankment are required to increase the factor of safety by 30%? Developing a computer program, find the minimum factor of safety for both the unreinforced and reinforced conditions.

63. List advantages and disadvantages of geosynthetic clay liners over a compacted clay liner.

64. In your opinion, what are the advantages and disadvantages of geosynthetic clay liners over a geomembrane liner?

65. Which type of geomembrane is the most chemically resistant of all geomembranes?

66. How does the seepage pattern through a geomembrane liner differ from the patterns through clay soil and composite soil liners?

67. What is the primary engineering assignment in designing, constructing, and operating landfills?
68. Explain the importance of 'advection' and 'diffusion' in landfill design and construction?
69. How will you make the stability checks for geomembrane lined slopes?
70. Describe some general design indications concerning the use of geosynthetics in ponds, reservoirs and canals.
71. Write the basic steps for the design of earth dams with geosynthetics.
72. Write the typical geosynthetic properties required in tunnel applications.

Chapter 6
Application guidelines

6.1 Introduction

In all the field applications of geosynthetics, the common objective is to install the correct geosynthetic in the correct location without having its properties impaired during the construction process. Many general and specific guidelines have been suggested to meet this common objective by the authors in the past (John, 1987; Ingold and Miller, 1988; Ingold, 1994; Van Santvoort, 1994, 1995; Holtz *et al.*, 1997; Pilarczyk, 2000; Shukla, 2002c,e; Koerner, 2005). Basically, the objectives of the application guidelines are to assist users to exercise their professional judgement and experience in developing site-specific recommendations and to promote the use of best practices in civil engineering constructions with geosynthetics.

Keeping the scope of this book in view, some general and specific geosynthetic application guidelines are given in the present chapter that may be followed while working with geosynthetics during the construction or the maintenance stage. It should be noted that no two projects are identical; unique site conditions may dictate different requirements, techniques and guidelines. The guidelines contained in the present chapter, therefore, may not be universally applicable to all geosynthetics under all field conditions. The project-specific guidelines will always supersede the general guidelines. Users can also find some information/ guidelines on the application of specific geosynthetics from the concerned manufacturers.

6.2 General guidelines

6.2.1 Care and consideration

In many projects, environmental factors during on-site storage and mechanical stresses during construction and initial operation place the most severe conditions on a geosynthetic during its projected lifetime. The successful installation of a geosynthetic is, therefore, largely dependent on the construction technique and the management of construction activities. Thus, the installation of geosynthetics in practice requires a degree of care and consideration.

In the past, most of the geosynthetic-related failures were reported to be construction-related and a few design-related. The construction-related failures were caused mainly by the following problems:

- loss of strength due to UV exposure,
- lack of proper overlap,
- high installation stresses.

Although the general nature of the installation-induced damages to geosynthetics, for example cuts, tears, splits, and perforations can be assessed by site trials, no test methods have yet been derived by which the same nature and degree of damage can be reproduced consistently in the laboratory. However, the strength reduction due to damage during installation might be partially or completely avoided by considering carefully the following elements, where the damage is found to be most severe:

1. firm, rock or frozen subgrades,
2. thin lift thickness using heavy equipment,
3. large particle size, poorly graded cover soil,
4. light weight, low strength, geosynthetics.

If the type of subgrade cannot be changed, the options remain of changing construction practice or modifying the geosynthetic being used for a specific application. However, one may attempt to do both by recommending less severe construction practices and adopting a set of criteria on the geosynthetic strength, such as reductions in the values of strength and strain to be taken into account when assessing the design tensile capacity of the geosynthetics.

When geosynthetics are applied, the following aspects are also taken into account:

- temperature during placement and service life,
- possibility of leaching of UV stabilizers, resulting in subsoil pollution,
- possibility of materials in the surroundings of the geosynthetic, which can act as a catalyst in degradation process.

Due care should also be taken during spreading and compaction of the fill materials on geosynthetic layers, particularly for very soft subgrades and/or very coarse fill materials (stones, rockfill, etc.), in order to avoid or minimize the mechanical damages to geosynthetics.

The relationship between any geosynthetic and the environment in which it is to be placed should be carefully considered.

6.2.2 Geosynthetic selection

Good geosynthetic specifications are essential for the success of any project. Due to a wide range of applications and the tremendous variety of geosynthetics available, the selection for a particular geosynthetic with specific properties is a critical decision. The selection of a geosynthetic is generally done keeping in view the general objective of its use. For example, if the selected geosynthetic is being used to function as a reinforcement, it will have to increase the stability of soil (bearing capacity, slope stability and resistance to erosion) and to reduce its deformation (settlement and lateral deformation). In order to provide stability, the geosynthetic has to have adequate strength; and to control deformation; it has to have suitable force-elongation characteristics, measured in terms of modulus (the slope of the force versus elongation curve) as explained in Sec. 3.3. Woven geotextiles and geogrids are preferred in most reinforcement applications.

When a geosynthetic has to function for filtration/drainage applications, the most suitable product is usually a thick nonwoven needle-punched geotextile with an appropriate

AOS (Apparent Opening Size). This is because of the higher permittivity and transmissivity of these nonwoven geotextiles, which is a primary requirement in such applications (Shukla, 2003b).

Methods of transport, storage and placement also govern the selection of geosynthetics. The selected geosynthetic should have a certain minimum strength, thickness and stiffness so that it will be fit enough to survive the effects of placement on the ground and the loads imposed by equipment and personnel during installation. In other words, the construction engineers should consider the field survivability/workability, transmissivity and permeability requirements of geosynthetics during their selection. These requirements can be expressed in terms of grab strength, puncture strength, burst strength, impact strength, tearing strength, permeability, transmissivity, etc. The actual values of these survivability properties of geosynthetics should be decided on the basis of the expected degree of damage (low, moderate, high or very high) on their installation in the specific field application.

In some geosynthetic applications, the colour and the surface feature of the geosynthetics should also be considered in their selection. For example, in the lining of ponds, reservoirs and canals, the white-surfaced textured HDPE geomembranes are preferred mainly because of lower heat absorption and, consequently, less expansion and contraction. White-surfaced geomembranes also allow for easier damage detection. As the geomembranes are to be left uncovered over the service life, less dramatic temperature changes inherent in the white-surfaced geomembrane are considered to be beneficial for the containment and conveyance projects. The textured geomembrane is selected to provide a suitable working base for the installation crew (Ivy and Narejo, 2003).

Many times, cost and availability of geosynthetics may also govern their selection.

6.2.3 Identification and inspection

Upon receipt, each shipment of geosynthetic rolls should be inspected for conformance to product specifications and contract documents and checked for damages. A construction quality assurance (CQA) representative should be present, whenever possible, to observe material delivery and unloading on site. Before storing or unrolling geosynthetic rolls, or both, the individual roll identification should be verified and should be compared with the packing list. Irregularities should be noted and reported. Upon delivery of the rolls of geosynthetic materials, the CQA consultant should ensure that conformance test samples are obtained. These samples should then be forwarded to the geosynthetic quality assurance laboratory for testing to ensure conformance with the site-specific specifications. Geosynthetic rolls not in compliance with the accepted material specifications may be rejected. The damaged, deformed or crushed geosynthetic rolls should be rejected and removed from the project site.

6.2.4 Sampling and test methods

The samples of geosynthetics must be cut from the product roll supplied by the manufacturer as per the standard sampling procedures to provide a statistically valid sample for the selection of coupons and test specimens (Fig. 6.1). At least one sample is generally taken for 5000 m^2 or less area of geosynthetic. Each roll selected should look undamaged and the wrapping, if any, should be intact. The first two turns of the roll should not be used for sampling. The sample should be cut from the roll, over its full width, perpendicular to the

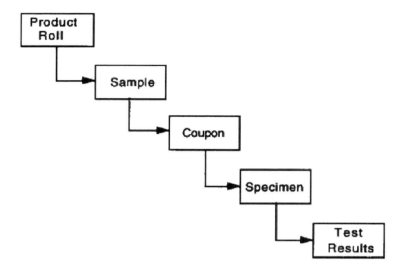

Figure 6.1 Relationship among roll, sample, coupon, and specimen test (Reprinted, with permission, from ASTM D 6213-97: *Standard Practice for Tests to Evaluate the Chemical Resistance of Geogrids to Liquid*, copyright ASTM International, 100 Barr Harbor Drive, West Conshohocken, PA 19428).

machine direction. A mark (e.g. arrow) should be used to indicate the machine direction of the sample. When two faces of the geosynthetic are significantly different, the sample should be marked to show which face was inside or which was outside the turn of the roll. The sample must be marked for identification purposes with the following information:

- brand /producer/supplier,
- type description,
- roll number,
- date of sampling.

The samples should be stored in a dry, dark place, free from dust at ambient temperature and protected against chemical and physical damage. The sample may be rolled up but preferably not folded. Sampling may be required for three purposes: one for manufacturer's quality control (MQC) testing, one for manufacturer's quality assurance (MQA) testing and a third for purchaser's specification conformance testing.

For each type of test, the required number of specimens should be cut from positions evenly distributed over the full width and length of the sample but not closer than 100 mm to the edge. Test specimens should not contain any dirt, irregular areas, or other defects, and they should be conditioned as per the requirement of the test. For atmospheric conditioning, test specimens should preferably be hung or laid flat, singly on open wire shelves allowing free access of air to all surfaces for at least 2 h. For dry conditioning, the test specimens should be placed in a desiccator until constant mass is attained. For wet conditioning, the test specimens should be immersed in water at a temperature of $20 \pm 5°C$ for a minimum of 24 h. For most of the tests on geosynthetics, air is maintained at a relative humidity between

50% and 70% and a temperature of 21 ± 2°C. If the test methods for determining the geosynthetic properties are not completely field simulated, the test values must be adjusted as discussed in Sec. 3.6. On any given project, the minimum average roll value (MARV) must meet or exceed the designer's specified value for the geosynthetic to be acceptable. With an understanding of the failure mechanism and exposure environment, the user can select appropriate test methods to best simulate the geosynthetic behaviour.

6.2.5 Protection before installation

Geosynthetics must be handled and stored properly to ensure that the specified properties are retained to perform their intended function as per the project needs. A proper choice of material and a careful handling of the geosynthetic can prevent mechanical damages during transport, storage and placement. When delivered, all the rolls of geosynthetics should be wrapped in a protective layer of plastic to avoid any damage/deterioration during transportation.

Storage areas should be located as close as possible to the point of end use, in order to minimize subsequent handling and transportation. It is usually adequate to stack the rolls with undamaged protective outer wrapper directly on the ground by covering with a waterproof tarpaulin or plastic sheet, provided that this is level, dry, well-drained, stable and free from sharp projections such as rock pieces, stumps of trees or bushes. The storage area must protect the geosynthetics from precipitation, standing water, UV radiation, chemicals (strong acids/bases), open flames and welding sparks, temperatures in excess of about 70°C, vandalism, animals and any other environmental condition that could damage geosynthetics before end use.

Enclosed indoor storage is preferred if the geosynthetic rolls are to be stored for a long period. However, if the rolls are to be stored outside for a long period of time, some form of shading is required with elevated base, unless the wrapper is of opaque material, to give protection against UV light attack. Acceptable limits of exposure to UV light depend upon the site environmental conditions such as temperature, latitude, time of year, wind, etc. and the assumptions used by the engineer during design. At no time the geosynthetics should be exposed to UV light for a period exceeding two weeks. If the wrapper gets damaged and it is beyond repair, the roll should be stored by making a suitable arrangement to prevent ingress of water. Without this, the geotextiles, particularly the nonwovens, can absorb water up to three times their weight, thereby causing handling and installation problems. In cold regions, it is nearly impossible to unroll wet, frozen geotextile without first allowing it to thaw; therefore geosynthetics should be protected from freezing. Note that that geosynthetics are generally hydrophobic (i.e. they repel water), so there is no wicking action in them. Where geosynthetics are to be used as filters, it is important to keep the wrapper intact to give protection against ingress of dust and mud. If geosynthetic rolls become wet, it must be allowed for a few days of exposure to wind after removing the wrapper in order to dry the geosynthetic.

The geosynthetic rolls may be stacked upon one another, provided they are placed in a manner that prevents them from sliding or rolling from the stack. The height of the stack should generally be three rolls. In fact, the height of the stack should be limited to that at which safe access is provided to equipment and labourers, and at which roll cores at the bottom of the stack are not distorted or crushed.

Good practice dictates that on site, the geosynthetic should be stored properly and handled according to the manufacturer's recommendations. In the absence of good, documented procedures, the guidelines given in this section may be used for general purpose.

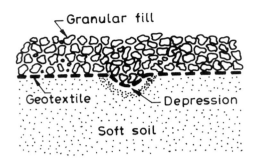

Figure 6.2 Effect of depression in the subgrade soil layer on the geosynthetic.

6.2.6 Site preparation

The original ground level may be required to be graded to some predetermined formation level. During site preparation the sharp objects, such as boulders, stumps of trees or bushes, which might puncture or tear the geosynthetic, should be removed if they are lying on the site. All protrusions extending more than 12 mm from the subgrade surface should be removed, crushed or pushed into the surface with a smooth-drum compactor. Disturbance of the subgrade should be minimized where soil structure, roots in the ground and light vegetation may provide additional bearing strength. All depressions and cavities must be filled with compacted material; otherwise the geosynthetic may bridge and get torn when the fill is placed (Fig. 6.2). For critical applications, the depressions can be lined with geotextiles before filling them with granular material. In brief, it is suggested that the subgrade be prepared well, since a geosynthetic between two stones is like a grain being ground in a mill – it cannot sustain the applied load very long. If any equipment causes rutting of the subgrade, the subgrade must be restored to its originally accepted condition before the placement of geosynthetic continues.

6.2.7 Geosynthetic installation

Geosynthetic material installation includes the placement and the attachment of the recommended geosynthetic. Geosynthetic properties are only one factor in the successful installation involving geosynthetics. Proper construction and installation techniques are essential in order to ensure that the intended function of the geosynthetic is fulfilled. Placing geosynthetics is thus the single most important step in the performance of the geosynthetic-reinforced soil systems. While handling the rolls manually or by some mechanical means at any stage of installation, the load, if any, should not be taken directly by the geosynthetic. It should be rolled/unrolled into place rather than dragging. The entire geosynthetic should be placed and rolled out as smoothly as possible. A temporary geosynthetic subgrade covering commonly known as a *slip sheet* or *rub sheet* may be used to reduce friction damage during placement.

Since the geotextile opening size in some applications, such as filtration and drainage applications, is chosen with a high degree of accuracy in the design stage, it is important to observe during installation stage that abrasion and excess straining must not enlarge the openings or even create holes before the final service state.

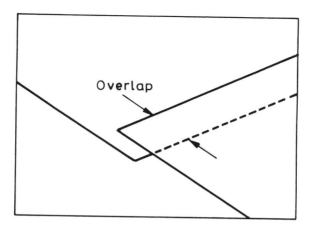

Figure 6.3 A simple overlap.

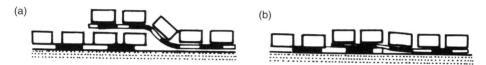

Figure 6.4 Overlap constructions: (a) wrong; (b) correct (after Pilarczyk, 2000).

An overlap between adjacent sheets must be provided while unrolling the geosynthetic into position after site preparation (Fig. 6.3). The overlap used is generally a minimum of 30 cm, but in applications where the geosynthetic is subject to tensile stresses, the overlap must be increased or the sheets of geosynthetic sewn/bonded as explained in the following section. If possible, the overlap should not be located at the place where the transition or edges of the cover layer may take place (Fig. 6.4).

It has been observed that a misunderstanding of expected imposed loads or unforeseen stresses arising from poor construction practices are the main reasons for the damages, particularly mechanical, during installation processes. Also with a careless installation, the parts of geosynthetics may get scattered into the surroundings resulting in a harmful influence on the environment. Therefore, the installer, that is, the party who installs, or facilitates installation of geosynthetics should consider the involved processes necessary for the perfect solution. Installation effects are often out of the designer's hands, so specifications, inspections, and protective measures must be agreed upon at each individual site.

6.2.8 Joints/seams

Geosynthetics are finite and therefore where geosynthetic widths or lengths greater than those supplied on one roll are required, it becomes necessary to make joints or overlaps. Since joints and overlaps are the weakest link in geosynthetic-reinforced soil structures, they have to be as limited as possible.

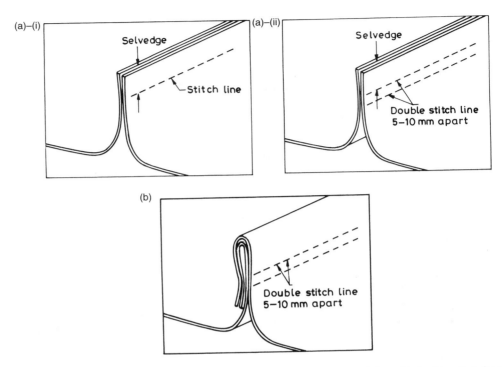

Figure 6.5 Sewn seams: (a) face to face ('Prayer') seams – (i) single stitch line, (ii) double stitch line; (b) Lapped ('J') seam.

When two sheets of similar or dissimilar geosynthetics (or related materials) are attached to each other by a suitable means, the junction so formed is known as a *joint*, and when a geosynthetic sheet is physically linked to, or cast into another material (e.g. the facing panel of a retaining wall), this is known as a *connection*. When no physical attachment is involved between two geosynthetic sheets or a geosynthetic sheet and another material, this is known as an *overlap*. However, sometimes overlap is also considered as a type of joint, called *overlapped joint*.

There are several jointing methods, such as *overlapping, sewing, stapling, gluing, thermal bonding*, etc. Different joints, presently in use, may be classified into *prefabricated joints* and *field joints* that are basically made during applications on field sites. In the vast majority of cases, the geosynthetic width or length is extended simply by overlapping, which is usually found to be the easiest field method for jointing (Fig. 6.3). Overlapping by 0.3–1 m may be employed if relatively small tensile forces are developed in geosynthetic layers to be jointed. Relatively more overlaps are required if the geosynthetic is placed under water. The overlaps involve considerable wastage of material and if not carried out with care they can be ineffective.

Geotextiles may be jointed mechanically by sewing or stapling, or chemically by means of adhesives. The term 'seam' generally refers to a series of stitches joining two or more separate pieces of geosynthetics (or related materials); however, it is also being used as a synonym for 'joint'. Figure 6.5(a) shows the most suitable seam configuration known as *prayer seams*. Another type of seam known as lapped ('J') seam is shown in Figure 6.5(b), which is reliably

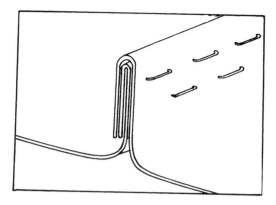

Figure 6.6 Stapled seam.

Figure 6.7 A bodkin joint.

soil tight even for fine-graded soils. Depending on the critical nature of the construction, either a single or double stitch is used. Several types of threads are available (nylon, high performance polymers, etc.) depending on the type of geotextiles and type of field applications. AASHTO M 288-00 recommends that the threads used in joining geotextiles by sewing should consist of high strength polypropylene, or polyester. Nylon thread should not be used.

The sewn joints must be projected up so that every stitch can be inspected. High-strength geosynthetics, employed for their reinforcing potential, should normally be sewn. For jointing the geotextiles by stapling method, corrosion resistant staples should be used. Figure 6.6 shows the stapled seam configuration. Stapling may be used with geotextiles to make the temporary joints. They should never be used for structural jointing. It is to be noted that sewn seams are most reliable and can be carried out on site using portable sewing equipment. Heat bonded or glued seams are generally not used.

For geosynthetics such as geonets, and geogrids, a *bodkin joint* may be employed whereby two overlapping sections are coupled together using a bar passed through the apertures (Fig. 6.7). Geogrids can also be sewn using a robust cord threaded through the grid apertures. Hog rings, staples, threaded loops, wires, etc. are also used for jointing geosynthetics.

The types of jointing/seaming techniques used to construct seams in plastomeric geomembranes (made from PE, PP, PET, PVC, etc.) include the following: extrusion, hot air, fusion (hot wedge/knife), ultrasonic, and electric welding methods. The *extrusion technique* encompasses extruding molten resin between two geomembranes or at the edge of two overlapped geomembranes to form a continuous bond. The *hot air technique* introduces high-temperature air or gas between two geomembrane surfaces to facilitate melting.

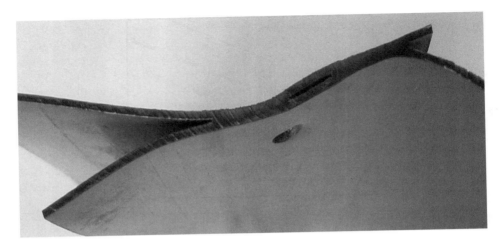

Figure 6.8 A typical dual hot wedge seam.

Pressure is applied to the top or bottom geomembrane, forcing together the two surfaces to form a continuous bond. *Hot wedge welding technique* consists of placing a heated wedge, mounted on a self-propelled vehicular unit, between two overlapped geomembrane sheets such that the surface of both sheets are heated above the melting point of the polymer. Pressure is applied to the top or bottom geomembrane, or both, to form a continuous bond. Fusion (hot wedge) welding is generally used for long seams. Extrusion welding is used for capping and patch repairs, and for joining panels at locations where fusion welding is not practical due to joint configuration. Hot air welding is also possible, but it is very sensitive to workmanship and is only used when other methods are not applicable.

Some seams are made by fusion (hot wedge) with dual bond tracks separated by a non-bonded gap, sometimes referred to as *dual hot wedge seams* or *double track seams* (Fig. 6.8). Some typical seams in geomembranes are shown in Figure 6.9. In general, the wedge welding with a dual-track wedge welder provides the best quality seams and is used as state of practice in environmental lining applications. Once finished, each air channel of the dual seam is tested by inflation. It should withstand an air pressure of 5 bars for 10 minutes without noticeable loss of pressure.

Elastomeric geomembranes are made from rubbers of various types as the barrier component. They require seaming by means of solution or adhesives.

A geomembrane seam, in service, must maintain its leak free condition. Metal hog rings should never be used when geonets are used in conjunction with geomembranes. The most important aspect of construction with geomembranes is the seam. Without proper seams, the whole concept of using geomembrane liners as a fluid barrier is foolish. A very important factor affecting seaming quality is the weather condition during seaming. Normal weather conditions for seaming are as follows (Qian *et al.*, 2002):

1   temperature between 4.5°C and 40°C,
2   dry conditions (i.e. no precipitation or other excessive moisture, such as fog or dew),
3   winds less than of 32 km/h.

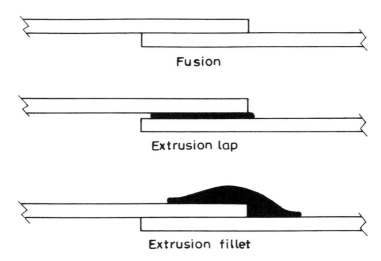

Figure 6.9 Some typical seams in geomembranes (after Giroud, 1994).

Geosynthetic clay liners are jointed by the application of bentonite at the panel joints.

It is important to understand the criterion for assessing the joint performance. This is expressed in terms of the load transmission between the two pieces of the geosynthetic. In some applications, it may be essential that the load transfer capability be equal to that of the parent material. For other situations, a more important criterion may be the magnitude of the deformation of the joint under load. Data on tensile strength of seams/joints are necessary for all functions if the geosynthetic is to be mechanically jointed and if load is transferred across the seams and joints.

The *seam/joint strength* is the maximum tensile resistance (that indicates the load-transfer capability), measured in kilonewtons per metre, of the junction formed by joining together two or more sheets of geosynthetics by any method (e.g. sewing or thermal bonding). The seam/joint efficiency (E) of a seam/joint between two sheets of geosynthetic is the ratio, expressed as a percentage, of seam/joint strength to the tensile strength of unseamed/unjointed geosynthetic sheet evaluated in the same direction. It is thus given as follows:

$$E = \left(\frac{T_s}{T_u} \times 100\right)\%, \tag{6.1}$$

where T_s is the seam/joint strength, expressed in kN/m and T_u is the tensile strength of unseamed/unjointed geosynthetic sheet, expressed in kN/m.

Ideally, the seam/joint should be stronger than the geosynthetic being jointed and should thus never fail in tension. In practice, in the field, high efficiencies are rarely obtained. Publications generally mention that laboratory obtained efficiencies are usually higher than field efficiencies. Thus, this is of little help to the field engineer trying to meet a consultant's specification. However, efforts should be made to make the seam efficiency near 100%. As the geosynthetic strength becomes higher, seams become less efficient. Above the

274 Application guidelines

geosynthetic strength of 50 kN/m, even the best of seams have efficiency less than 100%, and beyond 200–250 kN/m, the best ones can have approximately 50% efficiency. AASHTO M 288-00 recommends that when sewn seams are required, the seam strength, as measured in accordance with ASTM D4632, should be equal to or greater than 90% of the specified grab strength.

Murray et al. (1986) undertook research work into the seam strengths obtained from both sewn and adhesive bonded seams. Their work was comprehensive and stated that 100% efficiency could be obtained using adhesives. With sewn joints, they described efficiencies up to 90%, but they drew attention to the large deformations that were experienced. The technique of jointing geogrids by means of a bodkin joint proved to be an effective procedure whereby load carrying efficiencies of about 90% were obtained. Rankilor and Heiremans (1996) reported that the use of adhesives can reduce seam extension dramatically.

Since geomembranes are generally used to restrict fluid migration from one location to another in soil and rock, the integrity of their seams must be properly determined. The quality assurance engineer must analyse seam bonding shear strength and peal strength data obtained from tests with tensile instrumentation on geosynthetic seam shear and peel specimens (Fig. 6.10) to evaluate seam quality.

If geosynthetics are applied in reinforcement applications, overlaps and seams at right angles to the direction of the leading force are unacceptable. The termination of a geosynthetic sheet as well as its connection to another part of the structural elements needs special attention.

6.2.9 Cutting of geosynthetics

It is a labour-intensive, time-consuming operation, which in most cases can be avoided by forward planning. Although the total width of an area to be covered will rarely be an exact

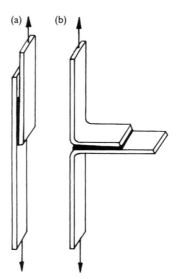

Figure 6.10 Geosynthetic seam specimens for (a) bonding shear strength test; (b) bonding peel strength test.

multiple of available widths. The maximum geosynthetic width is generally 5.3 m. There is less wastage of time and money if slightly larger overlaps or wraparound are allowed to take up the excess width, than if the geosynthetic is cut on site. In the case of walls and steep-sided embankments, the wraparound may enhance compaction at the edges and also helps to reduce erosion and may assist in the establishment of vegetation.

6.2.10 Protection during construction and service life

The damage due to UV light exposure can usually be avoided by not laying more geosynthetic in a day than can be covered by fill on same day. Unused portions of rolls must be rerolled and protected immediately. It is to be noted that when the geosynthetic is UV-stabilized; the degradation is largely reduced, but not entirely eliminated. Efforts should be made to cover the geosynthetic within 48 hours after its placement. A geosynthetic, which has not been tested for resistance to weathering, must be covered on the day of installation.

Protection of the exposed geosynthetic wall face against degradation due to UV light exposure and, to some extent, against vandalism can be provided by covering the geosynthetic with gunite (shotcrete), asphalt emulsion, asphalt products, or other coatings. A wire mesh anchored to the geosynthetic may be necessary to keep the coating on the face of the wall. In the case of walls constructed from geogrids, vegetation can easily grow through (or be placed behind) the large openings, and UV degradation of the relatively thick ribs is significantly lower. Thus the need to cover the geogrid wall face is not as compelling as with geotextiles, and the front of geogrid walls is sometimes left exposed.

The chemical resistance of the geosynthetic liner to the contained liquid must be considered for the entire service life of its installation. The minimum thickness of the geomembrane liner usually recommended is 20 mils (0.50 mm) irrespective of design calculations; however this lower limit may be 80 mils (2.0 mm) in the case of hazardous materials containment. When the secondary liner is also a geomembrane, it must be of the same thickness as the primary liner.

Geosynthetic clay liners (GCLs) are extremely sensitive to damage during and after construction owing to their small thickness and small mass of bentonite. So great care is required in applications with GCLs.

Before the placement of a granular fill on the geosynthetic, the condition of the geosynthetic should be observed by a qualified engineer to determine that no holes or rips exist in the geosynthetic. All such occurrences need repair. All wrinkles and folds in the geosynthetic should be removed. The following actions may result in puncturing, abrasion, or excessive straining that can lead to a loss of strength or reduction in the serviceability of the geosynthetic product, and therefore these actions must be avoided.

- dropping fill material from an unknown height,
- wheels passing over a relatively thin cover layer,
- compactors acting on the cover layer.

It may be noted that loads due to above actions may act on the geosynthetic during installation that will never occur again.

In road and railway construction, the geosynthetic damage from the impact of dropped fill material usually is not significant, unless the geosynthetic is very light and thin. Traffic or compaction loads cause severe harm than fill placement (Brau, 1996).

When placing fill on sloped surfaces, it is always advisable to start from the base of the slope. If there is no way to avoid having stones roll downslope, their weight should not exceed 50 kg. In general, when releasing stones that weigh less than 120 kg on geotextiles over well-prepared surfaces, the drop height should be less than 0.3 m. If the geotextile is covered by a cushion layer, the drop height may be up to 1 m. Larger stones should be placed without free fall. Compared to dumping in dry conditions, the falling height of cover material is often much larger under water but has less impact because it falls through water, rather than air. As a rule of thumb, the impact energy when falling through water of some depth is a mere 15% of that generated by an object falling from the same height in dry conditions (Heibaum, 1998).

No equipment that could damage the geosynthetic should be allowed to travel directly on the geosynthetic. In fact, once the geosynthetic is laid, it should not be used until an adequate layer of fill is placed over it, thus affording some protection; otherwise the geosynthetic may fail. One exception to this rule is where a heavyweight geosynthetic is used, which is specifically designed to be directly used by vehicles, but the principle 'thicker fill is better' is valid at every site. In road, railway and embankment construction, the first layer of fill material on the geosynthetic should have a minimum thickness of 200–300 mm, depending on aggregate size and weight of trucks/rollers. Exact answer will only come from the site tests. Maximum lift thickness may be imposed in order to control the size of the mud wave (bearing failure) ahead of the dumping due to excess fill weight. It has been observed that when more than 0.6–0.9 m of fill is placed, the geosynthetic sustains no significant damage from trucks or vibratory rollers (U.S. Department of the Interior, 1992).

No turning of construction equipment should be allowed on the first lift. Construction vehicles should be limited in size and weight to limit initial lift rutting to 75 mm. If rut depths exceed 75 mm, decrease the construction vehicle size and/or weight. For the initial stages of construction, low ground pressure and small dump trucks should therefore be used. For very soft formations, it is necessary to use special low bearing pressure tracked vehicles for spreading the fill over the geosynthetic layer. During filling operations, the blades or buckets of the construction plant must not be allowed to come into contact with the geosynthetic. A further lift may be placed after consolidation of the subgrade has increased its strength. Compaction of the first granular lift is usually achieved by tracking with construction equipment. Smooth-drum or rubber-tyred rollers may also be considered for compaction of the first lift. A continued buildup of cover material will allow vibratory rollers to be used. If localized liquefied conditions occur, the vibratory roller should not be used. Proof rolling by a heavy rubber-tyred vehicle may provide pretensioning of the geosynthetic by creating initial ruts, which should be subsequently refilled and levelled.

Proper care must be given during compaction of the top layers of wraparound reinforced walls and steep slopes. This is required because very high compaction results in very high stresses in the geosynthetic reinforcement due to movement of the fills in the wraparound sections as shown in Figure 6.11, and such situations are not desirable. All vehicles and construction equipments weighing more than 1500 kg should be kept at least 2 m away from the faces of the walls or steep slopes. The fill within 2 m of the face of the wall or steep slope should be compacted using vibro-tamper, vibrating plate compactor having a mass not exceeding 1000 kg or by vibrating roller having a total mass not exceeding 1500 kg.

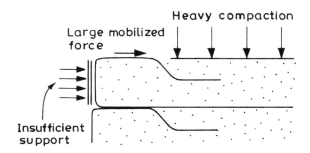

Figure 6.11 Effects of heavy compaction (after Voskamp et al., 1990).

If it is necessary to use poorly graded aggregate fill, and heavy construction equipment for placement and compaction, it may be prudent to place a cushioning layer of sand above the geosynthetic.

If a geosynthetic is used in conjunction with bituminous material, care must be taken to ensure that the temperature of the bituminous material is well below that of the melting point of geosynthetics. Tack-coat application quantity requires attention. The absence of adequate tack coat means the loss of paving fabric system benefits and can lead to damage to the overlay. Wet geosynthetic should not be used in such applications because it creates steam, which may cause the bitumen (asphalt cement) to be stripped from the geosynthetic because of a poor bond.

In the case of liquid containment pond, to shield the geomembrane liner from ozone, UV light, temperature extremes, ice damage, wind stresses, accidental damage, and vandalism, a soil cover of at least 30 cm thickness may be provided. For a proper design of containment, a geotextile should be used beneath the geomembrane, placed directly on the prepared soil subgrade before liner placement. The cover soils over geomembranes, installed on sloping ground, can unravel and slump very easily, even under static conditions. To alleviate this situation somewhat, it is a common practice to taper the cover soil, laying it thicker at the bottom and gradually thinner going toward the top. Due to the water pressure caused by the groundwater, the geomembrane may be ballasted to prevent its floating.

6.2.11 Damage assessment and correction

The ability to maintain design function (e.g. reinforcement, separation, filtration, etc.) and/or design properties (e.g. tensile strength, tensile modulus, chemical resistance, etc.) of a geosynthetic may be affected by damages to the physical structure of the geosynthetic during field installation operations. Therefore, before covering the geosynthetic with soil, the engineer should examine it for defects such as holes, tears, abrasions, etc. A test section may be used to assess the worst case installation techniques (e.g. overcompaction, thin lift heights, greater drop heights, etc.) and fill materials. The damages to geosynthetics from installation operations may be quantified by evaluating specimens cut from samples exhumed from representative field installation sites. Damage evaluation may be performed with visual examination and/or laboratory testing of specimens cut from the control (un-installed/original) and exhumed samples. Laboratory tests to perform will vary with type and function of the geosynthetic, and the project requirements. Note that the control sample should be a direct continuation of the exhumed sample so as to minimize differences in control and exhumed specimen properties due to inherent product variability. The positions of the test specimens on the control sample, relative to roll edge, must correspond identically with the positions of the exhumed sample. The amount, or area, of control sample to be retrieved should be equal to the area of exhumed sample. Details of the techniques for assessing the amount of damage, and documentation of installation and retrieval techniques used, may be found in the works of Bonaparte et al. (1988), Bush and Swan (1988), Paulson (1990), Allen (1991), Rainey and Barksdale (1993), and Sandri et al. (1993).

For more critical structures, such as reinforced soil walls and embankments, it is safest to remove the damaged section of geosynthetic, if any, entirely and replace it with undamaged geosynthetic. In these applications, a certain degree of damage may be acceptable, provided that this has been allowed for at the design stage. In low risk applications, where the geosynthetic is not subject to significant tensile stress or dynamic water loading, it is permissible to patch the damaged area by placing a new layer of geosynthetic extending beyond the defect in all directions at a distance equal to the minimum overlap (generally 300 mm) required for adjacent rolls.

Exposed geosynthetics, such as geomembranes used in lining of liquid containment and conveyance systems can be damaged by animals, construction equipment, vandalism or other elements of nature. Thus, it is imperative that a meaningful maintenance plan should be in place throughout the geomembrane service life. The maintenance plan may include occasional seaming and anchor trench soil cover maintenance.

6.2.12 Anchorage

To maintain the position of a geosynthetic sheet before covering with soil/fill, the edges of the sheet must be weighted or anchored in trenches (Fig. 6.12), thereby providing the significant pullout resistance. Anchorage selection depends on site conditions. In case of unpaved roads, the geosynthetic should be anchored on each side of the road. The bond length (typically around 1.0–1.5 m) can be achieved by extending the geosynthetic beyond the required running width of the road (Fig. 6.13(a)) or by providing an equivalent bond length by burying the geosynthetic in shallow trenches (Fig. 6.13(b)) or by wraparound (Fig. 6.13(c)). Similar approaches can also be adopted in other applications.

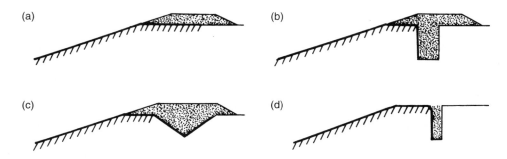

Figure 6.12 (a) Simple run-out; and anchor trenches: (b) rectangular trench; (c) V trench; (d) narrow trench (after Hullings and Sansone, 1997).

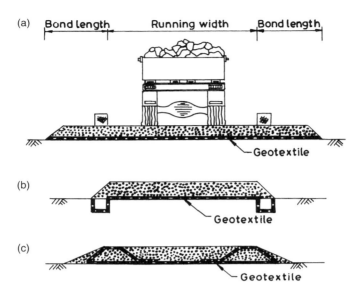

Figure 6.13 Use of geosynthetics in unpaved road construction (after Ingold and Miller, 1988).

6.2.13 Prestressing

Simple procedures such as prestressing the geosynthetic may enhance the reinforcement function in some applications. For example, to specifically add reinforcement for paved roads on firm subsoils, a geosynthetic prestressing system may be required. By prestressing the geosynthetic, the aggregate base will be placed in compression, thereby providing lateral confinement and will effectively increase its modulus over the unreinforced case.

6.2.14 Maintenance

All geosynthetic-reinforced soil structures should be subjected to a regular programme of inspection and maintenance. A habit should be developed to keep records of the inspections and any maintenance carried out.

6.2.15 Certification

Certification provides a benchmark for quality and thus it is a measure of assurance of success of geosynthetic-related projects. A quality certificate ensures that the geosynthetic delivered meets the design requirements. The initiative for certifying has to be taken by the manufacturer of the geosynthetic. It is the manufacturer's responsibility to perform through quality control testing for all properties requiring certification. The same system of quality assurance should be valid for the applications of geosynthetics by the concerned contractor by issuing certificates with complete description.

In the ponds, reservoirs, and any other containment of drinking water, the applied geosynthetics need a certificate, indicating that the geosynthetic concerned has been tested on aspects of health and has been approved. In view of the increasing demand on prevention of pollution, it is recommended to inquire the potential environmental effects. It may be noted that there is no danger of emission of toxic materials from the geosynthetics to the environment, except from some kinds of PVC.

6.2.16 Handling the refuse of geosynthetics

Geosynthetics, which become available after site clearing and demolition of a construction can be dumped on a landfill, burned or recycled. Special measures must be taken to prevent emission to the environment.

6.3 Specific guidelines

The applications of geosynthetics in the fields require many specific construction guidelines, as briefly described in this section for major civil engineering applications of geosynthetics. The readers and users of geosynthetics can have more details from the relevant standards and specialized codes of practice on specific geosynthetic applications, some of which are listed in Appendix B. Many geosynthetic manufacturers have developed their own design charts and the graphs as well as the construction guidelines for geosynthetic-reinforced structures. If a specific geosynthetic product is to be used, these guidelines can be considered. However, it should be noted that these guidelines are product-specific in their assumptions regarding allowable strength, factor of safety, etc. For important constructions and before application of newly developed geosynthetics, a test study in practice is recommended to study the collapsing mechanism and the chance of failing. Here also the execution-aspects, where there is a possibility of damages, should be studied. It must be underlined that proper equipments and well-planned construction techniques are important factors for geosynthetic applications to be successful in the fields.

6.3.1 Retaining walls

In the actual construction, geosynthetic (geotextile/geogrid)-reinforced soil retaining walls have continued to demonstrate excellent performance characteristics and exhibit many advantages over conventional retaining walls. To achieve better performance, the following points must be considered on site-specific basis.

1. Any unsuitable foundation soils should be replaced with compacted granular backfill material.
2. The geosynthetic layer should be installed with its principal strength (warp strength) direction perpendicular to the wall face.
3. With the geosynthetic layer it is necessary to take care not to tear it in the direction parallel to the wall face because a partial tear of this type will reduce the amount of tensile force carried out by the geotextile layer.
4. The overlap along the edges of the geosynthetic layer should generally exceed 200 mm. If there is possibility of large foundation settlements, then sewn or other suitable joints may be recommended between adjacent geosynthetic layers.
5. There should not be any wrinkles or slack in the geosynthetic layer as they can result in differential movement.

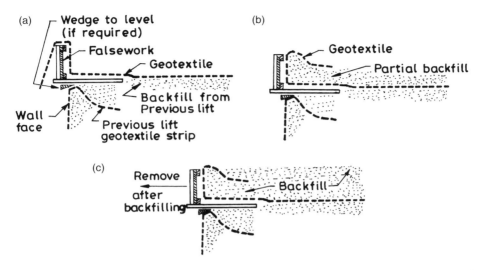

Figure 6.14 Lift construction sequence for geotextile reinforced soil walls: (a) (b) (c) (after Steward et al., 1977).

6   Granular backfill soil should generally be compacted to at least 95% of the standard Proctor maximum dry unit weight. Compacted lift thickness should vary from 200 to 300 mm. Efforts should be made to compact uniformly to avoid differential settlement.
7   Backfill soil should be compacted, taking care not to get the compactor very close to the facing element, so that it is not highly stressed, resulting in pullout or excessive lateral displacement of the wall face. It is therefore recommended to use lightweight hand-vibratory compactors within 1 m of the wall face.
8   Wraparound geosynthetic facing can be constructed using a temporary formwork as shown in Figure 6.14. The lap length should be generally not less than 1 m.
9   A construction system for the permanent geosynthetic-reinforced soil retaining wall (GRS-RW), widely used in Japan, can be adopted. This system uses a full-height rigid (FHR) facing that is cast in place using staged construction procedures (Fig. 6.15). This system has several special features such as use of relatively short reinforcement and use of low-quality on site soil as the backfill.
10 Geogrid-reinforced retaining walls can be constructed with geotextile filters near the face. Major construction steps are shown in Figure 6.16.
11 In the case of facing made with segmental or modular concrete blocks (MCBs), full-height precast concrete panels, welded wire panels, gabion baskets, or treated timber panels, the facing connections should be made prior to placing the backfill and must be carefully checked as per the design guidelines because the success of a geosynthetic-reinforced retaining wall is highly dependent on the facing connections.
12 It is necessary to have tight construction specifications and quality inspection to insure that the wall face is constructed properly; otherwise an unattractive wall face, or a wall face failure, could result.

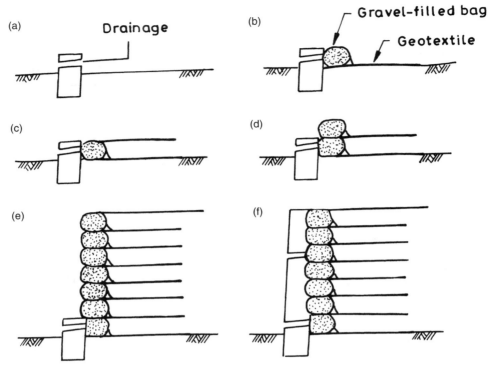

Figure 6.15 Standard staged construction procedure for a geosynthetic reinforced soil retaining wall (GRS-RW): (a) concrete base; (b) geotextile and gravel-filled bag placement; (c) backfill and compaction; (d) placement of the second layer of geotextile and gravel-filled bag; (e) all layers constructed; (f) concrete facing constructed (after Tatsuoka et al., 1997).

6.3.2 Embankments

For the construction of an embankment with a basal layer over very soft foundation soils, a specific construction sequence must be followed to avoid any possibility of failures (geosynthetic damage, non-uniform settlements, embankment failure, etc.) during construction. The following guidelines may help achieve this objective in practice.

1 A geosynthetic layer is placed over the foundation soil, generally with minimal disturbance of the existing materials. Small vegetative cover, such as grass and reeds, should not be removed during subgrade preparation. There can be several alternatives with regard to the installation of the geosynthetic layer inside the embankment. Some of them are as follows:

    a   a geosynthetic layer inside the embankment (Fig. 6.17(a));
    b   several geosynthetic layers along the embankment height (Fig. 6.17(b));
    c   a geocell at the base of the embankment (Fig. 6.17(c));
    d   a geosynthetic layer at the base of the embankment with folded ends (Fig. 6.17(d));
    e   combination of a geosynthetic layer with berms (Fig. 6.17(e));
    f   a geosynthetic layer (or layers) with vertical piles. (Fig. 6.17 (f)).

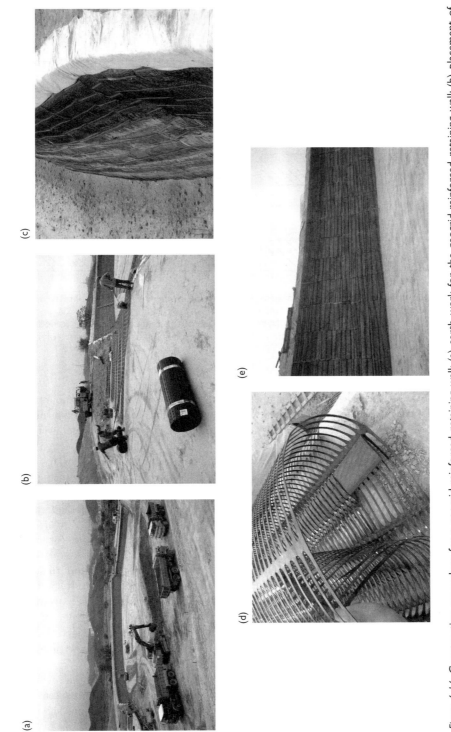

Figure 6.16 Construction procedure for a geogrid-reinforced retaining wall: (a) earth work for the geogrid-reinforced retaining wall; (b) placement of geogrid layers; (c) placement of geotextile filter layer near the face of the retaining wall; (d) connection between folded geogrid sheet and the next geogrid sheet; (e) a view of the completed retaining wall.

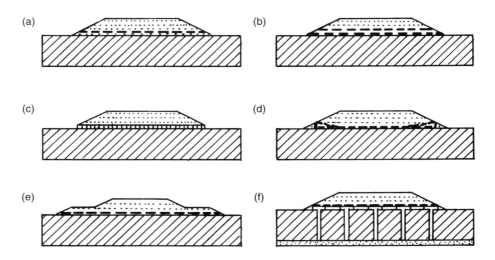

Figure 6.17 Geosynthetic installation: (a) a geosynthetic layer inside the embankment; (b) several geosynthetic layers along the embankment height; (c) a geocell at the base of the embankment; (d) a geosynthetic layer at the base of the embankment with folded ends; (e) combination of a geosynthetic layer with berms; (f) a geosynthetic layer (or layers) with vertical piles.

Each alternative has its own advantages. A geosynthetic layer inside the embankment (Fig. 6.17(a)), rather than along the interface between the fill material and the foundation soil, favours a better reinforcement anchorage length, particularly for geogrids due to interlocking effect. If different functions are to be achieved, then several geosynthetic layers of different types can be installed along the embankment height (Fig. 6.17(b)). Combination of geosynthetic layers creates a stiffer mass, which tends to reduce differential settlements. This effect can also be achieved with the use of geocells filled with embankment material (Fig. 6.17(c)). The increase of geosynthetic anchorage can be achieved by the use of folded edges (Fig. 6.17(d) and/or berms (Fig. 6.17(e)). If the settlements of the embankment are to be limited, then the geosynthetic layer (or layers) can be installed along with vertical piles (Fig. 6.17(f)).

2 The geosynthetic layer is usually placed with its strong/warp direction (machine direction) perpendicular to the centreline of the embankment (Fig. 6.18). It should be unrolled as smoothly as possible, without dragging, transverse to the centreline of the embankment. Additional reinforcement with its strong direction oriented parallel to the centreline may also be required at the ends of the embankment. The geosynthetic layer should be pulled taut to remove wrinkles, if any. Lifting by wind can be prevented by putting weights (sand bags, stone pieces, etc.).

3 Seams should be avoided perpendicular to the major principal stress direction, which is generally along the width of the embankment (Fig. 6.18). Since for surcharge/areal fills, a major principal stress direction cannot be defined, in such situations, seams should be made by sewing.

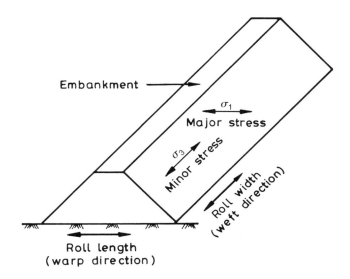

Figure 6.18 Geosynthetic orientation for linear embankments.

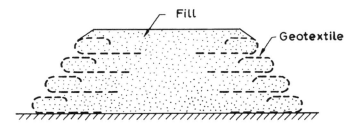

Figure 6.19 Embankment with wraparound side slopes.

4   Narrow horizontal geosynthetic strips may be placed along the side slopes with wraparound to enhance compaction at the edges (Fig. 6.19). Edge geosynthetic strips also help to reduce erosion and may assist in the establishment of vegetation.
5   The embankment should be built using low ground pressure construction equipments.
6   When possible, the first few lifts of fill material (0.5 to 1 m) just above the geosynthetic should be free draining granular materials; then the rest of the embankment can be constructed to grade with any locally available materials. This is required to have the best frictional interaction between the geosynthetic and fill, as well as drainage layer for excess pore water dissipation of the underlying foundation soils.
7   In the case of extremely soft foundations (when a mud wave forms), the geosynthetic-reinforced embankments should be constructed as per the sequence of construction shown in Figure 6.20. A perimeter berm system can be constructed to contain mudwave.
8   The first lift should be compacted only by tracking in place with dozers or end-loaders. Once the embankment is at least 60 cm above the original ground, subsequent lifts can

286 Application guidelines

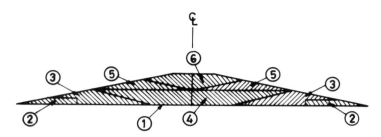

1 Lay geosynthetic in continuous transverse strips, sew strips together
2 End dump access roads
3 Construct outside sections to anchor geosynthetic
4 Construct interior sections to set geosynthetic
5 Construct interior sections to tension geosynthetic
6 Construct final centre section

Figure 6.20 Construction sequence for geosynthetic reinforced embankments over soft foundation soils (after Haliburton et al., 1977).

be compacted with a smooth drum vibratory roller or other suitable compactor. Traffic on the first lift should be parallel to the centreline of the embankment.

9 A minimum number of instruments, such as piezometers, settlement plates and inclinometers, can be installed in order to verify design assumptions and control construction. If piezometer indicates excessive pore water pressure, construction should be halted until the pressures drop to a predetermined safe value. Settlement plates installed at the geosynthetic level can help monitoring settlement during construction and thus adjusting the fill requirements appropriately. Inclinometers should be considered at the embankment toes to monitor lateral displacement.

6.3.3 Shallow foundations

The guidelines for geosynthetic installation and compaction of granular fills are governed by the type of the footing, the applied load and the foundation soil characteristics. In most of the shallow foundation applications, the geosynthetic layer(s) will be installed at the base of the foundation trench followed by the placement of a compacted granular fill.

6.3.4 Unpaved roads

A geosynthetic layer, generally a geotextile layer, is typically placed directly on the soil subgrade followed by placement and compaction of an adequate depth of granular layer. The construction practices, being adopted, must ensure that the geosynthetic will survive installation, and the construction sequencing will not lead to failure of the existing soil subgrade. Some specific guidelines for the field installation of geosynthetic layers are given below.

1 The designated site should be prepared by clearing topsoil and vegetation. The area for placement of the geosynthetic layer must be as smooth and as clean as possible to prevent geosynthetic damage and permit uniform granular layer thickness.

Table 6.1 Overlap requirement of geotextile for different CBR values (after AASHTO, 2000)

Soil CBR	Minimum overlap
Greater than 3	300–450 mm
1–3	0.6–1 m
0.5–1	1 m or sewn joint
Less than 0.5	Overlap is not recommended. Sewn joint should be provided
All roll ends	1 m or sewn joint

2 Prior to placement of the geosynthetic layer, the prepared subgrade surface, that is, the formation level must be provided an appropriate cross slope.
3 During geosynthetic placement, care must be taken not to damage the material or disturb the prepared subgrade. Care must also be taken to minimize wrinkles and folds in the geosynthetic.
4 Parallel rolls of geosynthetic should be overlapped or sewn as required. Recommended overlaps are given in Table 6.1. Overlaps of parallel rolls should occur at the centreline and at the shoulder. Overlaps should not be made along the anticipated main wheel path locations. Overlaps at the end of the rolls should be in the direction of the granular fill placement with the previously placed roll on top. Continuous visual inspection of all field seams and overlaps should be done throughout the installation to ensure that there are no voids in the seam or overlap area. Repairs that may be required during installation can be accomplished by patching by taking a piece of the primary geosynthetic that extends approximately 30 cm beyond each edge of the area to be repaired.
5 For very weak soil subgrades, the granular fill thickness should be limited to prevent construction-induced failure. The first lift of granular fill should be graded down to a thickness of 300 mm or the maximum design thickness. All remaining lifts of granular fill should be placed in lifts not exceeding 220 mm loose thickness. The maximum granular particle size in the initial lift should be limited to less than 1/4 of the lift thickness.
6 At no time should the equipment be allowed on the geosynthetic with less than 150 mm of granular fill between the wheels and the geosynthetic. Construction vehicles should be limited in size and weight such that rutting in the initial lift is less than 7.5 cm. The turning of construction vehicles should not be permitted on the first granular lift over the geosynthetic.
7 Care should always be taken to push the granular cover material over the top of the geosynthetic, rather than into the overlaps. Also, the material is pushed in such a way that the geosynthetic is not pulled, and wrinkles are not caused in front of the cover material. The compaction of granular fill by vibratory roller must be closely monitored where the subgrade is susceptible to liquefaction.
8 The ruts that develop during construction must be filled in with additional granular material; otherwise the geosynthetic may eventually become exposed at the crown between the ruts. In no case, should ruts be bladed down.
9 It must be ensured that there should not be any contamination of the granular layer from the fine-grained subgrade soils during construction and service life. It has been observed that the addition of 10–20% clayey fines to clean gravel reduces the bearing capacity of the gravel to that of the clay. It must also be ensured that the subgrade

must be free to drain as it consolidates under the traffic and roadway induced stresses. Thus, the geosynthetic must act as a filter to prevent the movement of soil fines into the granular subbase/base while allowing water to drain from the subgrade.

10 Edges of the geosynthetic layer must be attached to the side drains for allowing the water at the subgrade level to join the drain while moving along the plane of the geosynthetic layer.

6.3.5 Paved roads

A paving fabric interlayer system is looked upon as an economical tool, which effectively solves general pavement distress problems. It is easy to install and readily complements any paving operation. The ideal time to place a paving fabric interlayer system is in the very early stages of hairline cracking in a pavement. It is also appropriate in new pavement construction to provide a waterproof pavement from day one.

The installation of a paving fabric generally follows the same pattern wherever it is used. There are four basic steps in the proper installation of an overlay system with a geosynthetic interlayer. Surface preparation is followed by the application of tack coat, installation of the geosynthetic and finally the placement of the overlay. These steps along with general guidelines are described below.

Step 1 – Surface preparation: The site surface is prepared by removing loose material and sharp/angular protrusions, and sealing cracks, as necessary. The prepared surface should be levelled, dry, and free of dirt, oil and loose materials. Cracks, 3 mm wide or greater, should be cleaned with pressurized air or brooms and filled with a liquid asphalt crack sealant. This will prevent the tack coat from entering the cracks and reducing available tack for saturation of the fabric. Very large cracks should be filled with a hot or cold asphalt mix. Commercial crack filler can also be used. Cracks should be level with the pavement surface and not overfilled. If the quality of the existing pavement is poor, a levelling course of asphalt concrete is placed over it prior to the placement of the paving fabric interlayer system. On existing cement concrete pavements, a layer of asphalt concrete should be provided before laying the fabric. The surface on which a moisture barrier interlayer is placed must have a grade which will drain water off the pavement.

Step 2 – Tack coat application: Proper application of tack coat is crucial; mistakes can lead to early failure of the overlay. Straight paving-grade bitumen is the best and the most economical choice for paving fabric tack coat. Cutbacks and emulsions which contain solvents should not be used for tack coat; if they are used, they must be applied at a higher rate and allowed to cure completely. Minimum air and pavement temperature should be at least 10°C or more for placement of tack coat (IRC: SP: 59–2002). The temperature of tack coat should be sufficiently high to permit a uniform spray pattern. It should be spread at between 140°C and 160°C, to permit uniform spray and to prevent damage to the paving fabric. The target width of tack coat application should be equal to the paving fabric width plus 150 mm. Tack coat should be restricted to the area of immediate fabric lay-down.

Besides proper quantity, uniformity of the sprayed bitumen tack coat is of great importance. Application of hot bitumen should be done preferably by means of a calibrated distributor spray bar for better uniformity. Hand spraying and brush application may be used in locations of fabric overlap. When hand spraying, close attention must be paid to spraying a uniform tack coat.

Step 3 – Geosynthetic placement: The paving fabric is placed prior to the tack coat cooling and loosing tackiness. The paving fabric is placed onto the tack coat with its fuzzy side down leaving the smooth side up using a mechanical or manual lay-down equipment capable of providing a smooth installation without wrinkling or folding. Today most paving fabric is applied using tractor-mounted rigs. Slight tension can be applied during paving fabric installation to minimize wrinkling. However, stretching is not recommended, because it will reduce the thickness, changing the bitumen retention properties of the fabric. Too little elongation may result in wrinkles. Too much elongation produces excessive stretch, thinning the geosynthetic so it may not be thick enough to absorb the tack coat, leaving excess that may later bleed through the bituminous concrete on a hot day. Wrinkles and overlaps can cause cracks in the new overlay if not properly handled during construction process.

Overlaps and all overlapped wrinkles for fabric and grid composites should have an additional tack coat placed. Tack coat must be sufficient to saturate the two layers and make a bond. If not done correctly, a slip plane may exist at each overlapped joint, resulting in a possible crack of the asphalt from the fabric. Overlaps should be no more than 150 mm on longitudinal and transverse joints. This is different for grids, and each manufacturer has its own recommendations for overlaps. Paving multiple lanes has inherent installation problems. It is best to install in one lane and pave it for traffic prior to installing in another lane. 150 mm of fabric should be left unpaved for overlap on the adjacent panel of fabric to be installed.

A paving reinforcement geogrid is installed into a light asphalt binder or it may be attached to the existing pavement by mechanical means (nailing) or by adhesives, preventing the geogrid from being lifted by paving equipment passing over it. When a composite of geogrid and geotextile is installed, the tack coat is applied in the same way as in paving a geotextile alone.

Installing geosynthetic around curves without producing excessive wrinkles is the most difficult task for installers of paving synthetics. However, with the proper procedures, it can be accomplished with ease. Attempt should not be made to roll the geosynthetic around a curve by hand. It will wrinkle too much. Placing the fabric around a limited curve with machinery is preferable. Some minor wrinkles may occur. Grids have low elongation and thus do not stretch in curves. In most cases, the grid will need to be installed by hand or in short sections by machine to avoid wrinkles (Barazone, 2000).

Excess tack coat, which bleeds through the paving fabric, is removed by spreading hot mix, or by spreading sand over it. Any traffic on the geosynthetic should be carefully controlled. Sharp turning and braking may damage the fabric. For safety reasons, only construction traffic should be allowed on the installed paving fabric.

Step 4 – Overlay placement: All areas with paving geosynthetic placed are paved on the same day. In fact, asphalt concrete overlay construction should be done immediately after the placement of paving geosynthetic. Asphalt can be placed by any conventional means. Compaction should take place immediately after dumping in order to ensure that the different layers are bonded together.

The temperature of asphalt concrete overlay should not exceed about 160°C to avoid damage to the paving fabric. Overlays should not be attempted with temperature less than 120°C and air temperature less than 10°C. Adequate overlay thickness, if used, generates enough heat to draw the tack coat up, into and through the paving fabric, thus making a bond. In fact, the heat of the overlay and the pressure applied by its compaction force the tack coat into the paving fabric and complete the process. If sufficient residual heat after compaction is not present, the bonding process is disrupted, the results being slippage and eventual overlay failure. Thickness of the asphalt overlay should not be less than 40 mm. Compacting the asphalt concrete immediately after placement helps to concentrate the heat and supply pressure to start the process of the bitumen moving up into and through the fabric. This is very important when using a thinner overlay as it cools more rapidly. In cold weather, a thicker overlay may be necessary to achieve the same objective.

A paving fabric interlayer can also be used beneath seal coat or other thin surface applications. In such applications, sufficient heat is not there to reactivate the asphalt sealant. Therefore, the installed paving fabric must be trafficked or rolled with a pneumatic roller to push the fabric completely into the asphalt sealant. Sand can be applied lightly to avoid bitumen tackiness during trafficking. Once the paving fabric has absorbed the asphalt sealant, the seal surface treatment is applied exactly as it would be over any road surface.

It is suggested that the first-time users of paving fabric interlayer should obtain help from the paving fabric manufacturers and installers, keeping in view the site and material variables.

6.3.6 Railway tracks

For optimum performance of a geotextile layer, it must be properly installed. The geotextile can be installed under existing tracks in a number of ways, but is usually placed in conjunction with undercutting, plowing, or sledding operations as described by Walls and Newby (1993). In some instances, track sections are removed by crane during rehabilitation of the trackbed, with geotextiles being installed at the same time.

The following important points must be considered during installation (Raymond, 1986b, 1999; Tan, 2002):

1. The surface, over which the geotextile is being placed, should be prepared and contoured to remove debris and roadbed irregularities, with cross-fall gradients to facilitate drainage of water from track centreline to adjacent ditches and drains.
2. When joining geotextiles, an overlap of at least 0.5 m is recommended.
3. Geotextile should be placed so that water entering the geotextile can drain away from the track.
4. Geotextiles should be installed at a depth of not less than 200 mm, and preferably 300 mm, below the tie (sleeper) base and that tamping should not be permitted until that depth of ballast is in place. This is to prevent damage from normal tamping operations as shown in Figure 6.21. Ensuring sufficient ballast depth can be obtained through the use of sand bags filled with ballast.
5. Geotextile installations should be provided with day-lighted French drains on both sides of the track with inverts 15 mm below the load-bearing surface and the geotextile edges should be turned down into the French drains. If an outlet is not provided for internal drainage, the geotextile will drain water into the load-bearing area resulting in a worse condition than that obtained when the geotextile has not been used at all.
6. In order to allow smooth transition from the rehabilitated track with geotextile to the unrehabilitated areas, it is recommended to provide a transition zone of about 6 m where the track is undercut and not provided with a geotextile. This will decrease the suddenness of the track modulus change and reduce the associated dynamic traffic loadings.

Figure 6.21 Installation of geotextile and sledding of subballast (after Tan, 2002).

6.3.7 Filters and drains

Geosynthetics are used as a filter and/or a drain in many applications; the guidelines described below can be useful in most of such applications.

1 The geosynthetic filters and drains should be well protected to prevent any degradation, and care should be taken to avoid their contamination.
2 The geosynthetic, particularly a woven geotextile, should be placed with the machine direction following the direction of water flow.
3 The geotextile filters should be placed in intimate contact with the soil to be retained. Intimate contact between filter and soil is one of the major demands for all geotextile filters. If it is lacking, the adjacent soil can become suspended as water percolates towards the filter. Filtering a suspension flow is more difficult for geotextiles, since this condition hinders the forming of a secondary filter. Usually the geotextile acts as a kind of 'catalyst' for the developing filtration in the adjacent soil. Suspended soils cannot build a secondary filter, so clogging and/or blinding will result. If the construction process makes it difficult to establish close contact, a sand fill can be used between the geotextile and the base soil to provide a granular filter, which otherwise would be created in the soil itself.
4 The ends of subsequent rolls and parallel rolls of geotextile should be overlapped a minimum of 0.3–0.6 m. The overlaps may be increased for high hydraulic flow conditions and heavy construction. Joints, if provided, must prevent the infiltration of soil particles.
5 The overlying stones should be placed without damaging the geotextile. A cushion layer may be provided over the geotextile. This will reduce the impact (drop and abrasion) of large elements on the geotextile. The thickness of such a cushion layer should be

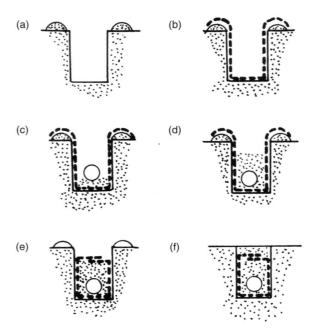

Figure 6.22 Construction procedure for geotextile-wrapped underdrains: (a) excavate trench; (b) place geotextile; (c) add bedding and pipe; (d) place/compact drainage material; (e) wrap geotextile over top; (f) compact backfill.

approximately equal to the diameter of the stones to be placed on it, up to about 0.4 m (Heibaum, 1998). During the initial period after installation, the large fill elements will agitate due to traffic, dynamic hydraulic loads or deformations of the subsoil caused by the new load until an equilibrium is found. Thin and very light geotextiles may not sustain this load, so the geotextile has to be chosen properly. When in doubt, some thickness should be added to the cushion layer.

6 Construction procedure, suggested by Holtz *et al.* (1997), can be adopted for geotextile-wrapped underdrains. The major construction steps are shown in Figure 6.22.

7 In the drainage system of retaining wall, the geosynthetic drain should be located away from the wall in an inclined orientation so that it can intercept seepage before it impinges on the back of the wall. Placement of a thin vertical drain directly against a retaining wall may actually increase seepage forces on the wall due to rainwater infiltration (Terzaghi and Peck, 1948; Cedergren, 1989).

8 Backfill materials should be placed by methods that will not disturb or damage the geosynthetic product. Backfill should be placed in a maximum of 150 mm lifts and compacted to a minimum of 90% standard Proctor density. Excessive compaction of backfill directly against the drain or filter should be avoided; otherwise loading during compaction may cause damage or deformation to the filter or the drain materials and their joints, if any. Coarse-grained, clean materials such as crushed stone, gravel and

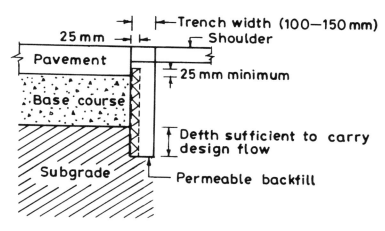

Figure 6.23 A typical type and arrangement of geocomposite buried drain.

sand are more readily compacted using vibratory equipment. Fine materials with high plasticity should not be used as a backfill material.

9   For installation of geocomposite pavement edge drains, trenches should be excavated with stable, straight and smooth sidewalls. If the sidewall of the pavement side trench is not reasonably straight and smooth, or if undercutting or sidewall sloughing occurs, the geocomposite drain material should not be installed against the sidewall of the pavement side trench. In such cases, the product may be installed in the centre of the trench or against the shoulder sidewall. If installed in the centre of the trench, the product must be supported during installation and backfill in such a way as to keep it straight, vertical, and stable. Also trench width may be increased to a minimum of 150 mm plus the product thickness. Figure 6.23 shows a typical type and arrangement of prefabricated geocomposite drain. The trench depth and width should be sufficient to carry the design flow below the base course – subgrade interface such that water will not be retained in the structural pavement section. Trench width is also governed by the required space for adequate compaction of the backfill using compaction equipment without damaging the geocomposite drainage panel. Geocomposite pavement drains should not be laid in standing or flowing water.

Prefabricated geosynthetic drains should be properly tied into the collector and outlet drainage systems. Outlet pipe, generally laid at a slope of 3%, should have sufficient diameter to remove the collected water from the geocomposite drain at a rate equal to or greater than the flow capacity of the geocomposite drain.

10 Since the function of the geotextile in a silt fence, as shown in Figure 6.24, is to filter and allow settlement of soil particles from sediment-laden overland water flow, the geotextile at the bottom of the fence should be buried in a 'J' configuration to a minimum depth of 150 mm in a trench so that no flow can pass under the silt fence. The trench should be backfilled and the soil compacted over the geotextile. The geotextile should be spliced together with a sewn seam only at a support post, or two sections of fence may be overlapped instead (AASHTO M 288-00).

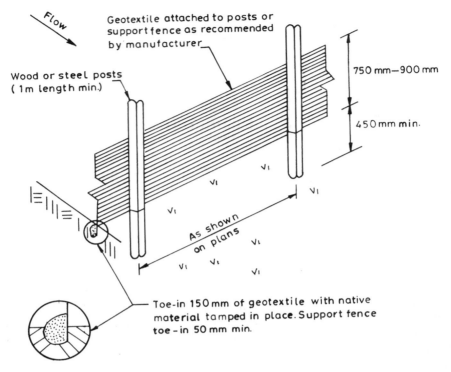

Figure 6.24 Typical silt fence detail (after AASHTO, 2000).

6.3.8 Slopes – erosion control

Erosion control systems such as riprap revetments with geosynthetic filters and mattress revetments may require the attention of the following guidelines.

1. Depressions in the slope should be filled to avoid geotextile bridging and possible tearing when cover materials are placed. A well-compacted slope is important in order to produce a smooth surface and thus ensure that there is a good connection between the revetment and the subsurface.
2. The geotextile sheet should be placed with warp direction (a.k.a. machine direction) (in case of a woven geotextile) in the direction of water flow, which is normally parallel to the slope for erosion control runoff and wave action, and parallel to the stream or channel in the case of streambank and channel protection. It should be placed in intimate contact with the subsoil without wrinkles or folds. There should be 1.5 m minimum offset between adjacent ends.
3. Jointing systems, which are without strain, can be made with an overlap of 0.5–1.0 m or with a lapped ('J') seam. Joints under stress must be avoided as much as possible. In particular cases where heavy forces occur in the main direction as well as at right angles to the main direction, it is usual to apply two layers, one in each direction. When overlapping, successive sheets of the geotextile should be overlapped upstream

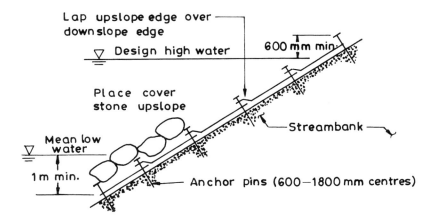

Figure 6.25 Cross section of stream-bank slopes with revetment (after AASHTO, 2000).

over downstream, and/or upslope over downslope. In cases where wave action or multidirectional flow is anticipated, all seams perpendicular to the direction of flow should be sewn. Overlaps should be made along the slope parallel to the direction of water flow. Overlapped seams of roll ends should be a minimum of 300 mm except where placed under water. In such instances the overlap should be a minimum of 1 m. Overlaps of adjacent rolls should be a minimum of 300 mm in all instances (AASHTO M 288-00). Geotextile sheets can be held in position by ballasting with sandbags or by pinning loosely with large-headed polymer pins (Fig. 6.25).

4   Place the armour (cover layer) over the geotextile as quickly as possible, preferably within 14 days. In underwater applications, the geotextile and the cover layer should be placed the same day.

5   The armour system placement should begin at the toe and proceed up the slope. Riprap and heavy stone filling should not be dropped from a height of more than 300 mm. The stone piece with a mass of more than about 100 kg should not be allowed to roll down the slope. The placing process should avoid any damage or stretching to the geotextile. It is a good practice to insist that the installation contractor demonstrates that his chosen placing method does not result in damage to the geotextile. Alternatively, the use of a cushion layer (granular sublayer) between the armour and the geotextile spreads the load and reduce the contact stress.

6   The geotextile must be checked for its temperature resistance at 130–140°C when used in conjunction with a bituminous armour placed in situ.

7   Once in place, the individual mattresses, if used, should be joined so that the edges cannot be lifted up under the action of water waves and currents. In addition, the top and bottom edges of the revetment including geotextile filter should be anchored as shown in Figure 6.26. In such a case, a toe structure may not be needed to stop revetments sliding. Keying the geotextile into the crest of the slope can be avoided, provided the geotextile can significantly be extended above the anticipated maximum high water level.

8   Bags for revetments should be filled and stacked against a prepared stable slope with their long axes parallel to the shoreline. While a stacked-bag revetment can be placed

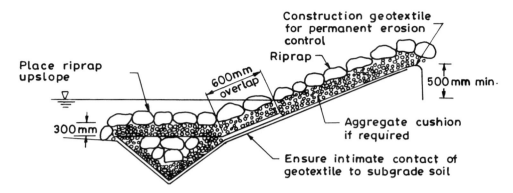

Figure 6.26 Key detail at top and toe of slope for geotextiles used for permanent erosion control (after AASHTO, 2000).

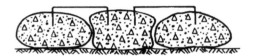

Figure 6.27 Reinforcement of concrete bags.

on a steeper slope, it should not exceed 1 vertical to 1.5 horizontal. A stacked-bag revetment should preferably be two bags thick, for example, with the outside layer concrete-filled and the interior bags sand-filled. Concrete-filled bags can be stabilized by steel rods driven through the bags (Fig. 6.27).

9 When the turf reinforcement mats (TRMs) are used, they should be installed first, then seeded and filled with soil. High strength TRMs provide sufficient thickness and void space to permit soil filling/retention and the development of vegetation within the matrix. Many erosion control revegetation mats (ECRMS) are generally installed prior to seeding.

10 If geonaturals (jute geotextile) are used aiming to speed up the vegetation growth, then straw or hay mulch must be placed beneath them to achieve optimum results. Anchoring trenches (450 mm deep and 300 mm wide) should be excavated at the top and toe of the slope along the length of the slope. Overlaps should be minimum 150 mm at sides and ends. The jute geotextile at the higher level on the slope should be placed over the portion to its next at a lower level. Side overlaps of jute geotextile piece should be placed over its next piece on one side and under the next piece on the other. It should be fixed in position by steel staples (usually of 11 gauge dia) as shown in Figure 6.28 or by split bamboo pegs. Stapling should be done normally at an interval of 1500 mm both in longitudinal and transverse directions. Seeds of vegetation or saplings of the appropriate plant species may be spread/planted at suitable intervals through the openings of the jute geotextile. Installation should be completed preferably before the monsoon to take advantage of the rains for quick germination of seeds. Watering/maintenance of the vegetation should be carried out as per specialist advice of agronomist/botanist.

11 Installation of erosion control mats and blankets can follow the manufacturer's recommended procedures. Figure 6.29 illustrates one such general installation procedure.

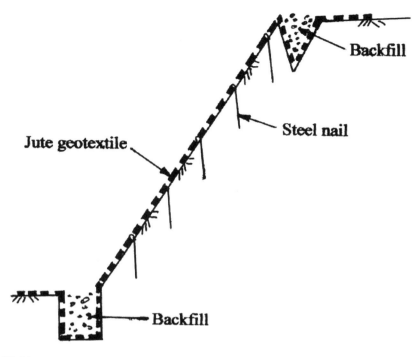

Figure 6.28 Use of a geonatural for erosion control of the sloping ground aiming to speed up the vegetation growth.

6.3.9 Slopes – stabilization

Like conventional soil slopes, reinforced slopes are generally constructed by compacting soil in layers while stepping the face of the slope back at an angle. Subsequently, the face is protected from erosion by vegetation or other protective systems. Additional geosynthetic elements are incorporated into reinforced steepened slopes to facilitate drainage, minimize groundwater seepage and to assure the stability of the steepened slope and the erosion resistance of the facing (Fig. 6.30).

In the present-day geosynthetic engineering, there are various slope stabilization methods in practice. However, the specific application guidelines are described only for a few popular methods of slope stabilization along with their basic description.

Geofabric-wrapped drain (GWD) method

This method was suggested by Broms and Wong (1986) and was used successfully in Singapore to stabilize a steep slope in residual soil and weathered rock. By this method, the stability of existing unfailed soil slopes can be increased, failed slopes can be stabilized or new steep slopes or high embankments can be constructed without exceeding the load-bearing capacity of soil. In these applications, the function of the geotextile, both as a tensile reinforcement and as a filter, is utilized.

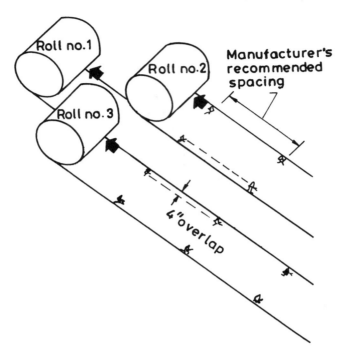

Figure 6.29 Typical installation instruction for erosion control mats and blankets (after Agnew, 1991).

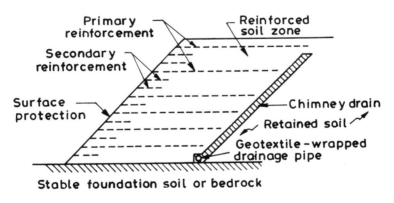

Figure 6.30 Components of a reinforced steepened slope.

In this method, the geotextile-wrapped drains consisting of granular materials are installed along the slopes as shown in Figure 6.31(a). The drains reduce the pore water pressure within the slopes during the rainy season and thereby the shear strength is increased. The geotextile layer acts as a filter around the drains, which prevents the migration of soil (internal erosion) within the slope into the drains. It also reinforces the soil along potential sliding zones or planes.

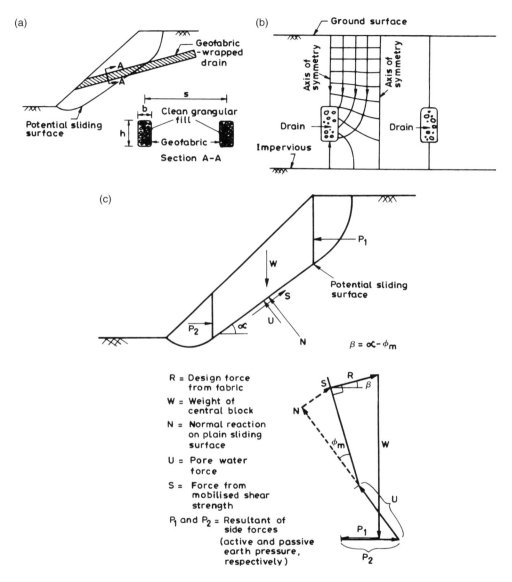

Figure 6.31 (a) Schematic of slope stabilization using geofabric-wrapped drains; (b) flownet showing steady state seepage; (c) computation of design tensile reinforcement to be provided by the geotextile (after Broms and Wong, 1986).

One additional advantage with this method is that the temporary decrease of the stability of the slope is only marginal during the construction of the deep trenches required for the drains. Here, only a limited width of the slope is affected. When concrete gravity or cantilever walls are used, the stability of the slope can be reduced considerably during the construction.

The required spacing of the drains wrapped in geotextile, as well as dimensions of the drains, depend on the pore water pressures in the slope which can be evaluated by means of a flownet (Fig. 6.31(b)). The granular material in the drains is considered to be infinitely pervious in relation to the slope material. The pore water pressure in the slope is reduced considerably by the drains both above and between the drains as can be seen from the flownet. For general situations, 0.5 m wide and 1.0 m high drains spaced 3.0 m apart would be reasonable.

The drains should be located deep enough so that they intersect potential slip surfaces in the soil. The required depth of the drains depends on the difficulties of excavating trenches along the slopes. The maximum depth is about 4 m. For slopes in residual soils or weathered rocks, this depth is usually sufficient because most slope failures in these materials are shallow, having maximum depth of failure surface less than 3–4 m.

The required tensile strength of the geotextile can be calculated by considering the force polygon for the sliding soil mass above possible sliding surfaces in the soil (Fig. 6.31(c)). The sliding surface is often located at the contact between the completely weathered and the underlying partially weathered material.

For a planar sliding surface, the orientation of the geotextile-wrapped drains should be perpendicular to the resultant of the normal reaction force and the force that corresponds to the mobilized shear strength along the potential failure surface, as shown in Figure 6.31(c) in order to utilize the geotextile effectively.

The required number of layers (N) of the geotextile in each drain can be determined as follows:

$$N = \frac{F_s R s}{aT}, \tag{6.2}$$

where R is the force per unit width (kN/m) to be resisted by the geotextile; s is the drain spacing (m); T is the tensile strength per unit width (kN/m) of the geotextile; a is the effective perimeter of the drain (m); and F_s is the factor of safety.

The geotextiles available in the market generally require an elongation of 14–50%, before the ultimate tensile strength of the geotextile is mobilized. The strain required to mobilize the ultimate strength is much less for woven geotextiles than for non-woven geotextiles. Only woven geotextiles should therefore be used. In view of the large strain required at failure, a factor of safety of at least three should be used in the design.

The length L that is required to transfer the load in the geotextile to the surrounding soil can be calculated as follows:

$$L = \frac{Rs}{2(hK\sigma_v' + b\sigma_v') \tan \phi_a'}, \tag{6.3}$$

where σ_v' is the vertical effective stress at mid height (centre) of the drains; K is the lateral earth pressure coefficient for the compacted granular material in the drains; h is the height of the drains; b is the width of the drains; and ϕ_a' is the friction angle between the geotextile and the soil.

The deformation δ of the geotextile to mobilize the required tensile force can be calculated from the following equation:

$$\delta = L \times \frac{e}{100} \tag{6.4}$$

where, e is the per cent elongation needed to mobilize the required tensile resistance of the geotextile.

During the construction of the granular fill drains, it is important to compact the fill carefully. The compaction will increase the lateral earth pressure and therefore the friction between the geotextile and the soil results in reduced transfer length L. For a well-compacted fill, a value of K equal to at least 1.0 can be used in the calculation of transfer length. The lateral earth pressure is highly dependent on the degree of compaction of the granular fill.

A second important point, with respect to compaction of the granular fill drains, is that the compaction should be done in the downhill direction in order to pretension the geotextile. In this way, the elongation of the geotextile, which is necessary to mobilize the required tensile force as well as the required displacement of the slope, will be reduced.

Anchored geosynthetic system (AGS) method

This method was suggested by Koerner (1984) and Koerner and Robins (1986) and is also known as '*anchored spider netting*' method. It is an in situ slope stabilization method in which a geosynthetic material (geotextile, geogrid or geonet) or other porous material is placed directly on the unstable or questionable slope and anchored to it with long steel rod nails at discretely reinforced nodes, 1–2 m apart. These nails must be long enough to penetrate up to, and beyond, the actual or potential failure surface. Figure 6.32 shows the idealized cross-section of a slope stabilized by this method along with its conceptualization. When the rods are properly fastened, they begin pulling the surface netting into the soil placing the net in tension and the contained soil in compression. When suitably deployed, this method offers a number of advantages in arresting slope failures:

- the steel rods in penetrating the failure surface aid stability;
- the stress caused by netting at the ground surface aids stability;
- the surface netting stress mobilizes normal stress at the base of the failure surface, which aids stability;
- the entire system causes soil densification, which increases the shear strength parameters of soil.

It is important to recognize that the mechanism by which an AGS stabilizes a slope is different from that of reinforced earth or soil nailing. Both reinforced earth ties and soil nails are passive systems that rely on soil strains to mobilize pullout, bending and shear resistances of the inclusions. By contrast, the anchors in an AGS are actively tensioned during installation. The increase in stability of the slope thus does not rely on soil movement to mobilize the soil–anchor interaction, but rather the increased stresses on potential failure surfaces, imparted by the tensioned fabric, increase the stability of the slopes (Ghiassian *et al*., 1996).

In recognition of the above, the analysis of a slope stabilized by anchored geosynthetics follows a traditional limiting equilibrium stability analysis for slopes. The effects of the AGS are considered as additional forces acting on a potential failure surface. Bending and shearing resistances of the anchors are disregarded for several reasons. First, the bending and shearing resistances of the anchors are not mobilized. Second, the spacing of anchors is typically greater than the spacing in soil nailing, and thus a coherent soil mass may not develop. A slope may fail by erosion and the flow of soil around anchors; therefore, the

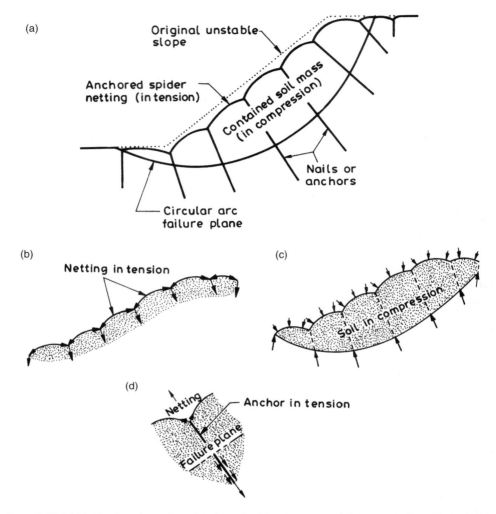

Figure 6.32 (a) Idealized cross-section of anchored spider netting in stabilizing a soil slope; (b) free-body diagram of netting; (c) free-body diagram of contained soil; (d) free-body diagram of anchor (after Koerner, 1984; Koerner and Robins, 1986).

bending and shearing resistances of the anchors may be irrelevant. Third, a variety of materials could be used for the anchors in an AGS, including cables with duck-billed anchors, which have essentially no bending resistance. Finally, the assumption is conservative.

Reinforced soil structures (RSS) method

Slopes can be stabilized with the construction of geosynthetic-reinforced soil structures. If reinforced soil structures have an inclination $\beta \leq 70°$, they are called *reinforced soil slopes*; otherwise they are called *reinforced retaining walls*, which have been addressed in

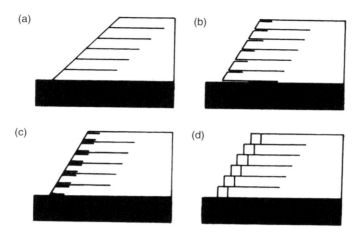

Figure 6.33 The construction schemes for green faced structures: (a) straight reinforcement; (b) wraparound reinforcement; (c) mixed scheme; (d) face blocks plus straight reinforcement (after Rimoldi and Jaecklin, 1996).

Sec. 6.3.1. In such stabilization methods, the failed soil can be used as a backfill material to make them economical.

Rimoldi and Jaecklin (1996) summarized the construction methods for green-faced reinforced soil walls and steep slopes in four main schemes:

1 Straight reinforcement (Fig. 6.33(a)): This type of reinforcement, made of geosynthetics only, is mainly used for shallow slopes ($\beta < 50°$). Generally, the face is left exposed, or covered by a geomat or biomat to prevent erosion. Therefore, the reinforcing geosynthetic is installed just at the face, without any wraparound. The installation is very easy. The reinforcement is laid down horizontally and straight, and then the soil is spread and compacted to the required height, smoothing the face with a vibrating table or with the bucket of a backhoe.

2 Reinforcement wrapped around the face (Fig. 6.33(b)): In this scheme, the geosynthetic is used both for reinforcement of the fill and for the face protection from soil washing and progressive erosion, by wrapping it around the face of the slope. This 'wraparound' technique has been the most widely used construction method in Europe. The 'wraparound' installation procedure can be used with or without formworks. The use of formworks is particularly suggested when it is necessary to have a smooth and uniform face finishing, as discussed in Sec. 6.3.1.

The most simple construction method with the wraparound technique is without any formwork – it consists in placing a geogrid layer; in laying down, spreading and compacting the fill soil; in smoothing and levelling the face of the slope at the desired angle with a vibrating table or with the bucket of backhoe; then the geogrid is wrapped around the face and fixed with a 'U' staple. This method provides a fast construction and affords good results if it is not necessary to obtain a perfectly smoothed face. In fact, 'bulging' of the face often occurs, with unpleasant aesthetic effect. Wraparound can also be made using movable formworks, or straight steel mesh or steel mesh shaped as an 'L' or a 'C'. There can be some other suitable means also. If steel meshes are used, they are left in place after the construction is terminated, which saves a lot of time and hence allows a very fast construction rate: a typical team of 4–5 workers well equipped and experienced enough, can install about 50 m² of wall face in one working day, but in particular situations 100 m² of face in one day can be achieved. The reinforcing geosynthetics can be connected to the steel meshes, but usually the two elements are independent.

3   Mixed scheme: straight reinforcement plus another geosynthetic wrapped around the face in 'C' shape (Fig. 6.33(c)): In this scheme, the two functions of reinforcement and face protection are played by two different geosynthetics. The reinforcing geosynthetic has high tensile strength and modulus, while the other one for face protection is lighter and is engineered to support the growing vegetation and to retain the soil, preventing washout and erosion.

4   Front blocks tied back by straight reinforcement (Fig. 6.33(d)): In this scheme a front block is used both to support the facing during construction and for providing the final face finishing. Blocks are usually made of compacted soil, encased in containers, made either of gabion baskets or of geosynthetics wrapped all around. Blocks are mechanically connected to straight reinforcing geosynthetics. This *front blocks method* has the advantage of not being dependent on weather situations. The prefabrication does not disturb any traffic and can be near the site. A standard excavator is used to place the face blocks quickly and the same excavator is also used for backfilling. No hydroseeding is needed because the grass seeds are already included inside the face blocks and grass starts growing immediately.

Over-steep geogrid reinforced slopes are usually associated with vegetation, and the facing of the slope is wrapped around by the geogrids or sometimes the facing is temporarily supported by a steel mesh allowing vegetation to grow through the mesh apertures. Hard facing is also in use with geogrid reinforced soil walls. Hard facing as opposed to soft facing refers to large precast concrete panels or the small modular concrete blocks (MCB). The MCB blocks are laid dry (i.e. without mortar) and the geogrid reinforcements are placed between the block courses and connected by means of insert keys or pins or by only the frictional interface between the courses. The footings for the geogrid reinforced modular concrete block wall systems (GRMCBWS) can be constructed from granular compacted materials or from cast-in-place concrete. The walls are usually constructed with stepped facing resulting in a batter ranging between 5° and 20°. The overall shape is equivalent to a steep slope as opposed to a vertical wall, and therefore analysis can be carried out using steep slope procedures. One advantage of GRMCBWSs is the simplicity of installation because the blocks are easily transportable. It is estimated that four persons can erect 30–40 sq. m of wall over an eight-hour working day. As for cost comparison, it is estimated that walls exceeding 1.0 m in height typically offer a 25–35% cost saving over conventional cast-in-place concrete retaining walls.

The following points may be followed during the construction of reinforced soil slopes:

1   A level subgrade is prepared by clearing the site and removing all slide debris.
2   Geosynthetic reinforcement should be placed with the principal strength direction perpendicular to the face of the slope.

3   Lightweight compaction equipment should be used near the slope face to help maintain face alignment.
4   A face wrap may not be required for slopes up to 1H:1V, if the reinforcement is maintained at close spacing, not greater than 400 mm.
5   Drainage layers, if required, should be constructed directly behind or on the sides of the reinforced section.

6.3.10 Containment facilities

In containment facilities (landfills, canals, ponds, reservoirs, dams, etc.), the installation of geomembrane basal liner is critical and relatively complicated, compared with installations of other geosynthetic products. The construction work involves the operation of heavy earth moving equipment as well as the minute handling of sensitive geosynthetic products. The construction personnel must be quality conscious. For effective installation of geomembrane and other geosynthetic materials, special care is required; Given below are general guidelines that should be followed during installation.

1   The subgrade/clay soil liner surface should be firm and unyielding with no abrupt elevation changes, voids and cracks, vegetation, roots, sharp-edged stones, construction debris, ice, standing water, and any other deleterious material that may cause damage to the geomembrane. If stones that could puncture the geomembrane exist in the clay soil liner/subgrade, they must be removed prior to the installation of geomembrane. Deviations from the theoretical plane surface should not exceed 20 mm over a distance of 4 m. The ruts of the compaction equipment may not be deeper than 5 mm. The subgrade should be protected from desiccation, flooding and freezing. If required, protection may consist of a thin plastic protective cover installed over the complete subgrade until the placement of the geomembrane liner begins. If the subgrade surface is too rough for the direct placement of the geomembrane, a nonwoven needle-punched geotextile can be placed on the subgrade prior to the placement of the geomembrane.
2   The method and equipment used to place panels must not damage the geomembrane or the supporting subgrade surface. Personnel working on the geomembrane must not wear shoes that can damage the geomembrane or engage in actions that could result in damage to the geomembrane. Adequate temporary loading and/or anchoring (i.e. sandbags, tires, etc.) must be done to prevent uplift of the geomembrane by wind.
3   There should be an intimate hydraulic contact between the geomembrane liner and the underlying soil subgrade/clay soil liner. To achieve intimate contact, the surface of the soil subgrade/clay liner on which the geomembrane is placed should be smooth-rolled with a steel-drum roller.
4   Since field seaming of the geomembrane panels is a critical aspect of their successful functioning as a barrier to liquid flow, it must be handled very carefully. In general, seams should be oriented parallel to the slope that is along, not across, the slope. All geomembrane sheets, regardless of type, should be seamed by thermal methods: fusion (hot wedge) welding and extrusion welding.

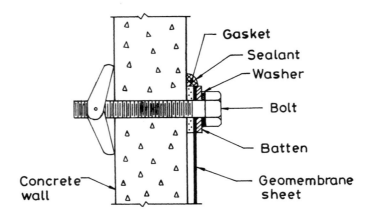

Figure 6.34 A typical attachment of the geomembrane sheet to a concrete wall.

5 The geomembrane liner must be attached properly to structures as shown in Figure 6.34. Welding or bonding the geomembrane to the structure may provide an attachment that has a lower possibility of leakage. If the geomembrane is attached directly to the structure, sealants are usually not required. The surface of the structure for which the geomembrane is to be attached should be constructed or formed without irregularities to limit damage to the geomembrane. Edges or corners of the structures should be rounded. If a structure cannot be constructed or formed without irregularities, then a cushion/sealant as a protective layer should be placed between the geomembrane and the structure.

6 The geomembrane should be placed and backfilled in a way that minimizes wrinkles. A field-deployed and seamed geomembrane must be backfilled with soil (generally drainage material with or without a geotextile protection layer) or covered with a subsequent layer of geosynthetics (generally a geonet or geocomposite drain) in a timely manner after its acceptance by the CQA personnel. Large voids under the geomembrane should be filled to stop the geomembrane from becoming overly stressed. Note that geonets are always covered with a geotextile, that is, they are never directly soil covered, since the soil particles would fill the apertures of the geonet rendering it useless.

7 The GCL should lie flat on the underlying surface, with no wrinkles or folds, especially at the exposed edges of the panels. Only as much geosynthetic clay liner should be deployed per working day as can be covered by suitable cover soils. The sealing between installed rolls of geosynthetic clay liners should be made by overlapping. The lengthwise seams should typically be overlapped a minimum 150 mm, and the widthwise seams a 500 mm. The geosynthetic clay liner should be placed so that seams are parallel to the direction of the slope. If a trench is used for anchoring the end of the geosynthetic clay liner, soil backfill should be placed in the trench to provide resistance against pullout. The geosynthetic clay liner should be sealed around structures embedded in the subgrade and pipe penetrations as shown in Figure 6.35.

(a)

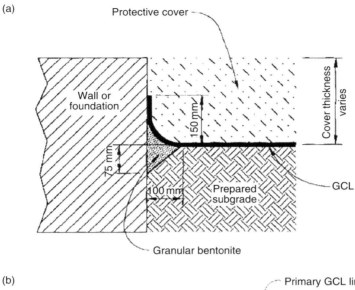

(b)

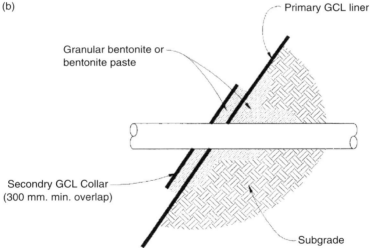

Figure 6.35 Geosynthetic clay liner sealing methods around: (a) a structure; (b) a pipe penetration (ASTM D 6102-97) test (Reprinted, with permission, from ASTM D6102-97: *Stanadard Guide for Installation of Geosynthetic Clay Liners*, copyright ASTM International, 100 Barr Harbor Drive, West Conshohocken, PA 19428).

8   Clay soil liners and covers should be compacted wet of optimum moisture content. Cover soils should be free of sharp edged stones or other foreign matter that could damage the geomembrane or the geosynthetic clay liner. Lift thickness should not be more than 6 inches (150 mm) after compaction. This differs from the lift thickness of 9–18 in. (230–300 mm) for embankments and other geotechnical applications. The cover

soil should be prevented from entering the geosynthetic overlap zones. Note that construction equipment should never be allowed to move directly on any deployed geomembrane.

9   Geomembranes and other involved geosynthetics should usually be terminated by a horizontal runout, an anchor trench or a combination thereof (Fig. 6.12). The runout and the anchor trench are covered with soil and suitably compacted to hold installed geosynthetics in place against applied loads. The holding capacity comes mainly from the frictional resistance between the geosynthetic and the soil, and it depends on several factors such as runout length, cover soil depth, shape and depth of anchor trench, the types of soil underlying and overlying the geosynthetic and the type of geosynthetic used.

10 All construction operations at a landfill site are sensitive to weather conditions. Obviously, the placement of a clay liner is impossible during heavy rain, snowfall or frost, and partly finished clay blankets must be protected against water and against desiccation due to dry wind and sunshine when the construction work is interrupted at weekends, due to bad weather or for any other reasons. For such a temporary protection, thin plastic membranes are used. The installation of GMBs requires favourable weather as well. It cannot be done in the rain. The minimum temperature for seaming polyethylene sheets is 5°C. Sufficient time has to be allocated to the placement of geomembranes to cope with unavoidable delays due to unfavourable weather.

6.3.11   Tunnels

Waterproofing of tunnels can be successfully carried out using geosynthetics. Even completely submerged tunnel can be waterproofed. Geomembrane sheet sealing with a protective nonwoven geotextile drainage layer has become the predominant sealing system worldwide. The staging of construction activities should be designed such that the installation of the geosynthetics becomes a separate and continuous operation. The geosynthetic installation process should minimize interruption of other tunnel construction activities. Based on the reported case studies (Benneton *et al.*, 1993; Davies, 1993; Posch and Werner, 1993), the major construction steps are summarized as follows:

- excavation of rock and/or soil;
- grouting to stop/minimize inflowing water, if present;
- supporting the exposed surface by shotcrete (gunite);
- fastening the thick needle-punched nonwoven geotextile, as protective screen as well as drainage medium, to the shotcrete by means of PVC plastic discs (plates) and fasteners (nails);
- fixing the geotextile to underdrains on each side of tunnel base;
- placement of a geomembrane (usually PVC) to PVC plastic discs by means of hot air welding;
- spot-bonding of a protective shield (3 mm thick PVC) to the geomembrane;
- placement of the concrete liner against the geomembrane;
- providing additional seals (consisting of an expansion product, for example butyl bentonite) at concrete restart points.

Self-evaluation questions

(Select the most appropriate answers to the multiple-choice questions from 1 to 20)

1. The construction-related failures of geosynthetic applications are caused mainly by

    (a) The loss of strength due to UV exposure.
    (b) The lack of proper overlap.
    (c) The high installation stresses.
    (d) All of the above.

2. The minimum number of test samples for 5000 m^2 or less area of geosynthetic is generally taken as

    (a) 1.
    (b) 2.
    (c) 3.
    (d) 5.

3. At no time should the geosynthetics generally be exposed to UV light for a period exceeding

    (a) One week.
    (b) Two weeks.
    (c) Three weeks.
    (d) One month.

4. The temperature at the geosynthetic storage site should not generally exceed

    (a) 21°C.
    (b) 27°C.
    (c) 70°C.
    (d) None of the above.

5. The geosynthetic overlap used is generally a minimum of

    (a) 15 cm.
    (b) 30 cm.
    (c) 1 m.
    (d) None of the above.

6. The temperature during geosynthetic seaming should be between

    (a) 0°C and 20°C.
    (b) 4.5°C and 20°C.
    (c) 0°C and 40°C.
    (d) 4.5°C and 40°C.

7. AASHTO M 288-00 recommends that when sewn seams are required, the seam strength, as measured in accordance with ASTM D4632, should be equal to or greater than

    (a) 50% of the specified grab tensile strength.
    (b) 70% of the specified grab tensile strength.
    (c) 90% of the specified grab tensile strength.
    (d) None of the above.

8. When placing fill on sloped surfaces, fill material should always be kept from the

    (a) Base of the slope.
    (b) Top of the slope.
    (c) Mid-point of the slope.
    (d) All of the above.

9. In general, when releasing stones that weigh less than 120 kg on geotextiles over well-prepared surfaces, the maximum drop height can be

    (a) 0.30 m.
    (b) 0.45 m.
    (c) 0.75 m.
    (d) 1.00 m.

10. In road, railway and embankment construction, the first layer of fill material on the geosynthetic should have a minimum thickness of

    (a) 200 mm.
    (b) 200 mm to 300 mm.
    (c) 300 mm.
    (d) 1 m.

11. All vehicles and construction equipments weighing more than 1500 kg should be kept away from the faces of the walls or steep slopes by at least

    (a) 1 m.
    (b) 2 m.
    (c) 5 m.
    (d) None of the above.

12. In the case of liquid containment ponds, to shield the geomembrane liners from ozone, UV light, temperature extremes, ice damage, wind stresses, accidental damage, and vandalism, a soil cover of at least

    (a) 15 cm thickness should be provided.
    (b) 30 cm thickness should be provided.
    (c) 75 cm thickness should be provided.
    (d) 1 m thickness should be provided.

13. In the geotextile-reinforced retaining wall, the geotextile layer should be installed with its principal strength direction (warp direction)

    (a) Perpendicular to the wall face.
    (b) Parallel to the wall face.
    (c) Inclined at 45° to the wall face.
    (d) None of the above.

14. In the geotextile-reinforced retaining wall with a wraparound facing, the minimum lap length should generally be

    (a) 0.5 m.
    (b) 1.0 m.
    (c) 1.5 m.
    (d) None of the above.

15. In unpaved roads, overlaps of parallel geosynthetic rolls should not occur at

    (a) The centreline of the roadway.
    (b) The shoulder of the roadway.
    (c) The anticipated main wheel path locations.
    (d) All of the above.

16. Which one of the following is the best and the most economical choice for paving fabric tack coat?

    (a) Paving-grade bitumen.
    (b) Cutback.
    (c) Emulsion.
    (d) None of the above.

17. To avoid damage to the paving fabric, the temperature of asphalt concrete overlay can be kept a maximum of

    (a) 50°C.
    (b) 100°C.
    (c) 160°C.
    (d) None of the above.

18. In railway tracks, when joining geotextiles, the minimum overlap generally recommended is

    (a) 0.3 m.
    (b) 0.5 m.
    (c) 1.0 m.
    (d) None of the above.

19. In the drainage system of retaining walls, the geosynthetic drain should be located away from the wall in

    (a) A parallel orientation.
    (b) A perpendicular orientation.
    (c) An inclined orientation.
    (d) None of the above.

20. Geofabric-wrapped drain (GWD) method of slope stabilization was suggested by

    (a) Broms and Wong (1986).
    (b) Koerner (1984).
    (c) Koerner and Robins (1986).
    (d) None of the above.

21. Why does the installation of geosynthetics in practice require a degree of care and consideration?
22. Give reasons why the white-surfaced textured HDPE geomembranes are preferred over the other types for their use in the lining of ponds, reservoirs and canals?
23. What are the general guidelines regarding the proper care and handling of geotextiles during the installation process?
24. How can you reduce the friction damages during geosynthetic installation?
25. What is the difference between a geosynthetic joint (seam) and a geosynthetic connection?
26. Describe seaming methods for geomembrane panels.

27. Under what type of weather conditions should field seaming be conducted?
28. List the factors influencing seaming quality.
29. Give the basic description of a bodkin joint.
30. What type of test would you recommend to evaluate the strength of a sewn seam?
31. Why is a uniform compaction of subgrade soils beneath the geosynthetic layer important?
32. What protection is required for geosynthetics during construction and service life?
33. What should be examined before covering the installed geosynthetic with soil?
34. What are the different approaches for providing the bond length to the installed geosynthetic?
35. What is the effect of prestressing the geosynthetic installed in a granular fill?
36. What is the purpose of certification in geosynthetic engineering?
37. What precautions should be taken during the construction of geotextile-reinforced retaining walls?
38. Describe a construction method for the permanent geosynthetic-reinforced soil retaining wall, widely used in Japan.
39. List the various alternatives with regard to the installation of a geosynthetic layer inside the embankment along with their advantages.
40. Draw a neat sketch to show the construction sequence for geosynthetic-reinforced embankments over soft foundation soils.
41. What is the purpose of placing narrow horizontal strips with wraparound along the side slopes of the embankment?
42. What should be the geosynthetic orientation for linear embankments?
43. What are the effects of contamination of the granular layer from the fine-grained subgrade soils on the performance of roadways?
44. Describe the basic steps in the proper installation of a pavement overlay system with a geosynthetic interlayer.
45. Would you recommend a smooth transition from the rehabilitated railway track with geotextile to the unrehabilitated tracks? If yes, why?
46. Intimate contact between the geosynthetic filter and the base soil is necessary. Is there any reason for this practice? Justify your answer.
47. Describe the major construction steps for the geotextile-wrapped underdrains.
48. Describe the typical arrangement of a geocomposite buried drain.
49. Although a geotextile functions as a filter in silt fence applications, consideration of its strength is also important in the selection of geotextiles? Do you agree with this statement? Justify your answer.
50. What is the role of a cushion layer (a granular layer) between the overlying stones and the underlying geotextile?
51. How will you install the turf reinforcement mat (TRM) for erosion control of slopes?
52. What is the typical size of an anchoring trench?
53. Describe briefly the geotextile-wrapped drain (GWD) method of slope stabilization. What are the special features of this method?
54. What are the components of reinforced steepened slope?
55. Illustrate the various mechanisms that soil nails in *anchored spider netting method* provide in soil slope stabilization.
56. In anchored spider netting method of slope stabilization, the geotextile is exposed on the surface of the slope. What advantages and disadvantages do you observe?

57. Describe the main schemes of construction methods for green-faced reinforced soil walls and steep slopes.
58. How will you attach the geomembrane liner to structures?
59. Describe the clay liner sealing methods around a structure and a pipe penetration.
60. What precautions are required while installing the geomembrane liner?
61. Describe the major steps for construction of waterproofed tunnels with geosynthetics.

Chapter 7

Quality and field performance monitoring

7.1 Introduction

The proper and intended functioning of a geosynthetic product or system in an engineered facility is strongly dependent on the quality of the material and construction, and it can be checked through a well-planned field performance monitoring programme. The users of geosynthetics must, therefore, be familiar with the quality evaluation and field performance monitoring techniques. This chapter introduces the reader to some aspects related to the quality evaluation and field performance monitoring.

7.2 Concepts of quality and its evaluation

Quality of a geosynthetic product or geosynthetic application is the confidence that can be placed in it, consistently meeting the numerically claimed variation limits in properties or functioning taken into account by the design engineer and extrapolated into the in situ conditions. Standards, test methods, testing frequencies, tolerances, and corrective actions are the means by which the quality can be measured and controlled. The purpose of quality evaluation is to facilitate continuous improvement of geosynthetic products and geosynthetic applications and to fully document the performance relative to a target.

In an ideal world quality considerations for geosynthetics would not be needed. A design engineer would properly design and specify a material, and the contractor would install the material in accordance with the design document, occasionally calling the engineer for a design clarification. Unfortunately, we do not leave in an ideal world, but in the world of the 'low bid' contractor. Low bid often means smallest profit margin, which potentially leads to attempts to cut corners and thus sacrifice quality. That is why a proper quality evaluation system is necessary for any construction project.

Quality control (QC) refers to a planned system of operational techniques and activities which sustain a quality of geosynthetic product, or geosynthetic application that will satisfy the needs as per the project plans, specifications and contractual and regulatory requirements; also the use of such a system. QC is provided by the manufacturers and the installers of the various components of the geosynthetic application. Geosynthetics must be properly manufactured and installed in a manner consistent with a minimum level of QC as determined by testing.

To achieve the QC, the manufacturer and the installer need a quality assurance system, of which QC is only a part. *Quality assurance* (QA) refers to a planed system of actions necessary to provide adequate confidence that a geosynthetic product, or geosynthetic application will satisfy the needs as per the project plans, specifications and contractual and regulatory requirements. QA is provided by an organization different from the organization that provides QC. An effective QA programme must define the route by which it will achieve its stated objectives. This will require an explanation of the qualifications, roles, responsibilities, authority, and interaction of all parties involved. The programme must identify and describe all QA activities and procedures (Menoff and Eith, 1990). The direct benefits of a QA programme to the facility owner/operator are to confirm that the project was constructed in accordance with the engineering design and specifications and to provide technical and legal documentation and certification of the quality of the work performed. The indirect benefit is perhaps of even greater importance. This is the knowledge that the product/facility has been manufactured/constructed with as great a degree of integrity as possible, thereby minimizing potential problems during operation and/or post-closure and ultimate liability.

Geosynthetics have four levels of quality management associated with them:

1   Manufacturing quality control (MQC)
2   Manufacturing quality assurance (MQA)
3   Construction quality control (CQC)
4   Construction quality assurance (CQA).

Manufacturing quality control (MQC), normally performed by the geosynthetic manufacturer, is necessary to ensure minimum (or maximum) specified values in the manufactured product (Koerner and Daniel, 1993). A factory production control scheme should be established and documented in a manual prior to a geosynthetic type being placed on the market. Subsequently, any fundamental changes in raw materials and additives, manufacturing procedures or the control scheme that affect the properties or use of a geosynthetic should be recorded in the manual. Additionally, manufacturing quality assurance (MQA) programme provides assurance that the geosynthetics were manufactured as specified in the certification documents and contract plans and specifications. MQA includes manufacturing facility inspections, verifications, audits, and evaluation of the raw materials and geosynthetic products to assess the quality of the manufactured materials.

Construction quality control (CQC), normally performed by the geosynthetic installer, is necessary to directly monitor and control the quality of a construction project in compliance with the plans and specifications. Construction quality assurance (CQA) programme provides assurance to the owner and regulatory authority (as applicable) that the structure was constructed in accordance with plans and design specifications. CQA includes inspections, verifications, audits, and evaluations of materials and workmanship necessary to determine and document the quality of the constructed facility. In fact, design, construction, and certification reporting are three phases in the life of a project during which a CQA monitor can have a beneficial role. The CQA monitor does not design the project. An experienced CQA monitor can, however, provide a significant role in the design phase by reviewing the design based on how a contractor would consider it (Thiel and Stewart, 1993).

Quite often, MQA and CQA are performed by the same organization. On the other hand, MQC and CQC are often performed by different organizations – the manufacturer and the installer. Of course, many of the larger manufacturers have their own installation crews (Qian *et al.*, 2002). Note that although MQA/CQA and MQC/CQC are separate activities, they have similar objectives and, in a smoothly running construction project, the processes will complement one another. Conversely, an effective MQA/CQA programme can lead to identification of deficiencies in the MQC/CQC process, but a MQA/CQA programme by itself, in complete absence of a MQC/CQC programme, is unlikely to lead to acceptable quality management. Quality is thus best ensured with effective MQC/CQC and MQA/CQA programmes. A major purpose of the MQA/CQA process is to provide documentation for those individuals who were unable to observe the entire construction process so that those individuals can make informed judgements about the quality of construction for a project. MQA/CQA procedures and results must be thoroughly documented.

Geosynthetic CQA monitors should have the following general qualifications: (Thiel and Stewart, 1993):

- familiarity with construction procedures and contract issues;
- experience of reviewing test results, quality control data, and contractor submittals;
- familiarity with the design issues regarding the type of construction project being monitored;
- ability to effectively communicate and prepare supporting documentation;
- geosynthetic monitoring experience gained under the supervision of more experienced individuals;
- experience of reporting, communicating, and resolving deficiencies and performing remediation activities.

The CQA monitor has to have certain qualities to effectively perform his/her job. The most important quality is probably assertiveness. When the monitor observes a problem, he or she has to act quickly, clearly identify the problem, recommend corrective action, verify that the corrective action is taken, and document the issue. The monitor does this knowing that the contractors' schedule and progress will be affected. If the monitor is working alone on a remote site, this assertiveness is essential for success and can require courage. CQA monitors who routinely monitor projects with geosynthetics should visit a geosynthetic manufacturing plant and a laboratory where geosynthetic tests are performed to gain a better understanding of the materials. Note that it is better to have a representative of the design engineer performing the CQA, because of liability issues and the need for understanding the design intent. In fact, the design engineer is in the best position to evaluate construction materials and methods because he/she understands the minute details and knows how the design elements are interconnected.

Figure 7.1 shows the usual interaction of the various elements in a total inspection programme. Note that the flow chart includes both the geosynthetic and the natural soil materials since both require similar concern and care.

Quality assurance and quality control are recognized as critical factors in many geosynthetic applications. Acceptance testing should be performed on a geosynthetic product to determine whether or not an individual lot of the product conforms to specified requirements. It can be done prior to geosynthetic shipment, directly after arrival of the geosynthetic roll at the site, and/or prior to geosynthetic installation. Irregularities should be noted and reported.

318 Quality and field performance monitoring

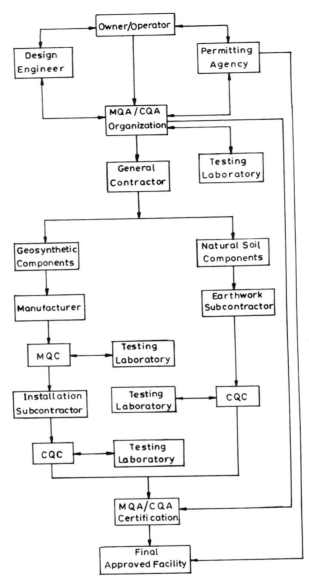

Figure 7.1 Organizational structure of MQC/MQA and CQC/CQA inspection activities (Koerner and Daniel, 1993).

Quality control on construction sites is done by index testing which has been discussed in Chapter 3. Index testing involves the use of very simple techniques, which do not give definitive design parameters for a geosynthetic, but do give reproducible results, suitable for QC and comparison of geosynthetic products. The users should always make at least a check for type and quantity of geosynthetics being delivered. To enable the user on site to identify the geosynthetic products as being identical to the products ordered, the following basic information should be indelibly marked on the outside wrapping of each roll so that it is clearly visible before the roll is opened:

- Name of the manufacturer and/or supplier and country of origin;
- Product name (brand name/commercial name);
- Product type (descriptive number or the code);
- Unit identification (number or other code given on each unit, for example roll, that allows the original manufacturer to trace at a later stage the production details, including place and date of production), for example batch number and date;
- Mass of unit, in kg;
- Unit dimensions, such as length × width or area (of material, not of package), in m;
- Mass per unit area, in g/m^2;
- Thickness of material, in mm;
- Major polymer type(s) for each component;
- Geosynthetic type/classification.

The above information should also be marked inside the core so as to be visible when the roll is partly used. More general and specific details are given on data sheet, brochures, technical sales literature, or similar documents.

Geosynthetics can be identified by simple means at the time of its installation, even if it is no longer in the original packaging. For the convenience of the user, the markings of the product name and type should be printed along the edge of the geosynthetic product at regular intervals of at most 5 m. A simple check on the mass per unit area may also be made using basic equipments such as a balance and a scale. In the case of high-risk applications, such as the use of geosynthetic filters in dams and geosynthetics as a soil reinforcement, testing of every roll, or at least every other roll should be performed. In such demanding applications, the most important property should be determined in addition to the basic index properties mentioned above. In the case of low risk applications, such as the use of geosynthetic as separator in unpaved roads, only the basic index tests need to be carried out for every 1 in 10 or 20 rolls. It is thus noted that the frequency and degree of QC testing are generally functions of application and the risk involved in that application. Note that any test that is carried out over a long period of time and is based on a complex procedure is not considered to be a routine QC test.

For adopting the QC in reality, it is absolutely essential to have competent and professional construction inspection. The field supervisor must be properly trained to observe every phase of the construction and to ensure that:

- the specified material is delivered to the project;
- the rolls of geosynthetics are properly stored;
- testing requirements are verified;
- the geosynthetic is not damaged during construction;
- the specified sequence of construction operations is explicitly followed;
- seam/joint integrity is verified on the basis of testing and evaluation.

A trained and knowledgeable supervisor/inspector should only be allowed to perform construction inspection. Efforts should be made to maintain a good documentation of construction activities.

7.3 Field performance monitoring

As with all geosynthetic applications, and especially with critical structures such as reinforced slopes and retaining walls, landfills, and dams, competent and professional field inspection is absolutely essential for successful construction. Field personnel must be properly trained to observe every phase of the construction. They must make sure that the specified geosynthetic is delivered to the project site, that the geosynthetic is not damaged during construction, and that the specified sequence of construction operations are explicitly followed. Other important details include the measures being adopted to minimize geosynthetic exposure to ultraviolet (UV). Field personnel should always review the checklist items for the project.

The in situ monitoring of geosynthetics- and geosynthetic-related systems usually has two goals. One addresses the integrity and safety of the system, whereas the other provides guidance and insight into the design process. It is to be noted that the purpose of the instrumentation in geosynthetic-related projects is not only for research but also to verify design assumptions and to control construction.

It is important to conceive and execute a monitoring plan with clear objectives in mind. Dunnicliff (1988) provides a methodology for organizing a monitoring programme in geotechnical instrumentation. The checklist of specific steps that are recommended follows:

1. Define project conditions.
2. Predict mechanism(s) that control behaviour.
3. Define the question(s) that need answering.
4. Define the purpose of the instrumentation.
5. Select the parameter(s) to be monitored.
6. Predict the magnitude(s) of change.
7. Devise remedial action.
8. Assign relevant tasks.
9. Select the instruments.
10. Select the instrument locations.
11. Plan for factors influencing the measured data.
12. Establish procedures for ensuring corrections.
13. List the purposes of each instrument.
14. Prepare a budget.
15. Write an instrument procurement specification.
16. Plan the installation.
17. Plan for regular calibration and maintenance.
18. Plan for data collection, processing, presentation, interpretation, reporting, and implementation.
19. Write the contractual arrangements for field services.
20. Update the budget as the project progresses.

Such a checklist should be considered in planning for the in situ monitoring of geosynthetic applications whenever permanent and/or critical installations are under consideration or are being otherwise challenged.

There are presently wide ranges of in situ monitoring methods/devices, which have generally resulted in reliable data. Table 7.1 provides the summary of the monitoring methods/devices

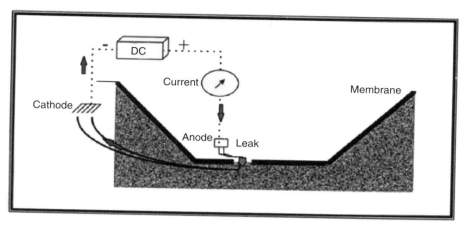

Figure 7.2 Schematic of electrical leak detection method (Reprinted, with permission, from ASTM D6747-02: *Standard Guide for Selection of Techniques for Electrical Detection of Potential Leak Paths in Geomembranes*, copyright ASTM International, 100 Barr Harbor Drive, West Conshohocken, PA 19428).

as presented by Koerner (1996). In this table, monitoring methods or devices are somewhat arbitrarily divided into recommended and optional categories. Table 7.2 gives a further description of the various methods/devices listed in Table 7.1. Since the monitoring is site-specific, its cost must be assessed on a case-by-case basis.

The geosynthetic product installation should be monitored by the engineer at a frequency appropriate to project requirements. Excessive post-installation and other distress should be monitored carefully by the engineer.

Experience shows that the installed geomembrane liner can have potential leak paths, such as holes, tears, cuts, seam defects and burned through zones. The damage to a geomembrane liner can be detected using electrical leak detection systems developed in the early 1980's. Such systems have been used successfully to locate leak paths in electrically insulating geomembranes such as polyethylene, polypropylene, polyvinyl chloride, chlorosulfonated polyethylene and bituminous geomembranes installed in ponds, reservoirs, canals, tanks and landfills. Types of potential leak paths have been related to the quality of the subgrade material, quality of the cover material, care in the cover material installation and quality of geomembrane installation.

The principle behind the electrical leak detection systems is to place a voltage across a geomembrane liner and then locate areas where electrical current flows through discontinuities in the liner as shown schematically in Figure 7.2. The liner must act as an insulator across which an electrical potential is applied. This electrical detection method of locating potential leak paths in a geomembrane liner can be performed on exposed liners, on liners covered with water, or on liners covered by a protective soil layer. This technique can locate very small leak paths, smaller than 1 mm. This technique cannot be used during stormy weather when the geomembrane is installed on a desiccated subgrade, or whenever conductive structures cannot be insulated or isolated.

A more recent survey by Nosko and Touze-Foltz (2000) summarized the results of electrical damage detection systems installed at more than 300 sites and covering more than

Table 7.1 Summary of monitoring methods/devices (after Koerner, 1996)

Geosynthetic type	Function or application	Recommended	Optional
Geotextiles	Separation	• water content measurements • pore water transducers	• level surveying • earth pressure cells • inductance gauges
	Reinforcement	• strain gauges • movement surveying • inclinometers • extensometers	• earth pressure cells • inductance gauges • pore water transducers • water content measurements • settlement plates • temperature
	Filtration	• water observation wells • pore water transducers	• flow meters • turbidity meters • probes for pH, conductivity and/or dissolved oxygen
	Drainage	(same as geotextile filtration)	
	Barrier (e.g. reflective cracking)	• surface deflections • level surveying • surface roughness measurements • profilometery (for rut depths)	• water content measurements
Geogrids	Walls	• strain gauges • inclinometers • extensometers • monument surveying	• earth pressure cells • piezometers • settlement plates • probes for pH • temperature

	Slopes	• strain gauges • inclinometers • extensometers
	Foundations	• strain gauges • level surveying • extensometers • flow meters • turbidity meters • earth pressure cells • piezometers • monument surveying • earth pressure cells • piezometers • settlement plates • probes for pH, conductivity and/or dissolved oxygen
Geonets	Drainage	• strain gauges • temperature measurement • flow meters • downgradient wells • piezometers
Geomembranes	Tensile stress Temperature Global leak monitoring	• flow meters • downgradient wells • extensometers • deformation telltales • turbidity meters • probes for pH, conductivity and/or dissolved oxygen • turbidity meters • probes for pH, conductivity and/or dissolved oxygen
Geosynthetic clay liners	Global leak monitoring Shear strength	• gypsum cylinders • fiberglass wafers • strain gauges (inductance coils) • level surveying
Geocomposites	Separation (e.g. erosion control) Reinforcement Drainage (e.g. edge drains) Barrier	• flow meters • turbidity meters (same as geotextiles and geogrids) • flow meters • turbidity meters (same as geotextiles, geomembranes and GCLs) • probes for pH, conductivity and/or dissolved oxygen

Table 7.2 Selected description and commentary on the methods and devices listed in Table 7.1 (after Koerner, 1996)

Category	Methods/device	Resulting value/information
Surveying	Monument surveying	Lateral movement of vertical face
	Level surveying	Vertical movement of surface
	Settlement plates	Vertical movement at depth
Deformation	Telltales	Measures movement of fixed rods or wires can accommodate any orientation
	Inclinometers	Measures vertical movement in a casing inclined movements up to 45°
	Extensometers	Measures changes between two points in a borehole
Strain measurement	Electrical resistance gauges • bonded foil • weldable	Measures strain of a material over gauge length, typically, 0.25 to 150 mm
	inductance gauges (coils) • static measurements • dynamic measurements	Measures movement between two embedded coils up to 1000 mm distance apart
	LVDT gauges	Measures movement between two fixed points 100 to 200 mm apart
Stress measurement	Earth pressure cells • diaphragm type • hydraulic type	Measures total stress acting on the cell, can be placed at any orientation, can also measure stress (pressure) against walls and structures
Soil moisture	Water observation wells	Measures stationary groundwater level
	Gypsum cylinders	Measures soil moisture content up to saturation
	Fibreglass wafers	Measures soil moisture content up to saturation
Groundwater pressure	Piezometers • hydraulic type • pneumatic type • vibrating wire type • electrical resistance type	Measures pore water pressures at any depth, can be installed as single point or in multiple point array, can be placed in any orientation
Temperature measurement	Bimetal thermometer	Measures temperature in adjacent area to ±1.0°C
	Thermocouple	Measures temperature at a point to ±0.5°C
	Thermistor	Measures temperature at a point to ±0.1°C
Liquid quantity	Tipping buckets	Measures flow rates (relatively low values)
	Automated weirs	Measures flow rates (relatively high values)
	Flowmeters	Measures flow rates (very high values)
Liquid quality	Turbidity meters	Measures suspended solids
	pH probes	Measures pH of liquid
	Conductivity probes	Measures conductivity of liquid

3,250,000 m² of geomembrane liners. This survey showed that the majority of the damages (71%) were caused by stones, followed by heavy equipment (16%) (Fig. 7.3(a)). Interestingly, most of the failures (78%) were found to be located in the flat areas of the liner (bottom liner); only 9% were found at the corners and edges of the landfills (Fig. 7.3(b)). It is interesting to note from the reported surveys that the bulk of the defects were related to mechanical damage caused by the placement of soil on top of the geomembrane. The

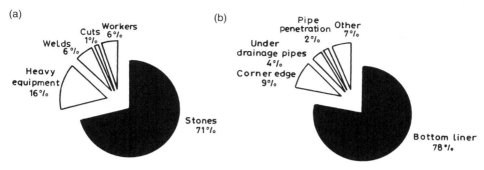

Figure 7.3 (a) Cause of damage in geomembrane liners; (b) Location of damage in geomembrane liners (modified from Nosko and Touze-Foltz, 2000; after Bouazza et al., 2002)

readers can refer to the recommendations made by Giroud (2000) to minimize geomembrane installation and post-installation defects.

To measure directly the extent of degradation on site it may be desirable to extract geosynthetic specimens at the following stages:

1. just after installation
2. after a certain fraction of design life
3. at the end of the design life.

To monitor the condition of the extracted geosynthetic, physical and chemical analysis methods are recommended in addition to normal index tests. In structures whose integrity is critical, the geosynthetic specimens should be placed such that they can be extracted after a certain fraction of design life and their condition can be compared with that predicted at the design stage. In this way, the user will obtain advanced warning of any degradation that is occurring. For the sake of comparison with the specimens extracted, the user should retain samples of pristine geosynthetic and a record of the original test results.

An important aspect of understanding the long-term performance of any material is to know the environmental conditions that it experiences during its service life. A key environmental condition, particularly for a geomembrane is its in situ temperature. Thermocouples can be installed for the purpose of long-term temperature monitoring. Accelerated ageing test can also be performed in the laboratory (Koerner, 2001).

Self-evaluation questions

(Select the most appropriate answers to the multiple-choice questions from 1 to 10)

1. Which one of the following programmes provides assurance that the geosynthetics were manufactured as specified in the certification documents and contract plans and specifications?

    (a) MQC.
    (b) MQA.

(c) CQC.
(d) CQA.

2. Manufacturing quality control (MQC) is normally performed by

    (a) The geosynthetic manufacturer.
    (b) The geosynthetic installer.
    (c) The owner of the engineering project.
    (d) The main contractor of the project.

3. Construction quality control (CQC) is normally performed by

    (a) The geosynthetic manufacturer.
    (b) The geosynthetic installer.
    (c) The owner of the engineering project.
    (d) The main contractor of the project.

4. A CQA monitor can have a beneficial role during

    (a) Design.
    (b) Construction.
    (c) Certification reporting.
    (d) All of the above.

5. 'MQA and CQA are performed by the same organization.' This statement is

    (a) Always true.
    (b) Quite often true.
    (c) Sometimes true.
    (d) Never true.

6. 'MQC and CQC are performed by different organizations.' This statement is

    (a) Always true.
    (b) Quite often true.
    (c) Sometimes true.
    (d) Never true.

7. The most important quality of a CQA monitor is

    (a) Ability to communicate effectively.
    (b) Familiarity with the design issues.
    (c) Experience of reviewing the test results.
    (d) Assertiveness.

8. Quality control on construction sites is generally done by

    (a) Personal judgement.
    (b) Index tests.
    (c) Field performance tests.
    (d) None of the above.

9. The electrical detection method of locating potential leak paths in a geomembrane liner can be performed on

(a) Exposed liners.
(b) Liners covered with water.
(c) Liners covered by a protective soil layer.
(d) All of the above.

10. A key environmental condition, particularly for a geomembrane is
    (a) In situ temperature.
    (b) In situ moisture.
    (c) In situ pressure.
    (d) None of the above.

11. What do you mean by the term 'quality'? List the purpose of quality evaluation for geosynthetic products and applications.
12. What is the difference between construction quality control (CQC) and construction quality assurance (CQA)?
13. What technical skills are required to perform CQA?
14. Can the MQA organization be the same as the CQA organization?
15. Do you agree with the statement that MQA/CQA and MQC/CQC, being separate activities, have similar objectives? Justify your answer.
16. Draw an organizational structure of MQC/MQA and CQC/CQA inspection activities.
17. What information should be marked on the outside wrapping of geosynthetic rolls for their complete identification?
18. How will you identify a geosynthetic by simple means at the time of its installation?
19. What do you mean by high-risk applications of geosynthetics? Give some practical examples. What special precautions are required in such applications?
20. What are the important roles of the field supervisor for geosynthetic applications?
21. What are the goals of in situ monitoring of geosynthetic-related systems?
22. What major inspections of geosynthetic-related structures are required by the field personnel?
23. Prepare a checklist of specific steps to be included in the performance monitoring plan for a canal lining project.
24. Why should the cost of monitoring for geosynthetic-related projects be assessed on a case-by-case basis?
25. What is the principle of an electrical leak detection system used to assess the damage to the geomembrane liners?
26. How would you estimate the field performance of geosynthetics in severe climatic conditions?

Chapter 8

Economic evaluation

8.1 Introduction

Geosynthetics are used in any application area to have technical benefits and/or the overall cost savings. Their use may result in lower initial cost and/or greater durability and longer life, thus reducing maintenance costs. The cost analysis of a geosynthetic-related project needs careful handling when taking decisions for the acceptance or the rejection of the option of using geosynthetics in the project just only on the basis of its cost. The current chapter deals with the fundamentals of cost analysis and related experiences from some completed geosynthetic-related projects and reported economic studies.

8.2 Concepts of cost analysis

The design engineer is usually confronted with an important task: whether a conventional solution or a geosynthetic-related solution should be preferred in a particular civil engineering project at a specific site. In order to give a rational decision, data related to the following aspects should be analysed carefully (Durukan and Tezcan, 1992):

- Relative economy
- Cost-performance efficiency (a.k.a. cost–benefit ratio)
- Factors of safety
- Feasibility
- Availability of materials
- Relative speed of construction.

The *rate of relative economy* (E_r) is defined as:

$$E_r = \left(\frac{C_c - C_r}{C_r} \times 100 \right) \%, \tag{8.1}$$

where C_c is the cost of conventional soil structure and C_r is the cost of geosynthetic-reinforced soil structure.

The cost of a geosynthetic-related structure or application should typically be presented as an engineering estimate of the capital, operational and maintenance costs. For having a general idea of the *cost-performance efficiency* (a.k.a. *cost–benefit (C/B) ratio*) of a geosynthetic or any other element of geosynthetic-related structure, it can be represented as the normalized cost (C_m). In the case of geosynthetic-reinforced soil retaining walls, C_m can be defined as:

$$C_m = \frac{C}{T}, \qquad (8.2)$$

where C_m is the normalized cost of geosynthetic reinforcement carrying a safe tensile load of 1 kN on a 1-m-run wall; C is the cost of 1 m² geosynthetic within a 1-m-run wall; and T is the safe tensile resistance of a one-layer geosynthetic for a 1-m-run wall. For any other reinforcing element of the structure, C_m can be defined similarly keeping in view the function served by that element.

It should be noted that the total cost of a geosynthetic-reinforced soil structure depends not only on the relative costs of individual elements, but also on the geometry of the reinforced soil structure and its site location. For the purpose of determining the relative economy as well as the cost efficiency of reinforced soil structures, a comprehensive cost analysis should be performed by taking into account the costs (both direct and indirect) of various elements of any application.

The cost–benefit analysis basically consists of a comparison of benefits and costs for a certain set of site-specific conditions. If the benefits exceed the costs for a certain set of conditions, the cost–benefit ratio is less than one. In the analysis, the preferred course of action is that which yields the smallest cost–benefit ratio (subject to being smaller than one). While making cost–benefit analysis of environmental applications, one should consider the reduction of human health and ecological risk as benefits of using geosynthetics.

In general, the benefits of geosynthetic-related structures are difficult to measure like the benefits of other public projects, whereas the costs are more easily determined. For simplicity, one can only attempt to quantify the primary benefits or effectiveness with which the application goals can be met in monetary or non-monetary measure by the extent to which that alternative, if implemented, will attain the desired objective. The preferred alternative is then either the one that produces the maximum effectiveness for a given level of cost or the minimum cost for a fixed level of effectiveness. In fact, the *cost-effectiveness method* allows us to compare alternative solutions on the basis of cost and non-monetary effectiveness measures.

The feasibility of a geosynthetic-related structure or application is strongly related to the type of problem at hand: an embankment on soft ground, a foundation, an unstable slope, a leaking dam or reservoir, a road or a railway track. Construction time available, speed of construction, availability of construction materials and equipments and the funds available are some of the factors which strongly govern the choice of geosynthetic-related structures or applications in a particular field project.

8.3 Experiences of cost analyses

In the case of reinforced soil walls, it is generally accepted that, under normal circumstances, and especially after a wall height of about 6 m, they become more economical, and also they are relatively easier and faster to build than their conventional counterparts (Ingold, 1982b). Reinforced soil retaining walls are almost indispensable when normal slopes may not be constructed due to property line constraints, high expropriation costs, existence of important structures or due to land being reserved for future structures. In most low to medium height retaining walls, the cost of the geosynthetic reinforcement was found to be less than 10% of the total wall cost. In these walls the greatest economy is obtained by using on site soils for backfill and an inexpensive facing. In order to arrive at a scientific conclusion, however, a comparative cost analysis must be performed.

It has been determined by a group of researchers in the UK that the rate of relative economy of reinforced soil walls increases steadily with the height of the wall, as shown in Figure 8.1. A similar cost-effectiveness study of reinforced soil embankments by Christie (1982) in the UK showed that when space restrictions or high land-acquisition costs necessitated steep walls, it was almost unavoidable to use soil reinforcement. Murray (1982) also reported that a repair project for a cutting using reinforced in situ soil saved about 40% when compared with conventional replacement techniques. It was reported by Bell et al. (1984) that the total cost of a series of geotextile-reinforced retaining walls varied between US $118 and US $134 per square metre of the wall surface. The average cost breakdown is shown in Table 8.1. In a blast protection embankment in London (UK), it was established by Paul (1984) that the geogrid-reinforced design was the most economical choice when compared with either the conventional reinforced concrete wall or the unreinforced soil wall. The relative costs for a 1-m-run of the wall are shown in Table 8.2.

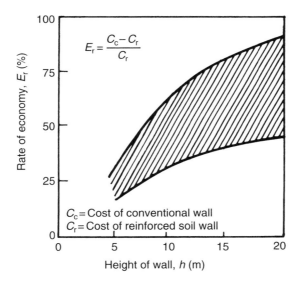

Figure 8.1 Rate of economy in reinforced soil walls (after Anon., 1979).

Table 8.1 Geotextile-reinforced soil walls (after Bell et al., 1984)

Item	Cost (US $/m²)	Share (%)
Geotextile	23	19
Labour	7	6
Equipment	7	6
Fill	53	44
Facing	30	25
Total	120	100

Table 8.2 Cost comparisons of an embankment (after Paul, 1984)

Wall type	Land width (m)	Cost (US $/m)
Reinforced concrete wall	18.9	2625
Geogrid-reinforced embankment	13.5	1775
Unreinforced embankment	32.5	1911

For evaluating the direct cost effect of geotextile applications in the context of India, four typical geotextile usages were examined by Ghoshal and Som (1993) for four different regions of the country. Costs of material, labour and land were collected for the metropolitan cities Mumbai, Bangalore, Delhi and Kolkata. For identical soil data and design parameters, the variation of cost with or without geotextiles for the selected functions was determined. An examination of the economic analysis reveals that the use of geotextiles depends significantly on the unit cost of different inputs. The apparent cost–benefit derived by using geotextiles is not uniquely determined on the basis of the cost of geotextiles alone. For example, where land cost is high, as in Mumbai, the economy of using geotextiles becomes more predominant in the slope stability function than in the separation function (see Tables 8.3 and 8.4). On the other hand, the separation function appears to give greater economy in Kolkata than in Mumbai because of the higher cost of stone aggregate in Kolkata.

For estimating the cost–benefit ratio of geosynthetic-reinforced unpaved roads, one must consider the future maintenance costs associated with both conventional and reinforced roadways. Such maintenance costs are dependent on the quality of the roadway initially constructed. In the case of geosynthetic-reinforced unpaved roads, the reduction of maintenance costs in excess of 25% has been reported (IFAI, 1992). For subgrades having California Bearing Ratio (CBR) values higher than 3, the geosynthetic-reinforced roadway may have an initial cost–benefit ratio near unity (that is, the additional cost of the geosynthetic is offset by a comparable savings in granular material), whereas it decreases as the strength of the existing soil subgrade decreases. This happens because for soil subgrades having CBR values lower than 3, geosynthetic layers provide an initial cost savings over alternative traditional stabilization methods such as excavation and replacement, preloading, staged construction to slowly increase the subgrade strength, lime or cement stabilization in addition to savings in granular material (Barenberg, 1992).

It is to be noted that in unpaved road construction the site preparation and the type of granular material greatly controls the selection, and therefore the cost of the geosynthetic based on the survivability during its installation. Increased site preparation increases construction costs but reduces the need for geosynthetics with higher survivability requirement. Similarly, the use of a granular soil with more angular particles produces a denser and stronger subbase/base but increases the need, and cost, for a more robust geosynthetic.

The cost effectiveness of a paving fabric interlayer system can be analysed by comparing its cost with the cost of additional bituminous/asphaltic concrete needed to produce the same pavement rehabilitating results. For the retardation of reflective cracking, the paving interlayer system has been proven to be equivalent to approximately 30 mm of additional bituminous concrete overlay thickness. This is significant since the installed cost of a paving fabric and bituminous concrete interlayer system is generally less than one half the cost of

Table 8.3 Separation function: comparison of cost, in Indian Rupees (INR), of an unpaved road with and without geotextile, base course thickness 800 mm (for 1 m² surface) (after Ghoshal and Som, 1993)

Place	Without geotextile		With geotextile		Ultimate saving (INR)
	Amount of loss per year (INR)	Amount of ultimate loss (INR)	Quantity of geotextile required (m²/m)	Cost of geotextile per m²/m (INR)	
Bangalore	8.80	26.40	1	50	−23.40
Mumbai	12	36	1	50	−14.00
Kolkata	32	96	1	50	46.00
Delhi	18.40	55.20	1	50	5.20

Notes
It has been assumed that a base course, 800 mm thick, will lose 10% of stone or metal per year and up to 30% of stone or metal will be lost on a long-term basis. The cost of geotextile has to be balanced against the cost of replenishment of stone or metals that will be required to maintain the yard in a usable condition. For comparison, the total cost of replenishment over a 3-year period has been considered.

Table 8.4 Stability of slope: comparison of cost, in Indian Rupees (INR), for an embankment with and without geotextile for one side slope (height of embankment = 8 m) (after Ghoshal and Som, 1993)

Place	Slope without geotextile		Slope with geotextile		Saving in land area per m run on one side (m²)	Cost of land saving (INR)	Saving per m run of embankment (INR)
	Angle of Slope (degree)	Cost (INR)	Angle of slope (degree)	Cost (INR)			
Bangalore	27	1701	45	3364	7.7	23,100	24,763
Mumbai	27	3465	45	4260	7.7	38,500	39,295
Kolkata	27	1575	45	3300	7.7	15,400	17,125
Delhi	27	3024	45	4036	7.7	30,800	31,812

the 25 mm of bituminous concrete. Thus, the pavement rehabilitation system with paving fabric interlayer system is economical. In addition, the paving fabric interlayer system provides the functions of moisture barrier and stress relief for the long-term fatigue resistance (IFAI, 1992).

For determining the cost effectiveness of geosynthetic versus conventional drainage systems, simply compare the cost of the geosynthetic with the cost of a conventional granular filter layer. It is to be noted that the use of geosynthetic can allow a considerable reduction in the physical dimensions of the drain without a decrease in flow capacity, resulting in reduction in the volume of excavation, the volume of granular material required, and the construction time necessary per unit length of drain.

The total cost of a riprap-geotextile revetment system will depend on the actual application and type of revetment selected. Additional cost for making special considerations while placing the geotextile below water level should be considered. Cost of overlapping and pins are also required. To determine cost effectiveness, these points should be considered and then cost–benefit ratios should be compared for the riprap-geotextile system versus conventional riprap-granular filter systems or other available alternatives of equal technical feasibility and operational practicality.

For many slope stabilization projects, when all financial aspects are considered, slope reinforcement is the most cost-effective technique to meet a change in grade. Issues include (Simac, 1992):

- Land acquisition costs
- Construction costs
- Soil-fill costs
- Financial return on usable area
- Future expansion.

However, since the largest cost-savings component comes from reducing land acquisition costs or optimizing the developable land from purchased property, slope reinforcement should be considered early in the project planning process to maximize the owner's economic benefits. It is considered that the slope reinforcement is justifiable when any of the following three site-specific factors apply to a project:

- High cost of real estate
- Steep topography
- Expensive, unsuitable or scarce soil-fill materials.

Slope reinforcement costs are offset by eliminating or minimizing the impact of these site-specific factors. Therefore, to limit evaluation of slope reinforcement only to construction and soil-fill costs may underestimate its value to a project owner. One should evaluate the technique based on its design features and total cost relative to other conventional solutions available. In fact, quite often, slope reinforcement presents both the best technical and cost-efficient solution for projects influenced by the key site-specific factors.

A reinforced slope will typically consist of three components; soil backfill, the geosynthetic reinforcement, and a surface erosion control system. The relative cost of these components will be a function of the height of the embankment and the slope of the face. In general the following trends have been noted (IFAI, 1992):

- The soil component will represent more than 80% of the reinforced slope cost.
- The per cent of total cost spent on reinforcement will increase with increasing slope height.
- The per cent of total cost spent on erosion control will increase with increasing slope angle of the reinforced slope face.

The true savings resulting from the use of a reinforced slope will depend on the value of the real estate saved and the potential use of locally available soil fill.

The life cycle cost of a landfill can be categorized as follows: construction, operation (including all monitoring), closure and long-term care (LTC). In many cases the cost of hauling is expected to be high because the landfills are located in remote areas. The cost of road construction for borrowed materials (e.g. clayey soil) must be included in the unit cost.

Table 8.5 Comparison of 34 canal lining test sections by the USBR (after Swihart and Haynes, 2002)

Type of lining	Construction cost ($/ft²)	Durability (years)	Maintenance cost (($/ft² -yr)	Effectiveness at seepage reduction (%)	Benefit/cost ratio
Fluid-applied membrane	1.40–4.33	10–15	0.010	90	0.2–1.5
Concrete alone	1.92–2.33	40–60	0.005	70	3.0–3.5
Exposed geomembrane	0.78–1.53	10–25	0.010	90	1.9–3.2
Geomembrane with concrete cover	2.43–2.54	40–60	0.005	95	3.5–3.7

Thus, enough money must be made available for not only constructing but also operating, maintaining and monitoring a landfill. Proper cost analysis must be done to ensure cash flow for performing all these tasks. Monitoring of a landfill may be required for 30–40 years after the closing of the last phase. The total cost of long-term maintenance and monitoring of a closed landfill can be higher than the cost of construction of a landfill. Methods for estimating these costs at a future date must take into account the inflation factor (Bagchi, 1994).

The United States Bureau of Reclamation (USBR) has constructed 34 canal lining sections in 11 irrigation districts in four northwestern states to assess the durability, cost and effectiveness of alternate lining technologies (Swihart and Haynes, 2002; Ivy and Narejo, 2003). The lining materials include combinations of geosynthetics, shotcrete, roller compacted concrete, grout mattresses, soil, elastomeric coatings and sprayed-in-place foam. Each test section typically covers 15,000 to 30,000 square feet. The test sections range in age from 1 to 10 years. Preliminary benefit/cost (B/C) ratios have been calculated based on initial construction costs, maintenance costs, durability (service life) and effectiveness (determined by preconstruction and post-construction ponding tests). Table 8.5 summarizes the performance of all 34 test sections dividing all 34 test sections into four generic categories. Exposed geomembranes were found to be 90% effective, that is, only 10% of the water was lost due to seepage. Notice also that concrete alone is only 70% effective in stopping the loss of water. Therefore, geomembranes are almost 20% more effective in stopping loss of water as compared to concrete. The benefit/cost ratio of geomembranes is approximately the same as that of concrete. The effectiveness of concrete at reducing seepage increases to 95% if underlined by a geomembrane. Thus, the geomembrane with concrete cover seems to offer the best long-term performance.

The geosynthetic construction quality assurance (CQA) cost is difficult to separate from the CQA costs for the total project. For landfill projects, it has been found that CQA costs range between 5% and 10% of the construction costs for the elements that are being monitored. Factors that influence the cost of a CQA programme are the project's duration, size, complexity, and work scope, ranging from complete construction management to performing CQA only for the geosynthetic elements (Thiel and Stewart, 1993).

ILLUSTRATIVE EXAMPLE   8.1

Make a group of the elements comprising the total cost of a geosynthetic-reinforced soil retaining wall. Past experiences available suggest an approximate breakdown of construction costs mainly in terms of materials, labour, plant and others.

Solution

The elements comprising the total cost of a geosynthetic-reinforced soil retaining wall may be grouped as follows:

1 Foundation soil improvement, if required.
2 Precast facing elements (including erection), if provided.
3 Soil backfill (including haul from borrow areas)
4 Compaction of the backfill soil.
5 Laboratory and in situ testing of soil.
6 Geosynthetic reinforcement and auxiliary parts.
7 Granular material for use around drainage.
8 Drainage pipes and cappings.
9 Transport of all materials.
10 Design.
11 Overhead and profit.

An approximate breakdown of construction costs can be given as follows:

1 Materials – 60%
2 Labour – 20%
3 Plant – 15%
4 Others – 5%

Answer

Self-evaluation questions

(Select the most appropriate answers to the multiple-choice questions from 1 to 8)

1. The cost of a geosynthetic-related structure or application should be typically presented as an engineering estimate of

    (a) Capital cost.
    (b) Operational cost.
    (c) Maintenance cost.
    (d) All of the above.

2. It is generally accepted that, under normal circumstances, geosynthetic-reinforced soil walls become more economical after a wall height of about

    (a) 3 m.
    (b) 6 m.
    (c) 9 m.
    (d) None of the above.

3. In most low to medium height retaining walls, the cost of the geosynthetic reinforcement has found be as high as

    (a) 5% of the total wall cost.
    (b) 10% of the total wall cost.
    (c) 15% of the total wall cost.
    (d) 20% of the total wall cost.

4. In the case of geosynthetic-reinforced unpaved roads, the reduction of maintenance costs has been found to be in excess of
   (a) 5%.
   (b) 10%.
   (c) 15%.
   (d) 25%.

5. The installed cost of a paving fabric and bituminous concrete interlayer system is generally
   (a) Less than one fourth the cost of the 25 mm of bituminous concrete.
   (b) Less than one half the cost of the 25 mm of bituminous concrete.
   (c) Less than the cost of the 25 mm of bituminous concrete.
   (d) Equal to the cost of the 25 mm of bituminous concrete.

6. Which of the following statements related to the cost of the three main components of a reinforced slope is incorrect?
   (a) The soil component generally represents more than 80% of the reinforced slope cost.
   (b) The per cent of total cost spent on reinforcement will increase with increasing slope height.
   (c) The per cent of total cost spent on surface erosion control will increase with increasing slope angle of the reinforced slope face.
   (d) None of the above.

7. The total cost of long-term maintenance and monitoring of a closed landfill can be
   (a) Higher than the cost of construction of the landfill.
   (b) Equal to the cost of construction of the landfill.
   (c) Lower than the cost of construction of the landfill.
   (d) All of the above.

8. Which one of the following types of canal lining is most effective in reducing seepage?
   (a) Concrete lining.
   (b) Geomembrane lining.
   (c) Geomembrane with concrete cover lining.
   (d) Any one of the above.

9. What do you mean by the term 'Relative Economy'? Explain this term by taking an example.

10. How will you assess the *cost-performance efficiency* of a geosynthetic or any other element of geosynthetic-related structure?

11. How does the cost–benefit analysis of environmental applications of geosynthetics differ from their other applications?

12. For the geosynthetic-reinforced roadways, the initial cost–benefit ratio decreases as the strength of the existing soil subgrade decreases. What are the possible reasons for this trend of variation?

13. How does the quality of site preparation affect the construction costs and the geosynthetic survivability requirement?

14. How will you analyse the cost-effectiveness of a paving fabric interlayer system?
15. What experiences are available on the cost-effectiveness of geosynthetic versus conventional drainage systems?
16. In the case of slope stabilization projects, what are the important issues that must be considered to make a cost–benefit analysis?
17. What are the components of the life cycle cost of a landfill project?
18. What are the factors that influence the cost of a CQA programme?
19. What difficulties do you expect in the economic evaluation of geosynthetic-related structures?
20. Make a group of the elements comprising the total cost of a geosynthetic-reinforced unpaved road. Based on the past experiences available in your state, suggest an approximate breakdown of construction costs mainly in terms of materials, labour, plant and others.

Chapter 9

Case studies

9.1 Introduction

Geosynthetics are available in a wide range of compositions appropriate to different applications and environments. Therefore, they have pervaded many areas of civil engineering, especially geotechnical engineering, environmental engineering, hydraulic engineering and transportation engineering. It is now no longer possible to work effectively without geosynthetics in these areas. There have been a number of case studies in the past four decades that support and verify the technical and economical feasibility of the construction techniques with geosynthetics. Geosynthetics have generally performed as expected, though relatively few installations have yet reached their designed service lives. In fact, the proven track record has resulted in the relatively quick acceptance of geosynthetics. Selected case studies are presented in the current chapter in order to develop the confidence of geosynthetics applications among engineering students, practising engineers and owners of the projects. It should be noted that as in other fields of engineering, confidence in the durability and better performance of geosynthetics can only be expected to develop as the technology matures and the results of long-term service experience accumulate.

9.2 Selected case studies

Case study 1 (retaining wall/steep-sided slope)

Geosynthetic-reinforced soil walls are gaining considerable attention as retaining structures and providing a valuable alternative to traditional concrete walls. No footing of any special kind is required in the case of retaining walls, and the lowest geosynthetic layer is placed directly on the foundation soil. They demonstrate the possibility of using soils of poor mechanical characteristics with an ample safety margin. Compared to the concrete walls, they present a low cost–benefit ratio and a low environmental impact.

Gourc and Risseeuw (1993) reported a case study on a geotextile-reinforced wraparound faced wall built in late 1982 in Prapoutel, France. Figure 9.1 shows the overall cross-section of the wall. Its length is 170 m and its height ranges between 2 and 9.6 m. Regarding the soils forming the embankment, the following characteristics were considered: $\gamma = 18$ kN/m^3, $\phi' = 30°$, $c' = 33$ kPa. The selected geotextile was Stabilenka 200, a woven polyester (PET) product, with a mass per unit area of 450 g/m^2, a tensile strength of $\sigma_t = 200$ kN/m and a strain at failure of $\varepsilon_t = 8\%$ (roll direction). Although the soil–geotextile friction ϕ_g measured in the laboratory was equivalent to the soil–soil friction, it was considered preferable to adopt the value $\phi_g = 0.8\phi'$. The design is based on an extremely steep embankment

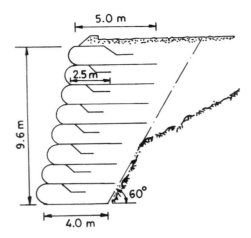

Figure 9.1 A cross-section of geotextile-reinforced wraparound faced wall, Prapoutel, France (after Gourc and Risseeuw, 1993).

(76% relative to the horizontal) using the anchor-tied wall design method proposed by Broms (1980) and considering a factor of safety of 1.3 in calculating the anchor force. The design principle assumes a uniform active earth pressure over the entire height of the embankment. It leads generally to a factor of safety less than those obtained with other design methods, which indicates that the method used was more conservative than others. The design calculation gave a vertical geotextile spacing of 1.20 m for an embankment height of 9.60 m. It is important to note that this wall was preferred to a reinforced concrete retaining wall or reinforced earth wall essentially for cost reasons (the reinforced earth wall was estimated to be 40% more expensive).

The embankment material available at the site was graded in layers 1.20 m thick, with successive layers being separated by a geotextile sheet. The geotextile reinforcement lengths adopted as per design were 5 m in the upper layers and 4 m in the lower layers. Specially made angle forms were used for placing the successive 1.20 m thick layers. The work rate was approximately 50 linear metre each layer per day. It was reported that 10 years after completion, the structure gave no cause for criticism regarding its overall stability. The only reservation concerns the facing of the embankment. Because of the steep slope, plant growth after hydraulic seeding proved to be difficult, and a bituminous protection with mulch was subsequently provided. However, it was reported to have some geotextile deterioration. At a few locations on the face, large stones punctured the geotextile, with no significant soil loss. A slight loss of soil from the facing occurred at certain joints between adjacent sheets. Joints were made simply by overlapping, probably over an insufficient width.

Case study 2 (retaining wall/steep-sided slope)
A geosynthetic-reinforced soil retaining wall with segmental facing panels was constructed on the Mumbai–Pune Expressway (Panvel bypass – package I) by the Maharashtra State Road development Corporation Limited, Mumbai, India. The height of the retaining wall varies from 2.5 to 13 m. The design was carried out using the Tieback wedge method

considering internal as well as external stability as per site conditions. Tensar 40 RE, 80 RE, 120 RE and 16 RE geogrids were used as reinforcements. Modular blocks (400 mm × 220 mm × 150 mm) as well as segmental panels (1400 mm × 650 mm × 180 mm) were used as facing elements. The extensive use of the Tensar connectors gave the perfect connection between the wall-facing panels and the Tensar geogrids. A nonwoven geotextile was used to wrap over the perforated pipe to allow free drainage. The construction sequences adopted were based on vast model experiments experiences and technical justifications. The construction work was completed in the year 2001. Figure 9.2 shows the details of the wall at one of its cross sections, along with soil and reinforcement characteristics. A portion of the wall during construction stage is shown in Figure 9.3. The retaining wall has been performing well without any noticeable problem since its completion.

Case study 3 (retaining wall/steep-sided slope)
Expansion of the City of Calgary's Light Rail Transit (LRT) system into northwest Calgary, Canada encroached into a hill side necessitating a retaining wall. The hill side which has an overall height of about 30 m at a 15° inclination consists of lacustrine sands and silts overlying glacial till with the water table lying near the interface. The wall was constructed in tiers with a maximum height of 10 m and four tiers (Fig. 9.4) with an overall slope of about 50°. Each tier consisted of full-height precast panels 2.4 m in width. The panels were reinforced with up to four layers of Tensar SR2 geogrid. The length of the geogrids varied, and, at some locations, the geogrids were tied into the shored excavation as shown in Figure 9.4. The number of geogrid layers, their length and vertical distribution were determined using the method suggested by Bonaparte *et al.* (1985). Backfill consisted of a well-graded free draining gravel compacted to a minimum of 95% standard Proctor dry density.

A drainage system consisting of a perforated pipe was provided behind the wall. A geodrain was provided at the joint between panels for additional drainage. The wall has an overall length of 155 m and a total area of 1280 m^2. Each panel was placed on concrete levelling pads and supported by inclined rakers on the outside face. The full length geogrid strip was then attached to the geogrid stub embedded in the concrete panel using a polymer bar threaded through the fingers of the connecting pieces. The geogrid was pretensioned to a predetermined value of 15–20% of design load using stretchers. The granular fill was then placed and compacted and the process was repeated at each geogrid layer. The project was constructed during the period April to June 1986 and the LRT line was put into service several months later. Follow-up inspection has shown no noticeable movement and the wall is functioning fully as intended.

Case study 4 (embankment on soft foundation soil)
The construction of a 3.5-km long embankment through the tidal area of Deep Bay in the New Territories district of Hong Kong was constructed between April and end November 1982 (see Fig. 9.5). The stability of the slopes of the 3.50-m high embankment was achieved by reinforcement of the base using a heavy-duty PET woven geotextile between the fill and the foundation mud. The undrained shear strength of the foundation soil was estimated to be as low as 5–10 kPa in the top 6 m of unconsolidated marine deposits, that is, a very soft clay. It was decided not to increase the base area for reducing the pressure on the ground because of the scarcity of land in Hong Kong. Also, the embankment went along existing fish ponds. The dikes of these ponds were not to be damaged in any way during the construction of the embankment. Additionally, the whole area was flooded twice a day due to tides.

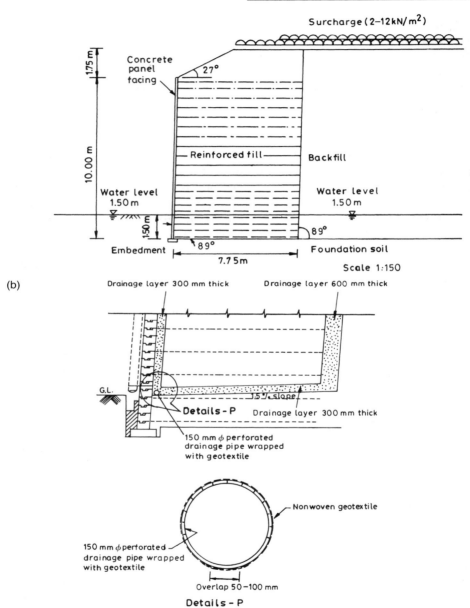

Figure 9.2 Geosynthetic-reinforced retaining wall built on the Mumbai–Pune Expressway (Panvel Bypass – package I), India: (a) cross-section with the details of soil and reinforcement; (b) details of the drainage system (courtesy of Netlon India, 2001).

Figure 9.3 A portion of retaining wall on the Mumbai–Pune Expressway (Panvel Bypass-Package I), India during its construction stage (courtesy of Netlon India, 2001).

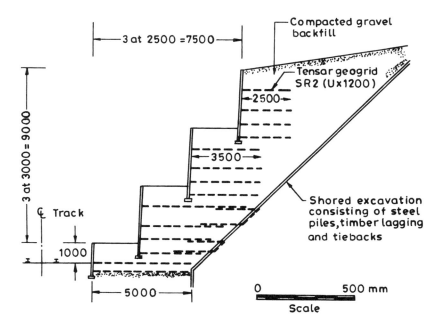

Figure 9.4 Typical four-tier section of geogrid-reinforced retaining wall for Calgary Northwest Light Rail Transit Project, Calgary, Canada (after Burwash and Lunder, 1993).

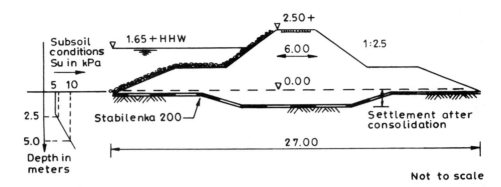

Figure 9.5 Cross section of the geotextile-reinforced embankment over tidal mud area of the Deep Bay, New Territories, Hong Kong (after Risseeuw and Voskamp, 1993).

Construction work started by installing 5-m wide geotextile (Stabilenka 200, manufactured and supplied by Akzo Industrial Systems). The reinforcing geotextiles were sewn together on site utilizing local labour. The muddy conditions presented the labourers with some difficulties as they struggled to carry and pull the 30-m lengths of geotextile into position. Once the geotextiles were positioned and sewn together, access was not so difficult. It was possible for the contractor to move in immediately. After a 1-m layer of fill was laid with the aid of light Komatsu dozers, the contractor was able to use heavily loaded 35-ton wheelbase trucks to dump the fill close to the front. Then, a fleet of large and small Komatsu swamp dozers was used to push the fill material into position. During the continued filling operation, the freshly deposited mud top layer (300–500 mm thick) squeezed out. The underlying stratum, however, was confined by the reinforcing geotextile. The average work output to bring the 3.5-km long embankment up to its 3.50-km height was 150 m per week. This was considered a good rate of construction, taking into account the fact that the installation of the geotextile and filling was restricted by the state of the tide.

The primary benefit of the geotextile reinforcement was to contribute to the short-term stability, whereas it was calculated that the embankment would be stable without the need for reinforcement after a prolonged period of subsoil consolidation. The field monitoring till 1993 indicated that no decrease of stability or excessive subsidence had been encountered.

Case study 5 (embankment on soft foundation soil)
A 100-year-old railway track from Magdeburg to Berlin had to be partially reconstructed for train velocities of 160 km/h. The railway line traverses deep deposits of soft organic soils. The organic soil consists, to a great extent, of peat with a water content of 300–600% well above the liquid limit, an unconfined compressive strength $c_u < 10$ kN/m^2 and a constrained modulus $E_s = 0.2$–0.8 MN/m^2 besides the old railway embankment. Below the old railway embankment the organic soil is preconsolidated and the constrained modulus is in the order of $E_s = 2$–6 MN/m^2 with the unconfined compressive strength $c_u > 15$ kN/m^2. The soft soil layers are not uniform. Apart from peat, sandy to clayey organic silts are encountered with varying amounts of plant residues and shells. The old railway tracks had suffered considerable settlement in the past, so it was necessary to improve the bearing capacity and deformational behaviour of the ground in this area.

Over a total length of 2100 m, the geogrid-reinforced structure was erected in several sections from 1994 to 1995. The structural system is sketched in Figure 9.6. It consists of the geogrid-reinforced embankment, the precast concrete pile caps, the ductile cast iron piles, the soft soil that prevents buckling of the piles, and, finally, the sand layer or glacial till at depth with sufficient bearing capacity to carry the total load. Based on the structural analyses, three layers of knitted PET geogrids were selected. Their short-term tensile strength in machine direction (MD) and in cross machine direction (CMD) is 150 kN/m each. Under design loading, the geogrids experience no more than 20% of their short-term tensile strength.

The section, shown in Figure 9.6, contains a sheet pile wall at the centre line of the embankment. This structural element served the purpose of securing half the embankment for railway traffic, while the other half was being reconstructed. The sheet piling was pulled after completion of the new embankment. Trains have been passing over the first section of the new structure sine May 1994, and the entire construction was completed in December 1995. The performance of the structure has been monitored by two extensive instrumented sections and in addition by conventional surveying. The data obtained over a period of more than seven years from displacement and strain measurements have demonstrated that the structure is safe and performs adequately with respect to serviceability. The railway personnel operating trains on the railway line have repeatedly reported that the trains are running more smoothly on the sections with the reinforced piled embankment than on sections with a conventional replacement of the soft organic soil with compacted sand (Zanzinger and Gartung, 2002b).

Case study 6 (foundation)
A section of Federal Highway B 180, more than 20 m long, at Neckendorf near Eisleben, Germany was destroyed across its entire width in 1987 by a sink-hole of diameter of about

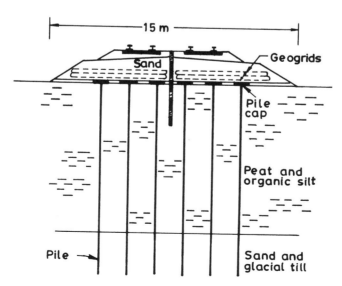

Figure 9.6 Geogrid-reinforced railway embankment on piles, Germany (after Zanzinger and Gartung, 2002b).

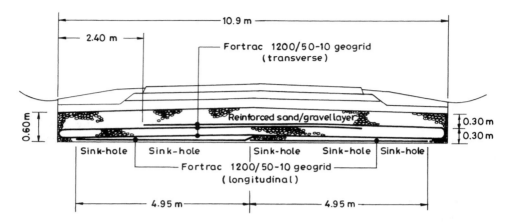

Figure 9.7 Geogrid-reinforced gravel layer bridges over sink-hole on Federal Highway B 180 near Eisleben, Germany (courtesy of HUESKER Synthetic GmbH & Co., Germany).

8 m located almost on the road axis below 30 m depth. Although the hole was filled with fill material, the danger of a new cave-in due to caverns deep underground still existed. To allow the roadway into operation, the opening had to be bridged over sufficiently to allow no more subsidence than 10 cm over 30 m of roadway even under heavy truck trafficking. The 20-m long weak section was bridged over with a geogrid-reinforced gravel/sand layer (see Fig. 9.7). The layer was about 60 cm thick by 60 m long and approximately 11 m wide. This layer supported the entire road surface.

The geogrid reinforcement was installed in three layers. The bottom layer consisted of two 5-m geogrid strips laid longitudinally side-by-side. The second layer consisted of a transverse geogrid strip, completely encapsulated and overlapped, resulting in a third layer. The design provided effective reinforcement against longitudinal and transverse deflection as well as torsion. The flexible *Fortrac 1200/50-10* geogrid is composed of very low elongation, low creep Aramid fibres with total tensile strength of 1200 kN/m and only 3% elongation. The mesh size is 10 × 10 mm. The reinforced layer was prepared within a few days in October 1993.

Case study 7 (unpaved roads)
The use of geosynthetics in unpaved roads on soft soils makes it possible to increase the load-bearing capacity of the soft soil and the granular fill. A geosynthetic layer in unpaved road allows the passage of heavily loaded vehicles over the granular fills of reduced thicknesses, placed on the soft soil subgrades. This, in turn, allows decreased consumption of materials, transport expenses and duration of construction.

One of the first roads in the USSR, where a domestic geotextile was first used, was a temporary road in Smolenskaya region (Kazarnovsky and Brantman, 1993). Construction of the road had to be accelerated to evacuate populated localities from areas to be flooded, when the reservoir of Vazuzskaya hydrosystem was being filled with water and also to allow for the movement of construction vehicles. A temporary road, about 20 km long, was to be constructed within the shortest period of time and with a minimum thickness of fill. The site was characterized by soft plastic loam soils, by a high groundwater level and by a prolonged stagnation of water above ground. To accelerate construction of the road, which at the same time provided a passage for construction vehicles delivering sand for the fill, the geotextile 'Dornit' ϕ-1 was used on several sections of the road where the soil consisted of a light silty loam and had a moisture content, at the time of construction, equal to 27–30%. The used geotextile, manufactured from waste synthetic fibres by needle-punching, was having the following characteristics: mass per unit area = 600 g/m², roll width = 1.7 m, nominal thickness = 4.5 mm, tensile strength = 12 kN/m (roll direction), and 6 kN/m (cross-roll direction), and strain at failure = 70% (roll direction) and 130% (cross-roll direction).

Construction procedures of the road sections on which geotextile was used included

- a rough grading of the soft subgrade by a bulldozer going back and forth with a lowered blade;
- unrolling the geotextile across the fill axis with 300 mm overlaps between adjacent rolls;
- filling and grading of 450–500-mm thick medium grained sand layer (containing gravel and 2% of silty and clayey particles), followed by compaction with a light weight roller.

The difference in driving conditions on sections with the geotextile and without it could be observed immediately after the installation of the geotextile. It was actually impossible to perform work after 8–10 passes of dump trucks along the same ruts in the section where the geotextile was not used and the sand fill thickness was limited to 400 mm for comparison. On the road section where the geotextile was placed under the sand fill, the rut depth did not exceed 100–120 mm and intermixing of the fill sand with the subgrade did not occur. On each working day, 300 vehicles, mostly dump trucks, travelled in both directions. After the road was used for several months, the ruts on the road section without geotextile had to be graded continually. Every morning, before the main traffic drove on the road, both the sections with and without the geotextile were graded with a bulldozer. Measurement of the ruts showed that, on the section without a geotextile, ruts from 200 to 250 mm deep were formed, and the traffic speed slowed to 5 km/h. On the road section with a geotextile, the rut depths were only 100–120 mm in spite of the fact that the dump trucks travelled along the same rut (Fig. 9.8). The traffic was able to maintain a speed of 25–30 km/h and the vehicles could pass each other using the whole width of the fill.

Case study 8 (paved road)

If the pavement of a road is only surface dressed or resurfaced, the deep-seated cracks, if any, will emerge to degrade the resurfacing. The crack resistance of pavement can be improved by installing a geosynthetic between the old and new surfacing. The geosynthetic may also prevent excessive moisture reaching and softening the subbase and subgrade, provided the geosynthetic used is having low permeability. In installing the geosynthetic, it is vital that a good bond be achieved between old and new works. This involves coating the old surfacing with a hot tar spray first, or an emulsion, before rolling out the geosynthetic. Once this is in place, the new surfacing is placed in the normal manner. A geosynthetic can also be used as a reinforcement to reduce the overlay thickness, if it is having high strength, high modulus, and low creep.

In 1995 it was observed that A587 Fleetwood Road, Cleveleys, Blackpool suffered severe fatigue failure with extensive cracking, localized rutting, depressions and pot holes. The overlay thickness was required to provide an extended service life of 15 years. This necessitated

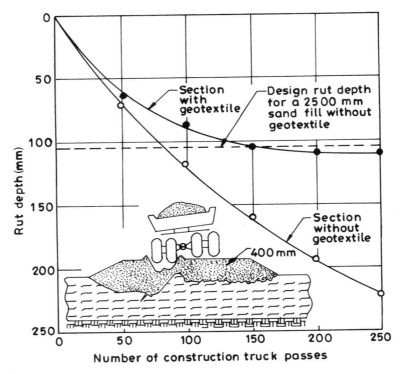

Figure 9.8 Change in the rut depths depending on the number of vehicles passing (sand fill thickness = 400 mm (after Kazarnovsky and Brantman, 1993).

the raising of the road by 160 mm. This thick overlay would require extensive kerb raising and alteration of side road access from commercial and private properties. Tensar International was asked to propose a design for a geosynthetic-reinforced overlay. The objective was to determine as thin an overlay as possible without reducing the design life of the road rehabilitation scheme. A long-term programme of research at the University of Nottingham on Tensar grids in asphalt showed that the fatigue life can be increased by up to a factor of 10. For a give design life, the thickness of asphalt can be reduced significantly. The Tensar design achieved a 36% reduction in overlay thickness with a combination of asphalt reinforcement and heavy-duty macadam base course (Tensar International, 2003).

After regulating the planned surface with 20 mm of rolled asphalt, the Tensar AR-G was installed on a bitumen bonding coat. Paving commenced as soon as the bonding coat had cured. Following the installation in 1995, the performance was observed and it has been reported that the road is performing satisfactorily at a time when it is approaching its half way design life.

Case study 9 (paved road)
If the water trapped at the interface between the concrete slab and the foundation soil is not drained off, it will be under pressure when vehicles pass ('beating' the concrete slab).

This may cause erosion of the foundation soil and the formation of cavities and deposits of fine soils, all factors of instability. Bordonado and Buffard (1993) reported that similar situations existed on cement concrete paved A26 motorway in Northern France. In Northern France, the foundation soil forming the subgrade is generally a fine, loamy soil, which is erodible. The drainage system initially selected for the A26 motorway involved placing, between the foundation soil and the concrete slab, a 100-mm thick drainage layer and using untreated limestone aggregates graded 5/40 mm or 6/20 mm. Although satisfactory results were previously obtained for this solution, the solution presented certain drawbacks: movement of construction equipments on this material was delicate, the solution was relatively costly, and its long-term performance was short of perfect. Therefore, in 1983, this drainage layer was replaced by a thick geotextile. The chosen product was designed to fulfil the following functions:

- low compressibility under traffic loads
- efficient long-term drainage
- freedom of movement of the concrete slab, without friction, under the effect of thermal gradients.

The project engineers selected a specific geotextile called *Drainatex*, a nonwoven, needle-punched polypropylene (PP) product supplied in 4-m wide rolls, with a draining part and a filter part bonded together in the manufacturing process. The draining part has a mass per unit area of 600 g/m^2 and is made up of short fibres, about 100 mm long, 60% of the fibres with 140 μm diameter and 40% with 30 μm diameter. The tensile strength of the geotextile is 3 kN/m and the strain at failure is 11% in the roll direction. A low tensile strength was not a handicap in the present case and the large strain facilitated adaptation of the geotextile to the deformation of the supporting soil. The filtration opening size of the geotextile is 80 μm on the foundation soil side.

A first experimental strip 400 m long was built in 1983 in the St Omer-Nordausques section of the motorway, with the thick concrete slab plus geotextile structure (Fig. 9.9). After one year of service, it was interesting to observe that the slab 'beating' values were lower with this technique than for sections built with a draining interface of granular materials. These encouraging results led to widespread adoption of this technique (110 km of carriageway completed with geotextiles by 1988). It is also to be noted that the winter of 1984–85 was particularly severe in Northern France (depth of frost = 0.70 m). Despite these

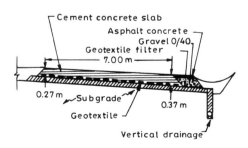

Figure 9.9 Standard cross-section of the concrete carriageway structure on A26 Motorway, Northern France, France (after Bordonado and Buffard, 1993).

conditions, no rise in carriageway level was observed in sections completed with geotextiles. Core sampling made down to the draining layer confirmed the satisfactory overall behaviour.

Case study 10 (paved road)

Shukla *et al.* (2004) described the use of a geotextile layer between soil subgrade and granular subbase course in the national paved roadway project, about 76 km long, being undertaken in India in the Varanasi zone. In the whole length of the highway, the bituminous pavement construction with nonwoven geotextile layer at the subgrade level is proposed. This project forms a part of the Golden Quadrilateral connecting four metropolitan cities of India, namely: Delhi, Mumbai, Chennai and Kolkatta with a total length of about 6000 km. The National Highway, NH-2, in Varanasi zone, lies broadly in flat to rolling terrain. The entire section traverses the flat flood plains of the Ganga and the Sone rivers. During normal monsoon drainage does not appear to pose any serious problem. In urbanized areas – where the road level is generally the same as the habitation on both sides and sometimes slightly lower – water flows on the pavement/shoulders. The problem is acute in all the urban areas. The area faces severe flood situations once in 10–15 years, resulting in the blockage of traffic. Such floods were experienced during the years 1971, 1978, 1987 and 1996. The project area has tropical climate. Mean annual rainfall in the area is 1500 mm, of which 80% falls during the monsoons (mid-June to end September). The project area is mainly covered by the Indo-Gangetic Alluvium. The soil subgrades consist mainly of fine-grained clayey/silty materials with soil class CL (silty clays with low compressibility) according to Indian Standard classification (A-4 to A-6 according to AASHTO classification system). The plasticity index of the soils is in the range of 0–22%. However, the major soil samples tested have a plasticity index in the range of 8–14%. Soils have negligible swelling characteristics.

The carriageway, in general, is two-lane with 8.75 m (including 1.5 m shoulder) width except when it passes through urban centres where the width is less than 8.75 m. The shoulders are unpaved for most of the length of the roadway and are of varying widths in the range 1.0–1.5 m. The height of embankment is generally less than 1 m. A level ground is mostly available in the urban sections.

The polypropylene continuous filament needle-punched nonwoven geotextile (Polyfelt TS50) was selected to be installed at the subgrade level (see Fig. 9.10) to function as a

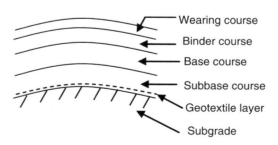

Figure 9.10 Typical cross-section of the asphaltic/bituminous pavement with a geotextile layer at the soil subgrade level.

separator and/or drainage medium. This geotextile has the following properties: mass per unit area = 212 g/m², thickness under pressure 2 kPa = 1.99 mm, tensile strength (MD/CMD) = 19.5/12.3 kN/m, Tensile elongation (MD/CMD) = 35/108%, CBR puncture resistance = 2465 N, Apparent opening size < 0.075 mm, permittivity = 0.0198 s^{-1}. The aim of geotextile application was to prevent intrusion/pumping of subgrade soil particles into the subbase/base course. It would also intercept and carry water in its plane to side drains on either side of the pavement. These two functions were intended to improve the overall performance of paved roadways and increase their operating life.

The pavement design was carried out for flexible pavements (bituminous pavements) without considering any reduction in granular layers due to the presence of a geotextile layer. The design was based on the design traffic calculated on the basis of the traffic and axle load surveys conducted in the feasibility phase. The values of vehicle damage factor (VDF) for single axles adopted for the design were 6.83 in the direction Varanasi – Aurangabad and 9.0 in the opposite direction.

The roadway was designed with the following typical main characteristics:

- design speed = 100 km/h
- number of two-lane carriageway = 2
- width of median between carriageways = 1.5–5 m
- lane width = 3.5 m
- width of paved shoulder = 1.5 m with the same pavement as the carriageway
- width of earthen shoulder outside the paved shoulder = 1 m
- cross-slope of paved areas = 2.5%
- height above the flood level = 1.5–2.0 m
- minimum radius of curve = 360 m (Almost all curves are have a radius of more than 500 m.)
- California Bearing Ratio (CBR) of subgrade = 5.0 %
- design load = 150 million equivalent standard axles (ESALs).

The construction started with excavation work at areas where the levels were higher than the subgrade levels. The unsuitable materials (debris, loose material, boulders, etc.), if found, were disposed off. Embankment works were carried out using crawler dozer, motor grader, vibration roller and water trucks. Prior to the filling works, batter peg markers were installed to indicate the limit of embankment/subgrade at regular intervals of 50 m as a guide. The suitable material obtained from excavation and approved borrow area was used at embankment areas by spreading the material in layers by using crawler dozer/motor grader not exceeding 200 mm and compacting it to the maximum dry density (MDD) of 95% compaction using a vibratory roller up to 500 mm below subgrade level. For upper portion, it was compacted up to MDD of 97% compaction. Upon completion of the embankment filling up to subgrade level, the side slopes were shaped and trimmed. The subgrade was provided a cross-slope of 2.5% prior to the placement of the geotextile layer. As per design calculations, four different layers were provided on the geotextile layer. The order of these layers from the bottom layer is as follows: 210-mm thick granular subbase course (GSBC), 250-mm thick wet mix macadam (WMM), 190-mm thick dense bituminous macadam (DBM) and 50-mm bituminous concrete (BC).

Figure 9.11(a) shows the installed geotextile layer along with the spreading operation of the granular subbase material over it by the motor grader. The compacted subgrade surface as well as sewn joints in geotextile layer can also be observed in this figure. Figure 9.11(b) shows the compaction of the granular subbase layer by the vibratory roller. The granular subbase course after final compaction by rubber-tyred roller is shown in Figure 9.11(c).

A portion of the roadway was completed in the end of March 2003 and opened to traffic (see Fig. 9.11(d)). It was exposed to the monsoons of 2003, 2004 and 2005. The roadway is presently functioning well, structurally as well as functionally without any noticeable problem. By using a nonwoven geotextile layer as separation and drainage layer, pavement layers were constructed conveniently and economically even over poor soil subgrades. Due to the permanent separation, settlements and the balance of earthworks quantity were predicted more accurately. The geotextile layer contributed to the construction speed of the granular lifts without any local shear failure of the soil subgrade.

Case study 11 (railway track)
Walls and Newby (1993) reported a case study on the use of geosynthetics for railway track rehabilitation in Alabama, USA. In May 1983, the existing track was inspected by railway personnel and consultants who identified several problems that needed to be corrected, namely: increase the load-bearing capacity of the soil subgrade, prevent the contamination of the ballast with subgrade fines and dissipate the high pore water pressures built up by cyclic train loading. Since geosynthetics have been used separately to solve each of the above problems, a combination of geotextiles and geogrids was considered to maximize the benefits of the two products in the following ways:

- geogrid to provide tensile reinforcement and shear resistance to increase the effective bearing capacity of the subgrade;
- geogrid to interlock with and confine the ballast, increasing its resistance to both vertical and lateral movement;
- geotextile for separation and filtration to prevent contamination of the ballast and providing quick relief of pore water pressures.

The design called for the removal of the fouled ballast to a depth of approximately 0.30 m below the base of the tie. Rather than disposing off the old ballast, it was to be placed along the edges of the embankment to widen the existing shoulders. Next a 380 g/m² nonwoven geotextile was to be placed on top of the remaining subballast followed by a geogrid placed directly on top of the geotextile and 0.30 m of new ballast. The specifications for the geotextile required that it should be a nonwoven needle-punched engineering grade fabric comprising PET fibres with a linear density of 0.8–1 tex per filament. The minimum fibre tenacity was to be 4 mN/tex and the minimum fibre length was to be 150 mm. The geotextile selected for this project was Quline 160 manufactured by Wellman Quline. The geogrid specified was Tensar SS2 geogrid manufactured by the Tensar Corporation. The important properties of the Tensar geogrids made it suitable for ballast reinforcement. These geogrids are having open grid geometry for interlock capability, high junction strength to resist lateral deformation of the grid cross members and high tensile modulus to reinforce at low strain levels.

Although the design thickness of ballast was selected arbitrarily at that time, recent developments in geogrid design technology for railway track bed indicate that about 600 mm of reinforced ballast and subballast would be required for a weak subgrade having a CBR

Figure 9.11 National Highway, NH-2, Varanasi zone, India: (a) a geotextile layer with the granular subbase material being spread over it (b) compaction of granular subbase material by the vibratory material; (c) granular subbase course after final compaction by the rubber-tyred roller; (d) the completed roadway opened to traffic (after shukla et al., 2004).

value of approximately 1% compared to 1 m of unreinforced ballast and subballast. However, since the actual depth of existing ballast and subballast was considerably greater than 300 mm, it was assumed that the remaining ballast and subballast provided enough support for the new reinforced ballast while fulfilling the total reinforced thickness requirement to prevent overstressing of the weak subgrade. Although other stabilization alternatives were proposed, it was decided to use the geotextile–geogrid combination because of its relatively low cost (just one sixth of the estimated total cost of the relocation alternative) and the proven performance of geogrids in other reinforcement applications. Construction of the reinforced track was carried out in December 1983. Following an initial observation period of approximately 3 months in which it was determined that the reinforced track structure was performing satisfactorily, it was decided to gradually increase the allowable speed up to 56 km/h and ultimately to 80 km/h. The rehabilitated track was reported to be in service for four years without any reccurrence of stability problems and only requiring routine track maintenance. This case study has supported the results of large-scale laboratory tests on tie-ballast loading that geogrids can increase the life of ballasted track between maintenance cycles, particularly over weak soil subgrades.

Case study 12 (slope erosion control)
Riverbanks and coastline erosion is counteracted by the protection of the surface to resist the forces generated by the flow and waves. The method widely used is to install a layer of stone pitching on the bank to stop the loss of soil. The rise and fall of the water level as well as wave action causes water to flow into the pitched bank and then drain away. This two-way flow, known as dynamic flow, is capable of dislodging and carrying away soil lying below the stone protection and ultimately causing the revetment to fail. Installing a filter between the stone layer and the soil may reduce the soil erosion. Traditionally, a granular layer is used as a filter, which allows the water to pass through freely but not the soil particles. The design and choice of a suitable granular material for this filter can be done, but it is not an easy task to achieve the function of filter accurately. The use of geosynthetic filters in such cases has proven to be an attractive alternative.

The Kolkata Port Trust authorities used jute geotextile as a revetment filter for riverbank protection at Nayachor Island, Haldia, India during 1992 (Sivaramakrishnan, 1994). The eroded site was first prepared to form a uniform slope 1:1. The bare jute geotextile was unrolled over the slope of the embankment, starting from the top of the bank. The geotextile was anchored at the top in a trench 1 m × 1 m and similarly at the sides. The overlapping portions were nailed with 10-in. long iron pegs at an interval of 1 m. The bottom portion of the jute geotextiles was fabricated in such a manner that it had multiple pockets to fill sand in. This was done to anchor the geotextile in its place and protect erosion by reverse current and eddies. After the entire area was covered with the jute geotextile, small laterite boulders were placed over the jute geotextile. The small laterite boulders were laid to provide a cushion effect to the geotextile. On the top of the small laterite boulders, big laterite boulders, weighing approximately 15–20 kg, were pitched to a height of 1.5 feet as shown in Figure 9.12. The entire operation was carried out during low tide. A good amount of siltation, up to a height of 600 mm, was observed after a period of eight months, which indicates that the jute geotextile was effective in protecting the slope.

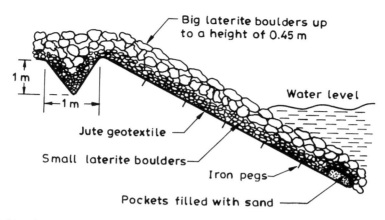

Figure 9.12 River bank protection at Nayachor Island, Haldia, India (after Sivaramakrishnan, 1994).

Case study 13 (slope erosion control)
Lee *et al.* (1996) carried out a full-scale-field experimentation of a new technique called *green coating* for protecting steep *mudstone* slopes in southwestern Taiwan. Mudstone is a weak sedimentary rock, formed during Miocene to Pliocene and Pleistocene. Many forms of geologic damage such as erosion, mud flow and slope failure were often seen in the mudstone area during the rainy season. The new technique consisted of three main elements: (1) cutting the natural mudstone slope into a multistage slope with a steep angle and a short height in each stage, (2) spraying RC-70 liquid asphalt on the slope and covering it with green geotextile sheets, and (3) placing concrete platforms on the top of each stage of slope for drainage and vegetation. The total surface area of the cut slopes treated with the *green coating* technique was about 630 m².

The construction began at the top of the hill (test site) and gradually worked towards the bottom. Immediately after each slope stage was completed, the waterproofing and drainage work were carried out. The first step in this work was to clean up the slope surface, removing loose rock and broken pieces. The clean surface was then sprayed with RC-70 asphalt. This asphalt coating serves two purposes: (1) preventing water from entering the mudstone and (2) providing adhesion between the mudstone and the geotextile sheet. It was observed that the sprayed asphalt firmly stuck to the surface of the newly excavated mudstone and thus was effective in preventing erosion. The drainage strips were next installed on the slope surface. Finally, the slope was covered by geotextile sheets, which had two layers: the inner layer was an asphalt coating and the outer layer was a geotextile. The width of each geotextile sheet was 1 m with 10 cm overlap with the next sheet. Steel nails were used to fasten the sheets to the mudstone surface, and waterproofing treatment of these nails by asphalt coating was done immediately. Meanwhile, excavation continued for the next stage of the slope. During the excavation, it was evident that many joints were present in the test hill.

To avoid the newly excavated slope surface from erosion by water (which could reduce the strength of mudstone and cause the failure of the slope), no excavation was allowed on rainy days. The construction of the designed slopes began in January 1992 and ended in March 1992. Two types of measurement were made to observe the movement and the inclination of the treated slopes. Based on the field observations and measurements, the treated slopes did not show any signs of significant erosion and movement.

Case study 14 (slope erosion control)
The research and applications of geonaturals, especially geojute, have been directed towards developing these materials as alternative to geosynthetics. Due to their biodegradability and short life, most target use areas have been their applications as soil saver for supporting vegetation growth and temporary filter and drainage layers. The design life of geonaturals may end when vegetation grows and soil layers consolidate to some extent.

Erosion of railway and road embankments and hill slopes is caused principally by rain and wind. Erosion of the top soil gradually destabilizes the earthen embankments. Denuded hill slopes are always vulnerable to erosive forces of rains particularly during the monsoon. Jute geotextile (geojute) when applied on an exposed soil surface acts as miniature check dams or micro terraces, reduces the kinetic energy of rain splashes, diminishes the intensity of surface runoff, prevents detachment and migration of soil particles and ultimately helps in the quick growth of vegetation on it by the formation of mulch. Jute geotextile therefore helps in controlling erosion in road and railway embankments and hill slopes naturally. In India, jute geotextiles have been used in several erosion control projects in the past. Some of them are described very briefly below (IS 14986:2001):

- sand dune stabilization (5000 m^2) with 500 g/m^2 jute geotextile in Digha Sea Beach, Midnapur, Forest Department, Government of West Bengal, India in 1988, 80% covered by vegetation after 6 months;
- control of top soil erosion (10,000 m^2) with 300 g/m^2 and 400 g/m^2 jute geotextile in Arcuttipur, T.E. Cachar, Assam, India in July 1995, 93–97% reduction in soil loss;
- erosion control in embankment (3000 m^2) in Valuka, Maldah Irrigation Department, Government of West Bengal, India in August 1996, no damage by rains in 1996 and 1997;
- afforestation and erosion control (2000 m^2) with 25 g/m^2 jute geotextile in Hijli and Porapara, Midnapore Forest Department, Government of West Bengal, India in August 1997, growth of trees in the treated area significantly higher, no sign of erosion.

Case study 15 (slope stabilization)
The method suggested by Broms and Wong (1986), may be called *Geofabric-Wrapped Drain (GWD) Method*, as described in Sec. 6.3.9, was adopted for the stabilization of a landslide on the campus of the Nanyang Technological Institute (NTI) in Singapore. The landslide occurred in early 1984, during a period of heavy rainfall, on the NTI campus in the western part of Singapore. One student dormitory, Block E, was located at the toe of the slope. Two other dormitories were at the crest. An existing rubble wall, which had been constructed along the whole length of slope with height varying from 1.70 to 3.50 m, failed during the landslide. The average height of the slope was about 7 m. The inclination of the slope was 37° prior to the failure. A scupper drain at the toe of the rubble wall was damaged and closed up as a result of the movement of the slope. The ground immediately in front of the displaced rubble wall heaved about 200 mm. The whole sliding mass continued to move

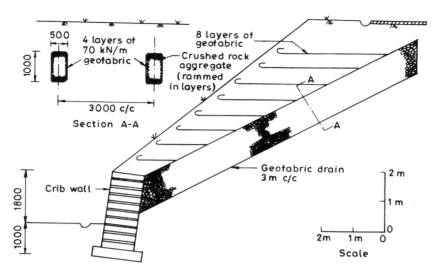

Figure 9.13 Stabilization scheme – Nanyang Technological Institute block E slide, Singapore (after Broms and Wong, 1986).

at a slow rate during the rest of 1984. Large cracks appeared on the displaced rubble wall. The total displacement of the wall was approximately 700 mm at the end of 1984. The toe had moved about 300 mm. The slope was composed of residual soil and highly and completely weathered sedimentary rocks.

The remedial stabilization works at the Block E slope consisted of the installation of eight fabric-wrapped crushed rock drains (Fig. 9.13). The drains, 0.5 m wide and 1.0 m high, were spaced 3.0 m apart. Based on the sliding surface and a residual friction angle of 18°, the required tensile force of the geofabric was 85 kN/m of the slope or 255 kN per drain. For each drain, four layers of 3.4-m wide PET fabric, with an ultimate tensile strength of 70 kN/m (238 kN per layer) at 14% elongation, were used. The fabric was wrapped around the two sides and the bottom of each drain. The drains were connected to the crib wall at the lower end of the slope for drainage. The crib wall was filled with crushed rock to allow discharge of the water from the transverse drains. Horizontal layers of the fabric were also placed in the slope between the ground surface and the transverse drains to increase the stability of the slope with respect to shallow slides above the transverse drains. The far end of each fabric strip was anchored in the crushed rock drain. Another layer of the fabric was placed along the drains between the horizontal strips as a filter to prevent the soil above from being washed (eroded) into the drains. No further movements of the slope were observed after the installation of the drains.

Case study 16 (slope stabilization)

The Anchored Spider Netting Method, as described in Sec. 6.3.9, has been used to stabilize a 4.5-m high clayey silt slope at a uniform slope angle of 25° (Koerner and Robins, 1986). The slope was in an active state of failure. The slope was hand cleared of vegetation and

358 Case studies

graded so as not to have any concave depressions. A knitted geonet made of bitumen coated nylon was used as netting. The anchors were 13-mm diameter steel rods in 1.2 m long sections which were threaded into one another during installation. Since the existing failure zone was a shallow slope failure, the rods were only 2.4 m long. This length easily penetrated the failure plane as it drew the net into the surface soil. Two adjacent widths of netting were used on this slope (each being 5.6 m wide) along with a total of 73 steel rod anchors. Upon completion of the anchored spider netting, the slope was seeded with a rapid growing rye grass and mulched. The grass grew within two weeks and had completely hidden the netting. The slope was reported to be in a stable condition.

Case study 17 (slope stabilization)

Dixon (1993) reported the geogrid-reinforced soil repair of a slope failure in clay on the North Circular Road, London, UK. Figure 9.14 shows the cross-section of the repair of the slip failure. The slope was a cut slope (side slope = 2H : 1V, maximum height = 8 m) formed in London clay in 1975. Some seven years after construction, slip failures began to occur along a 500-m length of cutting causing damage to fence lines and spillage onto the carriageway. Main earthworks began with the excavation and removal from the site of a 35-m long strip of slipped soil in September 1985. Excavation extended beyond the failure plane with benched steps cut into the undisturbed clay. To control any seepage, a 300-mm thick granular drainage layer was included over the excavated surface on the north side, where the forest slopes towards the cutting. The general sequence then adopted was to reinstate the first strip using fill excavated from an adjacent strip, thereby minimizing double handling. The second strip was then reinstated using the fill excavated from a third strip and so on. Fill was tipped from dump-truck, placed using the bulldozer and compacted to a maximum layer depth of 200 mm using the vibrating roller towed by the bulldozer. The 2.0 m width of Tensar SS1 secondary reinforcements were obtained by cutting the standard 4.0-m wide rolls into half on site with a disc cutter. Tensar SR2 rolls were cut to the required length and laid perpendicular to the slope alignment. Adjacent rolls were butt jointed.

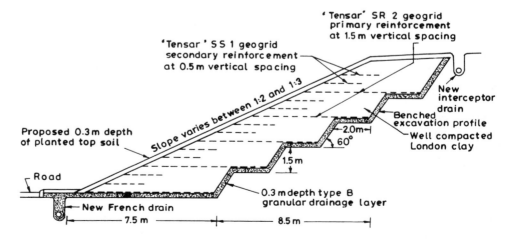

Figure 9.14 Cross-section of the repair of the slip failure of a slope (after Dixon, 1993).

The slope face was over filled and trimmed in conventional manner. Earthworks were completed in early February 1986. No special site equipment or expertise was required for installation, which was carried out using conventional plant and labour. The average construction time per 35-m long strip was about 3 days (typical strip quantities being: excavation – 1200 m³, fill – 800 m³, gravel drain – 350 m³, geogrids – 2000 m²). After construction no discernable movements were noted and the grass cover on the slope was reported to be in good condition with a pleasing appearance.

Case study 18 (slope stabilization)

Dikran and Rimoldi (1996) described a case study in which the geogrid-reinforced modular concrete block wall systems (GRMCBWS) was successfully used for facing steep cuttings for the approaches of a tunnel under the main A225 Tonbridge Road, Sevenoaks, Kent, UK. Two walls were designed and built at each end of the tunnel with a total length of 120 m and heights varying from 6.3 to 1.0 m. The geometry of the walls was chosen in a manner to give a pleasant aesthetic view and for adequate stability (Fig. 9.15). The blocks used in the project were of 'GEOBLOCK' type, and the construction sequence was as follows:

1. The foundation was prepared and the footings were cast using mass concrete.
2. The first course of the blocks was placed along the desired building line using standard rib units of Type 1.
3. The second course was built using the insert blocks of the Type 2 where the first layer of Tenax TT301 SAMP geogrid was required.

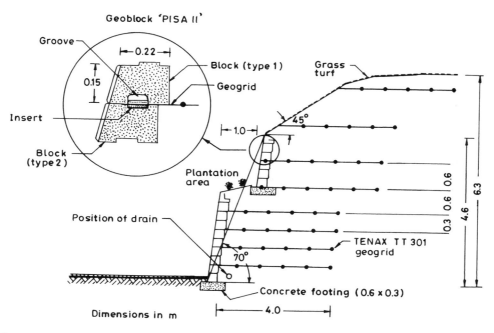

Figure 9.15 Geogrid-reinforced modular concrete block wall, Sevenoaks School, Kent (after Dikran and Rimoldi, 1996).

4   The inserts were placed in the groove on the top of the block with the narrow end of the finger pointing towards the face of the wall.
5   Tenax TT301 SAMP geogrids were cut to the required lengths and placed over the inserts so that every aperture in the grid was located over a finger of the inserts. The next row of blocks was placed over the insert and geogrid to hold the grid in place. The geogrid was then pulled from the back of the wall pulling the transverse rib of the grid back across the end of the fingers of the inserts.
6   Dug silty sand materials were used for the geogrid-reinforced fill. The compaction was carried out in 150-mm layers using a plate compactor in areas within 1.0 m from the face of the wall and a vibrating roller with a mass per metre width of 1300 kg for the remainder of the length of the reinforcement.
7   When the fill was placed and compacted up to the level of the next ground, the geogrid was laid down on the top of the fill and construction continued as in steps 2–6. The grass turf was placed over the back slopes.

Case study 19 (landfills)
Designing a constructible composite liner system for the side slopes of a landfill is a challenging task. To meet this challenge at the Lopez Canyon Sanitary Landfill, Los Angeles, USA, an entirely geosynthetic composite liner (GCL) and a leachate collection and removal system (LCRS) was developed in the year 1991 (Snow *et al.*, 1994). A schematic cross-section of the geosynthetic alternative developed for the side-slopes disposal area C is shown in Figure 9.16. The veneer of concrete was specified to have a

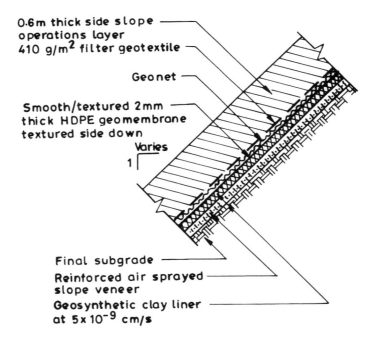

Figure 9.16 A geocomposite liner system for steep Canyon landfill side slopes in Los Angeles, USA (after Snow *et al.*, 1994).

compressive strength of 170–205 kPa and was sprayed on to the graded canyon side slopes to provide support and a smooth surface for the composite liner. A polyethylene (PE) geonet was used in lieu of granular soil to provide an LCRS on the side slopes. The primary advantages of the geonet are simple installation and a high drainage capacity resulting in a low liquid head on the composite liner. Construction of the geosynthetic side-slope liner system was subjected to large temperature variations, high winds and the steep slopes at the site. The familiarity of the geosynthetics installer with these conditions from his work on other landfills in the area was a significant benefit for the project. A total of about 15,500 m^2 and 77,000 m^2 of geosynthetic composite side-slope liner system was placed during Phase I and II of liner system construction, respectively. Phase I GCL joints were simply overlapped with no additional preparation, while the Phase II GCL joints were prepared by the addition of powdered bentonite at the rate of 1.5 kg/m^2 in the overlap areas. Performance of the liner system under dynamic loading was observed during the Northridge earthquake, of Richter magnitude 6.6, which struck Los Angeles on 17 January 1994. The Lopez Canyon site is located less than 15 km from the earthquake epicentre. Recording stations nearby measured horizontal peak ground accelerations of up to 0.44 g. Observations made that same day indicated that the geosynthetic side-slope liner system performed very well.

Case study 20 (reservoir)
The Gennevilliers (France) recreation reservoir was constructed in 1986 over an old landfill and gravel quarry and designed for recreational use such as sailing. It is still, to date, one of the largest geomembrane-lined artificial lakes in Europe (Ialynko and Gonin, 1993; Duquennoi, 2002). Figure 9.17 shows the typical section of the reservoir. Many technical difficulties arose from the fact that it was built on a landfill, which in itself represents a reference case. The results of preliminary geotechnical studies led to the design of a geomembrane lining system. The subgrade preparation consisted in bank consolidation, followed by the compacting and grading of a 30–40 cm thick sand layer over the entire surface of the reservoir. The eventuality of a water table rise, together with possible remaining waste gases, led to the design of a water and gas drainage system consisting of a 10-cm thick 10–25-mm gravel layer, enhanced by geocomposite drain strips. The gravel layer is connected to a central drain pipe and to peripheral gas vents. A 3-mm high density polyethylene (HDPE) geomembrane was selected regarding the following criteria: tensile properties to resist differential settlement, static and dynamic puncture resistance and roll width and length to minimize overall seam length. It has to be noted that the geomembrane was not associated with any geotextile in this project, neither as a underliner nor as a protection overliner. Geomembrane protection systems have been designed on the banks only. They generally consist in 10-cm thick reinforced concrete slabs which were poured in situ. Wherever landscaping purposes required specific bank works, such as gabions and vegetation, the concrete slabs were topped with gravel and topsoil. The geomembrane was unprotected at the bottom of the reservoir.

Case study 21 (canal)
The Mulhouse (France) canal was constructed in 1997 to convey water from the city of Mulhouse water treatment plant to the Hardt irrigation canal (Potié, 1999; Duquennoi, 2002). The canal is 9 km long with a trapezoidal section and a total output of 7 m^3/s

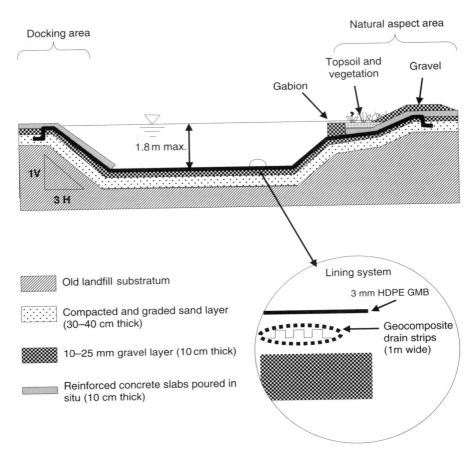

Figure 9.17 The Gennevilliers (France) recreation reservoir: typical cross-section and specific features (not to scale) (after Duquennoi, 2002).

(Fig. 9.18). The subgrade preparation consisted of alluvial gravelly soil excavation and grading. A 3.5-mm bituminous geomembrane factory-surfaced with an overliner geotextile was installed directly on the subgrade (Fig. 9.19). The geotextile was designed to drain eventual infiltration water under the extruded concrete cover. Since the concrete cover was to be poured with a sliding formwork machine, the eventuality of geotextile sliding by the machine had to be prevented by bonding geotextile and geomembrane together. The bituminous geomembrane was selected because of its higher puncture resistance and its ease of installation under harsh climatic conditions. Welding had to be done using 40-cm wide bituminous geomembrane strips because of the surface geotextile. It must be underlined that anchoring the geomembrane on top of the embankment was not necessary because of the ballasting effect of the concrete layer. A protecting layer of extruded concrete was laid on the geosynthetic lining system using a sliding formwork machine, specifically designed for this work. The rate of installation, including lining system installation and extruded concrete pouring, was 300 m/day.

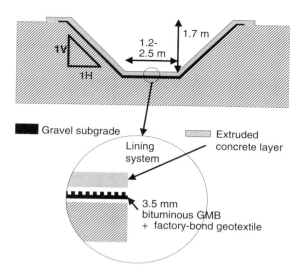

Figure 9.18 The Mulhouse (France) canal: typical cross-section and specific features (not to scale) (after Duquennoi, 2002).

Figure 9.19 The Mulhouse (France) canal: geomembrane–geotextile composite installation (after Duquennoi, 2002).

Case study 22 (canal)

The geomembranes provide an effective lining alternative for minimizing the loss of water through seepage. The type of geomembrane is an important factor as it affects project performance, longevity and cost. The United States Bureau of Reclamation (USBR) has constructed 34 canal lining sections in 11 irrigation districts in 4 northwestern states to assess durability, cost and effectiveness of alternate lining technology. A detailed report on various lining materials used in these evaluations has recently been published by the USBR (Swihart and Haynes, 2002).

Ivy and Narejo (2003) described the performance of white-surfaced textured HDPE geomembrane as the lining material in the Buffalo Rapids Main Canal Test Section. The general schematic of the lined section of the canal is provided in Figure 9.20. The total length of the lined section is 1490 m. The base width of the section varies from 3.5 to 4 m. The height of the lined embankment varies from 1.5 to 1.8 m with 0.6 m benches along both sides to accommodate the runout anchoring system for the liner. The width of the lined segment of the canal at the top is 11.5–12 m. White-surfaced textured HDPE geomembrane with thickness of 8 mils (2 mm) was manufactured by GSE Lining Technology Inc. The downstream 395 m of the test section was too rough to place a geomembrane directly. The section was first lined with a 335-g/m^2 nonwoven needle-punched geotextile.

An eight-man crew installed the geomembrane and geotextile on the prepared subgrade. The 4.5-m wide geotextile was installed lengthwise in the canal bed. The geomembrane rolls were 6.9 m wide. Rolls were unrolled across the canal by a trackhoe operating in the canal invert. The geomembrane was temporarily secured with sandbags and pins. Working from an access road, a second trackhoe then covered the anchor berm with 0.3–0.6 m of cover soil. The geomembrane panels were shingled downstream, overlapped 10–15 cm and hot-wedge welded. The cost–benefit ratio and effectiveness against seepage indicate that white-surfaced textured geomembranes provide a suitable alternative to traditional materials.

The effective service life of geomembranes, in general, was found to be 10–25 years. However, the USBR study predicts the service life for HDPE geomembranes to be 20–25 years. The service life of geomembranes varies depending on installation, service conditions and maintenance programmes.

Case study 23 (dam)

Dams are the most frightening and, potentially, the most dangerous structures. As a result, the designers of dams have traditionally been conservative, which explains why the use of

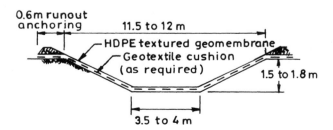

Figure 9.20 General schematic of the lining of Buffalo Rapids Canal, USA (after Ivy and Narejo, 2003).

geosynthetics has developed more slowly in dams than in other geotechnical applications (Giroud, 1992). In the past four decades, however, a wide variety of geosynthetics have been used in virtually all types of dams for new construction or rehabilitation.

A bituminous geomembrane was used in 2000 for the construction of the upstream waterproofing face of La Galaube Dam on Alzeau River in the south of France (Gautier et al., 2002). The dam is 380 m long at its ridge and the slopes have a gradient of 2 horizontal to 1 vertical, that is, an angle of about 26°. The maximum height above the foundations is 43 m.

The dam waterproofing is prolonged inside the foundation by an injection well through the upstream plinth. Construction works were divided into 4 independent contracts: earthworks and civil engineering, grouting, equipment and upstream waterproofing face. The upstream waterproofing face construction began in July 2000, with 23,000 m² to be waterproofed. The geomembrane is based upon a nonwoven PET geotextile impregnated with 100/40 blown bitumen. Its properties were reported as: tensile strength – 28 kN/m in longitudinal direction, 20 kN/m in transverse direction, together with more than 70% deformation at break; puncture strength – 500 N; and a good ageing behaviour based on more than 20-year-old references, including the field of dam waterproofing.

Rolls were lifted and unrolled with a specifically designed hydraulic beam carried by a track excavator. The geomembrane strips were laid side-by-side, with a 20-cm cover for the future weld. The high mass per unit area of geomembrane reduced the risks of uplifting by the wind and the creation of creases, which therefore eased up the welding operations. After removing the kraft paper that protected the edge of the strips, welds were carried out with a gas burner, which heated the bitumen until fusion of surfaces of both geomembranes. This operation was then followed, within 2 m from the flame, by the rolling of the welded strips, to ensure a proper contact between the two geomembranes. To avoid any risk of geomembrane slipping along the slope, it was temporarily anchored by steel pins and loaded by gravel material upon the crest of the dam. At the foot of the slope and along the whole periphery of the impervious face, the geomembrane was fastened to the reinforced concrete plinth, which is anchored into the rock foundation. The geomembrane was hot-welded on the concrete face, which has been previously covered with a tack-primer, then anchored with stainless steel plates bolted into the plinth. A 1-m wide strip of geomembrane was then welded above the steel plates. The protection to geomembrane was made of in situ cast concrete slabs (5 m × 10 m), on a nonwoven geotextile (see Fig. 9.21). The minimum thickness

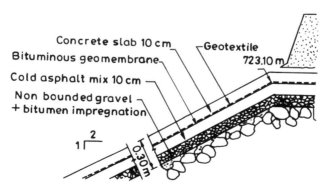

Figure 9.21 Typical cross-section of the waterproofing structure of La Galaube Dam on Alzeau River, France (after Gautier et al., 2002).

of the slabs was 10 cm, and PP fibres were added to the concrete to prevent cracking. The joints between the concrete slabs in the water variation range were filled with an elastomeric binder, added with solvents to allow a cold application on vertical surfaces, even under wet conditions.

Case study 24 (dam)
Lafleur and Pare (1991) described the experience gained from the placement of 400,000 m² of geotextiles in temporary and permanent water-retaining embankments of the James Bay hydroelectric project in North Quebec, Canada. Scarcity of suitable granular materials and ease of installation of geotextiles favoured their use in the earth and rockfill structures. Their main function was to act as filters in cofferdams between the coarse rockfills and the upstream cores made of cohesionless silty moraine. Cofferdams are temporary embankments permitting dewatering of the river bed during the period of construction of the permanent water-retaining structures. Their service life was limited to about 3 years and their maximum height was generally less than 15 m. The typical section, shown in Figure 9.22, indicates that they were generally composed of a rockfill embankment with a thick moraine blanket dumped upstream to provide imperviousness.

According to the conditions encountered at the James Bay project, the construction procedure for the cofferdams was generally the following: the rockfill was first dumped into water across the river, forcing the diversion tunnel to start operating. This rockfill was used as the working platform to place the remaining top part of the cofferdam in the dry using rolled lifts. The geotextile sheets, anchored at the top, were then laid and spread along the slope from top to bottom, using either boats or cranes. Due to the coarseness of the rockfill embankment, little hydraulic head was developed across the fill. To help sinking the geotextiles and ease stable overlapping (1.5 m), moraine was placed on the geotextile sheets using a backhoe. The joints were generally made by overlapping (1.5 m minimum) or sometimes, in cases of low and short cofferdams, by sewing. After this sinking operation, the moraine was dumped into the water over the geotextiles with final slopes in the bottom-submerged part of the embankment, equal to an angle of repose of about 10 horizontal to 1 vertical. In case of high cofferdams where the fill was above water, the rockfill, the geotextile, and the moraine were placed using conventional placement and compaction equipment. In this case, the geotextile sheets were installed parallel or perpendicular to the slope depending on the site conditions. The side slopes of the rockfill were kept at 1 vertical to 1.4 horizontal and the upstream slope of the compacted moraine core, protected by a proper

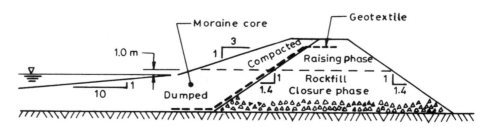

Figure 9.22 Typical cross-section of cofferdams in the James Bay Hydroelectric Project, North Quebec, Canada (after Lafleur and Pare, 1991).

riprap, was constructed at 1 vertical to 3 horizontal. Greater care was taken to check visually the continuity of the joint overlapping of the geotextiles.

Performance tests indicated that, for placement under water over dumped rockfill, heavy-duty nonwoven geotextiles had to be specified to avoid punching by large angular stones. In rolled embankments constructed above water, geotextiles with masses greater than 300 g/m² worked satisfactorily because of the possibility of working out a better bedding surface at the interface.

Case study 25 (tunnel)

The Arlberg Tunnel, Austria's outstanding and most important road tunnel, was constructed during 1974–78 (Posch and Werner, 1993). The design methodology was the New Austrian Tunnelling Method (NATM), using the following three different types of lining systems for the tunnel waterproofing over 13,972-m long section (Fig. 9.23):

1. a 300–400-mm thick concrete lining without any sealing measures or separation layers (34.3% of tunnel length);
2. a 300–400-mm thick concrete lining and PP needle-punched nonwoven geotextile of 700 g/m² with 3.6 kN puncture resistance (Polyfelt TS 008), functioning as a separation and drainage layer (12.4% of tunnel length);
3. a 300–400-mm thick concrete lining and waterproofing system, which is a combination of a 1.5-mm thick polyvinyl chloride (PVC) geomembrane and a 500 g/m² PP needle-punched nonwoven geotextile protection and drainage layer (53.3% of tunnel length).

Needle-punched PP nonwoven geotextiles were selected based on the mechanical resistance, chemical resistance, and in-plane water permeability criteria. The concrete lining was constructed in 12-m long ring sections of waterproof pumped concrete B30. The concrete consisted of 240–250 kg of Portland cement PZ 275, 40–60 kg of fly ash and aggregate with a maximum size of 32–60 mm.

Ten years after completion, this almost classic example of the NATM has revealed remarkable results in terms of reducing crack formation in the concrete lining. A 40% reduction in crack formation has been found in the geotextile and concrete lined tunnel sections. In this section the geotextile is the separation and drainage layer between the shotcrete and

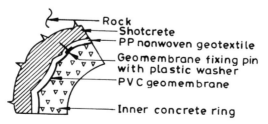

Figure 9.23 Typical cross-section of the tunnel lining in Arlberg Tunnel, Austria (after Posch and Werner, 1993).

inner concrete lining. Geotextile samples were removed 12 years after construction, and the following results were obtained:

- The mass per unit area has increased by 9% due to sedimentation of fine particles.
- The coefficient of normal permeability was reduced from 5.72 to 3.25 mm/s.
- There was no indication of clogging caused by mineral precipitation.

It should be noted that the use of a nonwoven geotextile increased the cost of the concrete lining by only 9%. The insignificant difference in the number of cracks in the geotextile lined section, compared with the geotextile–geomembrane lined section, makes this design an alternative to lined waterproofing. Particularly, the use of a geomembrane increases the liner costs by 70%. In the limited area where cracks did occur in the geotextile-lined section, effective drainage of moisture prevented seepage of water into the tunnel through the cracks. This was true even when drains at the floor of the tunnel showed seepage occurring behind the lining. It has been recommended that a geotextile should be used even in those sections which do not show any immediate ingress of rock water during excavation of the tunnel, because the water leakage path in the transition zone may change over time from the sealed to the unsealed area.

9.3 Concluding remarks

The knowledge developed during the past 3–4 decades allows engineers to safely design and construct with geosynthetics. The use of geosynthetics in civil engineering will go on increasing as more engineers learn about geosynthetics and become aware that using geosynthetics is beneficial, provided design and construction are adequate. Efforts should always be made to avoid misuse of geosynthetics in any application because of simple ignorance or negligence. In fact, geosynthetics must be treated as any other construction material, although geosynthetic producers are still improving upon the technology. Their strengths and weaknesses must be recognized and their properties should be evaluated through adequate testing. Engineers designing geosynthetic applications should not be afraid of using geosynthetics nor should they have unrealistic expectations about their capabilities. A rational approach is certainly the best way to ensure the development of proper applications of geosynthetics in civil engineering projects.

Self-evaluation questions

(Select the most appropriate answers to the multiple-choice type questions from 1 to 10)

1. Geosynthetics can be effectively installed in

    (a) Geotechnical engineering projects.
    (b) Environmental engineering projects.
    (c) Transportation engineering projects.
    (d) All of the above.

2. The primary benefit of the geotextile reinforcement is generally to contribute to the short-term stability of

    (a) Paved roadways.
    (b) Retaining walls.

(c) Embankments on soft foundation soils.
(d) None of the above.

3. The number of geosynthetic layers in foundation applications

   (a) Is one.
   (b) Is two.
   (c) Is three.
   (d) Can be greater than one.

4. The use of geosynthetic reinforcement in unpaved roadways may allow decreased

   (a) Consumption of granular materials.
   (b) Duration of construction.
   (c) Transport expenses.
   (d) All of the above.

5. Select the incorrect statement related to the application of geotextile–geogrid composites in railway tracks.

   (a) Geotextile performs its fluid barrier function to prevent contamination of the ballast.
   (b) Geogrid provides tensile reinforcement and shear resistance to increase the effective bearing capacity of the subgrade.
   (c) Geogrid confines the ballast through interlocking, thus increasing its resistance to both vertical and lateral movement.
   (d) None of the above.

6. Geotextiles are used mainly for filtration function in construction of

   (a) Dams.
   (b) Tunnels.
   (c) Canals.
   (d) All of the above.

7. Which of the following methods is not applicable to slope stabilization?

   (a) Geofabric-wrapped drain method (Broms and Wong, 1986).
   (b) Anchored spider netting method (Koerner and Ronbins, 1986).
   (c) Both (a) and (b).
   (d) Green coating method (Lee et al., 1996).

8. Which of the following standards deals with the guidelines for application of geojute (jute geotextile) for rain water erosion control applications?

   (a) IS 14986: 2001.
   (b) IS 14715: 2000.
   (c) IS 14324: 1995.
   (d) None of the above.

9. In tunnel lining projects, which of the following geosythetics is generally not used?

   (a) Geotextile.
   (b) Geomembrane.
   (c) Geosynthetic clay liner.
   (d) All of the above.

10. The development of proper applications of geosynthetics in civil engineering requires
    (a) Rational approach.
    (b) Empirical approach based on past field experiences.
    (c) Both (a) and (b).
    (d) None of the above.
11. What is the main purpose of preparing case studies related to geosynthetic applications?
12. What are the special features of the case study, presented by Gourc and Risseeuw (1993), on a geotextile-reinforced wraparound faced wall built in late 1982 in Prapoutel, France? Select the common features reported in all the case studies on geosynthetic-reinforced retaining walls/steep-sided slopes, presented in the current chapter.
13. What were the challenging tasks during the construction of a long embankment through the tidal area of Deep Bay in the New Territories district of Hong Kong, as reported by Risseeuw and Voskamp (1993)?
14. How will you bridge over a weak section of a roadway with a geogrid-reinforced gravel–sand layer? Give a step-wise description of the engineering involved.
15. Describe the steps of geotextile installations in unpaved roads, constructed in the USSR, as reported by Kazarnovsky and Brantman (1993). What were the improvements observed with geotextile installations?
16. Describe the geotextile installation steps at the subgrade level in paved roadways, as reported by Shukla et al. (2004). Can you suggest any other improved method of installation?
17. What are the functions served by the geosynthetics in railway track application, as reported by Walls and Newby (1993)?
18. Develop a method of erosion control with geosynthetics using the concepts presented in the case studies as described in the current chapter.
19. Do you think that the green coating technique of erosion control described by Lee et al. (1996) is applicable to any type of slope? In your opinion, what are the limitations of this technique?
20. What are the advantages and limitations of jute geotextiles (geojutes) for erosion control applications?
21. Describe the performance effectiveness of the following methods of slope stabilization based on the reported cased studies:
    (a) Geofabric-Wrapped Drain Method
    (b) Anchored Spider Netting Method.
22. Describe the construction sequence of geogrid-reinforced modular concrete block wall system (GRMCBWS), as reported by Dikran and Rimoldi (1996).
23. What special precautions are required for the construction of composite liner systems for the side slopes of a landfill?
24. What differences do you observe in the lining of a reservoir and a canal in the reported case studies?
25. Mention the locations in dam sections where geosynthetics have been used in the reported case studies. Briefly describe the functions served by the geosynthetics and their performance effectiveness.

26. Describe the lining system for tunnel waterproofing as reported by Posch and Werner (1993). How did geotextiles perform in tunnel applications?
27. List the geosynthetics, along with their properties, used in the case studies reported in the current chapter.
28. Prepare some case studies on geosynthetic applications available in your locality with the help of the engineers, contractors and owners of projects.
29. Can you think of some new geosynthetic applications? Translate your thought into action.

Appendix A

Answers to multiple choice type questions and selected numerical problems

Chapter 1

1. b
2. d
3. c
4. d
5. c
6. b
7. a
8. a
9. b
10. b
11. d
12. c
13. c
14. c
15. b
16. b

Chapter 2

1. b
2. d
3. b
4. a
5. b
6. d
7. c
8. a
9. d
10. d
11. c
12. b
13. d
14. c
15. d

Chapter 3

1. b
2. c
3. a
4. c
5. b
6. b
7. b
8. b
9. d
10. b
11. d
12. b
13. d
14. c
15. b
16. d
17. b
18. d
19. a
20. b
21. a
22. b
23. c
24. b

27. 1920 g/m²
42. Adhesion ≈ 15 kPa, Angle of interface shear resistance ≈ 15°
44. 90.8%
54. 1.44×10^{-3} m²/s
55. 0.538 s^{-1}, 1.13×10^{-3} m/s
56. 1.13×10^{-5} m²/s, 4.36×10^{-3} m/s
57. 2.6×10^{-5} m²/s, 0.01 m/s
72. 22.7 kN/m

(Note: Assumed reduction factors are: $f_{ID} = 1.1, f_{CR} = 2.0, f_{CD} = f_{BD}$)

73. 0.8 s^{-1}
    (Note: Assumed reduction factors are: $f_{CB} = 2.0, f_{CR} = f_{IN} = f_{CC} = f_{BC} = 1.0$)
74. 60.7 kN/m
    (Note: Assumed reduction factors are: $f_{ID} = 1.1, f_{CR} = 1.5, f_{CD} = f_{BD} = 1.0$)

Chapter 4

1. b
2. c
3. d
4. a
5. b
6. b
7. c
8. d
9. b
10. a
11. c
12. b
13. d
14. a
15. d
16. b
17. d
18. b
19. c
20. c
21. c
22. d

69. Permittivity $\geq 0.2 \text{ s}^{-1}$
    AOS ≤ 0.25 mm
    Grab strength ≥ 1100 N
    Sewn seam strength ≥ 990 N
    Tear strength ≥ 400 N
    Puncture strength ≥ 400 N
    UV stability $\geq$ (50% of 1100 N = 550 N) after 500 h of exposure
70. Grab strength ≥ 450 N
    Ultimate elongation $\geq 50\%$
    Mass per unit area ≥ 140 g/m^2
    Asphalt retention $\geq$ as recommended in the manufacturer certification
    Melting point $\geq 150°$C

Chapter 5

1. d
2. d
3. d
4. a
5. b
6. b
7. c
8. c
9. b
10. b
11. d
12. b
13. a
14. a
15. b
16. d

29. Geotextile strength, $T = 122$ kN/m; Geotextile modulus, $E = 813.3$ kN/m
52. Permittivity $\geq 0.1 \text{ s}^{-1}$
    AOS ≤ 0.22 mm
    Grab strength ≥ 700 N

Sewn seam strength ≥ 630 N
Tear strength ≥ 250 N
Puncture strength ≥ 250 N
UV stability ≥ (50% of 700 N = 350 N) after 500 hrs of exposure

53. Permittivity ≥ 0.02 s^{-1}
AOS ≤ 0.60 mm
Grab strength ≥ 900 N
Sewn seam strength ≥ 810 N
Tear strength ≥ 350 N
Puncture strength ≥ 350 N
UV stability ≥ (50% of 900 N = 450 N) after 500 hrs of exposure

54. $k_n > 1 \times 10^{-5}$ m/s
$\psi \geq 0.5$ s^{-1}
AOS = 0.88 mm
n ≥ 50% for nonwoven geotextile filter
POA ≥ 4% for woven geotextile filter

55. 0.0018 m²/s; No, a very thick or multilayered nonwoven geotextile or composite sheet drain will be necessary.
58. 0.36 m
59. 1.09 m

Chapter 6

1. d
2. a
3. b
4. c
5. b
6. d
7. c
8. a
9. a
10. b
11. b
12. b
13. a
14. b
15. c
16. a
17. c
18. b
19. c
20. a

Chapter 7

1. b
2. a
3. b
4. d
5. b
6. b
7. d
8. b
9. d
10. a

Chapter 8

1. d
2. b
3. b
4. d
5. b
6. d
7. d
8. c

Chapter 9

1. d
2. c
3. d
4. d
5. a
6. a
7. d
8. a
9. c
10. c

Appendix B

Standards and codes of practice

B.1 Introduction

The application of geosynthetics is an area where there has been improvement continuing in the basic concepts and developments related to raw materials, manufacturing processes and methods of analysis, design and construction for the last four decades. Preparation of new test standards and codes of practice and some revision works have also been continuing. In developed countries a large number of standards have been prepared on different aspects of geosynthetics and some standards are still under preparation/revision. The developing countries are also trying to have their own standards so that geosynthetics can be used on a scientific basis for economical solutions of civil engineering problems. In fact, a completely unified set of worldwide standards and test methods is currently not available.

B.2 General information

A *standard* is basically a set of rules that identifies what features or characteristics a specific product, service or process should have. In other words, a standard contains technical specifications or other precise criteria to be used consistently and guidelines or definitions of characteristics, to ensure that the product, service or process is fit for its purpose.

Standards affect every part of our life. They contribute to making life simpler, increasing the reliability and effectiveness of the products, services or processes. There are many standards helping to improve our life. We can also say that standards help our world work in harmony.

Most standards are voluntary. Manufacturers or suppliers usually are not required by law to make their products comply with standards, but that does not mean they can just ignore them. Customers often insist on them. Standards that have a role in protecting public safety and health may become mandatory through inclusion in laws and regulations by the Government.

The work of developing standards is performed by balanced committees of experts from industry, governments, consumers and other relevant sectors. It is a highly democratic process that ensures that proposed standards are acceptable to everyone affected by them. For example, the International Organization for Standardization (ISO) is a worldwide federation of national standards bodies (ISO member bodies). The work of preparing International Standards is normally carried out through ISO technical committees. Draft International Standards adopted by the technical committees are circulated to the member bodies for voting. Publication as an International Standard requires approval by at least 75% of the member bodies casting a vote.

Specification is a precise statement of a set of requirements to be satisfied by a material, product, system or service that indicates the procedures for determining whether each of the requirements is satisfied (ASTM D4439-02).

A *code of practice* contains material, which is both for the information and guidance of engineers, and facts, which form recommendations for good practice. Engineering judgement is essentially required to determine when the recommendations of the code should be followed and when they should not. It should be noted that a code of practice embodies the experience of engineers successfully engaged in the design and construction of the particular class of works. It is intended for the use of engineers with some knowledge of the subject as a basis for the design of similar works. In fact, any code of practice takes the form of guidance and recommendations, and it should not be quoted as if it were a specification.

A standard or code of practice reflects latest scientific and industrial experience available at the time it is prepared. Technical developments can render certain part(s) of the standard or code of practice obsolescent in time. Though standards are kept under continuous review after publication and are updated regularly to take account of changing technology; however, it is the responsibility of the engineers and users to remain conversant with the developments, which have taken place since publication of the standards or codes of practice.

Many organizations prepare and publish standards and codes of practice on geosynthetics. The following abbreviations are used to refer to the standards and codes of practice:

AS	Australian Standard
ASTM	American Society for Testing and Materials
BS	British Standard
CGSB	Canadian General Standards Board
IS	Indian Standards
ISO	International Organization for Standardization

B.3 Standards on test methods

Many test procedures have been laid down in standards. Some test standards are listed below. The manufacturers and users of geosynthetics can find the details of a specific test in one or more standards.

Sampling and preparation of test specimens of geosynthetics
   AS 3706.1-1990
   ASTM 4354-99
   BS EN 963-1995
   CAN/CGSB 148.1 No. 1–94
   IS 14706-1999
   ISO 9862-1990

Determination of mass per unit area of geosynthetics
   ASTM D5261-92 (Reapproved 1996)
   ASTM D5993-99
   ASTM D6566-00
   BS EN 965-1995

CAN/CGSB 148.1 No. 2-M85
IS 14716-1999
ISO 9864-1990

Determination of thickness of geosynthetics
ASTM D5199-01
ASTM D5994-98
ASTM D6525-00
CAN/CGSB 148.1 No. 3-M85
BS EN 964 (Part 1)-1995
IS 13162 (Part 3)-1992
ISO 9863 (Part 2)-1996

Determination of stiffness of geosynthetics
ASTM D6575-00
CAN/CGSB 148.1 No. 14–93

Determination of tensile properties of geosynthetics
AS 3706.2-2000
ASTM D4595-86 (Reapproved 2001)
ASTM D6637-01
ASTM D6768-02
ASTM D6818-02
IS 13325-1992
ISO 10319-1993

Determination of grab tensile strength of geosynthetics
ASTM D4632-91 (Reapproved 1996)
CAN/CGSB 148.1 No. 7.3–92

Determination of tensile properties of geosynthetics by multi-axial tensile test method
ASTM D5617-99

Determining the compression behaviour of geosynthetics
ASTM D6244-98
ASTM D6364-99

Determination of seam strength of geosynthetics
AS 3706.6-2000
ASTM D4884-96
IS 15060-2001
ISO 10321-1992

Determining the integrity of geomembrane seams
ASTM D6214-98
ASTM D6365-99
ASTM D6392-99

Determination of tearing strength of geosynthetics by trapezoidal method
AS 3706.3-2000
ASTM D4533-91 (Reapproved 1996)
IS 14293-1995

Determination of static puncture strength of geosynthetics by plunger method (or CBR method)
AS 3706.4-2001
ASTM D6241-99
ISO 12236-1996

Determination of impact strength (dynamic puncture strength) of geosynthetics by falling cone method
AS 3706.5-2000
BS EN 918-1996
IS 13162 (Part 4)-1992

Determination of bursting strength of geosynthetics
ASTM D751
ASTM D5617–99
CAN/CGSB 148.1 No. 6.1–93

Determination of soil–geosynthetic or geosynthetic–geosynthetic interface friction by direct shear method
ASTM D5321-02
BS 6906 (Part 8)-1991
IS 13326 (Part 1)-1992

Determination of geosynthetic pullout resistance in soil
ASTM D6706-01

Determination of pore size distribution of geotextiles by dry sieving method
AS 3706.7-2003
ASTM D4751-99a
BS 6906 (Part 2)-1989

Determination of in-plane water flow capacity of geotextiles and geotextile-related products
AS 3706.10.1-2001
ASTM D4716-01
ASTM D6574-00
BS 6906 (Part 7)-1990
ISO 12958-1999

Determination of cross-plane water flow capacity of geotextiles and geotextile-related products
AS 3706.9-2001
ASTM D4491-99a

ASTM D5493-93 (Reapproved 1998)
BS 6906 (Part 3)-1989
CAN/CGSB 148.1 No. 4–94
IS 14324: 1995
ISO 11058: 1999

Guide for selection of test methods to determine rate of fluid permeation through geomembranes for specific applications
ASTM D5886-95 (Reapproved 2001)

Evaluation of hydraulic properties of geosynthetic clay liners permeated with potentially incompatible liquids
ASTM D6766-02

Determination of abrasion resistance of geotextiles by sand paper/sliding block method
ASTM D4886-88 (Reapproved 2002)
IS 14714-1999
ISO 13427-1998

Evaluating the tensile creep behaviour of geosynthetics
ASTM D5262-02a
BS 6906 (Part 5)-1991
ISO 13431-1999

Determination of the clogging potential of the soil–geotextile systems by the gradient ratio test
ASTM D5101-01

Evaluating the filtration behaviour of soil–geotextile systems by hydraulic conductivity ratio test
ASTM D5567-94 (Reapproved 2001)

Determination of the potential for the biological clogging of geotextile or soil/geotextile filters
ASTM D1987-95 (Reapproved 2002)

Selecting test methods for experimental evaluation of geosynthetic durability
ASTM D5819-99

Evaluating the chemical resistance of geosynthetics to liquids
AS 3706.12-2001
ASTM D5322-98
ASTM D5496-98
ASTM D5747-95a (Reapproved 2002)
ASTM D6213-97
ASTM D6388-99
ASTM D6389-99

Evaluating resistance of geosynthetics to microbiological agents
    AS 3706.13-2000
    BS EN 12225-2000

Evaluation of the dispersion of carbon black in geosynthetics
    ASTM D5596-94

Deterioration of geotextiles by exposure to light, moisture and heat
    AS 3706.11-1990
    ASTM D4355-02
    IS 13162 (Part 2)-1991

Determining effects of temperature on geosynthetics
    ASTM D4594 -96
    ENV ISO 13438

General tests for evaluation of geosynthetics following durability
    BS EN 12226-2000

Evaluating the resistance of geosynthetics to hydrolysis
    BS EN 12447-2001
    ISO-13439

Determination of asphalt retention of paving fabrics
    ASTM D6140-00

B.4 Codes of practice

Identification, storage, and handling of geosynthetic rolls and samples
    AS 3705-1990
    ASTM D4873-02
    ASTM D5888-95 (Reapproved 2002)
    ISO 10320-1999

Strengthened/reinforced soils and other fills
    BS 8006-1995

Quality control of geosynthetic clay liners
    ASTM D5889-97

Installation of geocomposite pavement drains
    ASTM D6088-97 (Reapproved 2002)

Installation of geosynthetic clay liners
    ASTM D6102-97

Mechanical attachment of geomembrane to penetrations or structures
    ASTM D6497-02

Characteristics of geosynthetics required for use in the construction of roads and other trafficked areas (excluding railways and asphalt inclusion)
    BS EN 13249-2001

Characteristics of geosynthetics required for use in the construction of railways
    BS EN 13250-2001

Characteristics of geosynthetics required for use in earthworks, foundations and retaining structures
    BS EN 13251-2001

Characteristics of geosynthetics required for use in drainage systems
    BS EN 13252-2001

Characteristics of geosynthetics required for use in erosion control works (coastal protection, bank revetments)
    BS EN 13253-2001

Characteristics of geosynthetics required for use in the construction of reservoirs and dams
    BS EN 13254-2001

Characteristics of geosynthetics required for use in the construction of canals
    BS EN 13255-2001

Characteristics of geosynthetics required for use in the construction of tunnels and underground structures
    BS EN 13256-2001

Characteristics of geosynthetics required for use in solid waste disposals
    BS EN 13257-2001

Characteristics of geosynthetics required for use in liquid waste containment projects
    BS EN 13265-2001

Guidelines on durability of geosynthetics
    AS HB 154-2002

Application of jute geotextile for rainwater erosion control in road and railway embankments and hill slopes
    IS 14986: 2001

Appendix C

Some websites related to geosynthetics

Table C.1 Geosynthetic-related journals and magazines

Journals/magazines	Editor/Publisher	Website address
Geosynthetics International	Professsor T.S. Ingold	www.thomastelford.com
Geotextiles and Geomembranes	Professor R.K. Rowe	www.elsevier.com
Geotechnical Fabrics Report (GFR)	Mary Hennessy	www.gfrmagazine.info

Table C.2 Geosynthetic-related societies/associations

Societies/associations	Website address	Country/continent
International Geosynthetics Society (IGS)	www.geosyntheticssociety.org	USA
North American Geosynthetics Society (NAGS)	www.nagsigs.org	USA
Geosynthetic Materials Association (GMA)	www.gmanow.com	USA
Geosynthetic Institute	www.geosynthetic-institute.org	USA
International Erosion Control Association (IECA)	www.ieca.org	USA
Landfill Systems & Technologies Research Association of Japan (LSA, NPO)	www.npo-lsa.jp	Japan

Table C.3 Geosynthetic companies/business groups/testing laboratories

Companies/Business Groups/Testing Laboratories	Website address
Archana Structural Engineering (India) Pvt. Ltd, India	www.aseipl.com
Australian Lining Co. P/L, Australia	www.austliningco.com.au
BBA Fiberweb	www.bbafiberweb.com
Bidim Geosynthetics SA, France	www.bidim.com
Bradley Industrial Textiles, Inc., USA	www.bradley-geotextile.com
BTTG-British Textile Technology Group, UK	www.bttg.co.uk
Cemagref, France	www.cemagref.fr
Chemir/Polytech Laboratories Inc., USA	www.chemir.com
Colorado Lining International, USA	www.coloradolining.com
CSI Geosynthetics, USA	www.csigeo.com
DX2 Geosyntex Inc., USA	www.dx2.net
Engepol Ltda., Brazil	www.engelpol.com
Engineered Linings (Pty) Ltd, South Africa	www.engineered-linings.co.za
Environmental Fabrics, Inc., USA	www.environmentalfabrics.com
G and E Co. Ltd, China	www.g-and-e.com
Garware-Wall Ropes Ltd, India	www.garwareropes.com
Geomembranes Ltda, Colombia	www.geomembranes.com.co
Geosynthetics Limited, UK	www.geosyn.co.uk
Greenfix America, USA	www.greenfix.com
GSE Lining Technology Inc., USA	www.gseworld.com
H.C. Nutting Company, USA	www.hcnutting.com
K.K. Enterprises, India	www.florafab.com
Huesker Synthetic, GmbH & Co., Germany	www.huesker.com
LEISTER Process Technologies, Switzerland	www.leister.com
LGA Geosynthetic Institute, Germany	www.lga.de
Material Testing Inc., USA	www.materials-testing.com
Naue GmbH & Co. KG, Germany, UK	www.naue.com
	www.betofix.com
Netlon India, India	www.netlonindia.com
Netlon Limited, UK	www.netlon.co.uk
Nylex, Malaysia	www.nylex.com
Polyfelt, Germany	www.polyfelt.com
Precision Geosynthetic Laboratories, USA	www.precisionlabs.net
Permathene Ltd, New Zealand	www.permathene.com
Poly-Flex, Inc., USA	www.poly-flex.com
Propex Fabrics de Mexico, S.A. de C.V., Mexico	www.geotextile.com
Reed and Graham, Inc., USA	www.rginc.com
Sci-Lab Materials Testing Inc., Canada	www.sci-lab.com
SGI Testing Services, USA	www.interactionspecialists.com
SI Geosolutions, USA	www.sigeosolutions.com
STRATA Systems, Inc, USA	www.geogrid.com
Tenax Spa, Italy	www.tenax.net
Tensar International, UK	www.tensar.co.uk
Terratest Tecnicas Especiales S.A., Spain	www.terratest.es
Terrafix Environmental Technology, Inc., Canada	www.terrafixgeo.com
Terram Limited, UK	www.terram.com
Triumph Geo-Synthetics, Inc., USA	www.triumphgeo.com
Wolf Environmental Lining Systems, USA	www.wolfels.com

References

AASHTO (1993). *Guide for Design of Pavement Structures*, American Association of State Highway and Transportation Officials, Washington, DC.

AASHTO (2000). *Geotextile Specification for Highway Applications*, AASHTO Designation: M 288-00, American Association of State Highway and Transportation Officials, Washington, DC

Abramson, L.W., Lee, T.S., Sharma, S., and Boyce, G. (2002). *Slope Stability and Stabilization Methods*. John Wiley & Sons, Inc., New York.

ACZ (1990). *Application and Execution Aspects of Sand Tube Systems*. Internal Report of ACZ Marine Contractors B.V., Gorinchem, The Netherlands.

Adams, M.T. and Collin, J.G. (1997). Large model spread footing load tests on geosynthetic reinforced soil foundations. *Journal of Geotechnical and Geoenvironmental Engineering*, **123**, 1, 66–72.

Agnew, W. (1991). Erosion Control Product Selection. *Geotechnical Fabrics Report*, IFAI, Roseville, MN, USA, April, pp. 24–27.

Ahn, T.B., Cho, S.D., and Yang, S.C. (2002). Stabilization of soil slope using geosynthetic mulching mat. *Geotextiles and Geomembranes*, **20**, 2, 135–146.

Akagi, T. (2002). The last decade of geosynthetics in Japan. *Proceedings of the 7th International Conference on Geosynthetics*, France, pp. 1467–1470.

Ali, F.H. and Tee, H.E. (1990). Reinforced slopes: field behaviour and prediction. *Proceedings of the 4th International Conference on Geotextiles, Geomembranes and Related Products*. The Hague, The Netherlands, pp. 17–20.

Allen, T.M. (1991). Determination of long-term tensile strength of geosynthetics: a state-of-the-art review. *Proceedings of Geosynthetics '91*, IFAI, Roseville, MN, USA, pp. 351–379.

Almeida, M.S.S., Britto, A.M., and Parry, R.H.G. (1986). Numerical modeling of a centrifuged embankment on soft clay. *Canadian Geotechnical Journal*, **23**, 2, 103–114.

Andrawes, K.Z., McGown, A., and Murray, R.T. (1986). The load-strain-time-temperature behaviour of geotextiles and geogrids. *Proceedings of the 3rd International Conference on Geotextiles*. Vienna, Austria, pp. 707–712.

Anon. (1979). *Reinforced Earth and Other Composite Techniques*. Supplementary Report No. 457. Transportation and Roads Research Laboratory, London, UK.

Anthoine, A. (1989). Mixed modeling of reinforced soils within the framework of the yield design theory. *Computers and Geotechnics*, **7**, 1 and 2, 67–82.

AREA. (1985). *American Railway Engineering Association Manual*, AREA, USA.

AS 3706.3-2000. *Determination of Tearing Strength of Geotextiles – Trapezoidal Method*. Standards Australia International Ltd, Sydney, Australia.

AS 3706.4-2001. *Determination of Burst Strength of Geotextiles – California Bearing Ratio (CBR) – Plunger Method*. Standards Australia International Ltd, Sydney, Australia.

AS 3706.5-2000. *Determination of Impact Strength (Dynamic Puncture Strength) of Geosynthetics by Falling Cone Method*. Standards Australia International Ltd, Sydney, Australia.

AS 3706.9-2001. *Determination of Permittivity, Permeability and Flow Rate*. Standards Australia Australia International Ltd, Sydney, Australia.

ASTM D1987-95. (Reapproved 2002). *Standard Test Method for Biological Clogging of Geotextile or Soil/Geotextile Filters*. ASTM International, West Conshohoken, PA, USA.

ASTM D4533-91. (1996). *Standard Test Method for Trapezoid Tearing Strength of Geotextiles*. ASTM International, West Conshohoken, PA, USA.

ASTM D4632-91. (Reapproved 1996). *Standard Test Method for Grab Breaking Load and Elongation of Geotextiles*. ASTM International, West Conshohoken, PA, USA.

ASTM D5101-01. *Standard Test Method for Measuring the Soil-Geotextile System Clogging Potential by the Gradient Ratio*. ASTM International, West Conshohoken, PA, USA.

ASTM D5567-94. (Reapproved 2001). *Standard Test Method for Hydraulic Conductivity Ratio (HCR) Testing of Soil/Geotextile Systems*. ASTM International, West Conshohoken, PA, USA.

ASTM D5886-95. (Reapproved 2001). *Standard Guide for Selection of Test Methods to Determine Rate of Fluid Permeation Through Geomembrane for Specific Applications*. ASTM International, West Conshohoken, PA, USA.

ASTM D6102-97. *Stanadard Guide for Installation of Geosynthetic Clay Liners*. ASTM International, West Conshohoken, PA, USA.

ASTM D6213-97. *Standard Practice for Tests to Evaluate the Chemical Resistance of Geogrids to Liquids*. ASTM International, West Conshohoken, PA, USA.

ASTM D6364-99. *Standard Test Method for Determining the Short-Term Compression Behaviour of Geosynthetics*. ASTM International, West Conshohoken, PA, USA.

ASTM D6747-02. *Standard Guide for Selection of Techniques for Electrical Detection of Potential Leak Paths in Geomembranes*. ASTM International, West Conshohoken, PA, USA.

Bagchi, A. (1994). *Design, Construction, and Monitoring of Landfills*. John Wiley & Son, Inc., New York.

Baker, T. (1998). The Most Overlooked Factor in Paving-Fabric Installation. *Geotechnical Fabrics Report*, IFAI, Roseville, MN, USA, April, pp. 48–52.

Barazone, M. (2000). *Installing Paving Synthetics – Overview of Correct Installation Procedures*. Geotechnical Fabrics Report, May, pp. 17–20.

Barenberg, E.J. (1992). Subgrade stabilization. In *A Design Primer: Geotextiles and Related Materials*. Industrial Fabric Association International, Minnesota, USA, pp. 10–20.

Bassett, R.H. and Last, N.C. (1978). Reinforcing earth below footings and embankments. *Proceedings of the Symposium on Earth Reinforcement, ASCE*, New York, pp. 202–231.

Bathurst, R.J., Hatami, K., and Alfaro, M.C. (2002). Geosynthetic-reinforced soil walls and slopes – Seismic Aspects. In *Geosynthetics and Their Applications*, Shukla, S.K., Ed., Thomas Telford, London, pp. 327–392.

Beckham, W.K. and Mills, W.H. (1935). Cotton-fabric reinforced roads. *Engineering News Record*, pp. 453–455.

Bell, J.R., Barrett, R.K., and Ruckmann, A.C. (1984). Geotextile earth-reinforced retaining wall tests. *Transportaion Research Record*, No. 916, pp. 59–69.

Benneton, J.P., Mahuet, J.L., and Gourc, J.P. (1993). Geomembrane waterproofing of dry tunnel under two rivers, Lyon metro tunnel, France. In *Geosynthetics Case Histories*, Raymond, G.P. and Giroud, J.P., Eds, on behalf of ISSMFE Technical Committee TC9, Geotextiles and Geosynthetics, pp. 118–119.

Bhatia, S.K., Smith, J.L., and Christopher, B.R. (1994). Interrelationship between pore openings of geotextiles and methods of evaluation. *Proceedings of the 5th International Conference on Geotextiles, Geomembranes and Related Products*. Singapore, pp. 705–710.

Binquet, J. and Lee, K.L. (1975a). Bearing capacity tests on reinforced earth slabs. *Journal of Geotechnical Engineering Division, ASCE*, **101**, 12, 1241–1255.

Binquet, J. and Lee, K.L. (1975b). Bearing capacity analysis of reinforced earth slabs. *Journal of the Geotechnical Engineering Division, ASCE*, **101**, 12, 1257–1276.

Bonaparte, R. and Berg, R. (1987). Long term allowable tension for geosynthetic reinforcement. *Proceedings of the Geosynthetics '87*. New Orleans, Louisiana.

Bonaparte, R. and Christopher, B.R. (1987). Design and construction of reinforced embankments over weak foundations. *Proceedings of the Symposium on Reinforced Layered Systems. Transportation Research Record*, No. 1153, Transportation Research Board, Washington, DC, pp. 26–39.

Bonaparte, R., Holtz, R.D., and Giroud, J.P. (1985). *Soil Reinforcement Design Using Geotextiles and Geogrids*. ASTM Special Technical Publication No. 952, Geotextile Testing and the Design Engineer.

Bonaparte, R., Ah-Line, C., Charron, R., and Tisinger, L. (1988). Survivability and durability of a non-woven geotextile. *Proceedings of ASCE Symposium on Soil Improvement*, pp. 68–91.

Bordonado, G. and Buffard, J.Y. (1993). Geosynthetics for concrete paved motorway, *Proceedings on Geosynthetics Case Histories, ISSMFE*, TC9. A26 Motoway, Northern France, France, pp. 138–139.

Bouazza, A., Zornberg, J.G., and Adam, D. (2002). Geosynthetics in waste containment facilities. *Proceedings of the 7th International Conference on Geosynthetics*, France, pp. 445–514.

Bourdeau, P.L. (1989). Modeling of membrane action in a two-layer reinforced soil system. *Computers and Geotechnics*, **7**, 1–2, 19–36.

Bourdeau, P.L. and Ashmawy, A.K. (2002). Unpaved Roads. In *Geosynthetics and Their Applications*, Shukla, S.K., Ed., Thomas Telford, London, pp. 165–183.

Bourdeau, P.L., Harr, M.E., and Holtz, R.D. (1982). Soil-fabric interaction – an analytical model. *Proceedings of the 2nd International Conference on Geotextiles*. Las Vegas, USA, pp. 387–391.

Boynton, S.S. and Daniel, D.E. (1985). Hydraulic conductivity test on compacted clay. *Journal of Geotechnical Engineering, ASCE*, **111**, 4, 465–478.

Brau, G. (1996). Damage of geosynthetics during installation – experience from real sites and research work. *Proceedings of the 1st European Conference on Geosynthetics*. Eurogeo 1, Netherlands.

Broms, B.B. (1980). Design of fabric reinforced retaining structures. *Proceedings of the Symposium on Earth Reinforcement*, Pittsburg, USA.

Broms, B.B. and Wong, I.H. (1986). Stabilization of slopes in residual soils with geofabric. *Proceedings of the 3rd International Conference on Geotextiles*. Vienna, Austria, pp. 295–300.

Broms, B.B. and Wong, K.S. (1990). Landslides. In *Foundation Engineering Handbook*, Fang, H.Y., Ed., Van Nostrand Reinhold, New York.

Broms, B.B., Chu, J., and Chora, V. (1994). Measuring the discharge capacity of band drains by a new drain tester. *Proceedings of the 5th International Conference on Geotextiles, Geomembranes and Related Products*. Singapore, pp. 803–806.

BS 6906: Part 8: 1991. *Determination of Sand-Geotextile Frictional Behaviour by Direct Shear*. British Standards Institution, London.

BS 8006: 1995. *Code of Practice for Strengthened/Reinforced Soils and Other Fills*. British Standards Institution, London.

Burwash, W.J. and Lunder, D. (1993). Geogrid reinforced cantilever wall to retain slope light rail transit project, Calgary, Canada. In *Geosynthetics Case Histories*, Raymond, G.P. and Giroud, J.P., Eds, on behalf of ISSMFE Technical Committee TC9, Geotextiles and Geosynthetics, pp. 254–255.

Bush, D.I. and Swan, D.B.G. (1988). An assessment of the resistance of TENSAR SR2 to physical damage during the construction and testing of a reinforced soil wall. In *The Application of Polymeric Reinforcement in Soil Retaining Structures*, Jarrett, P.M. and McGown, A., Eds, NATO ASI Series, Kluwer Academic Publishers, Dordrecht, The Netherlands, pp. 173–180.

Bush, D.I., Jenner, C.G., and Bassett, R.H. (1990). The design and construction of geocell foundation mattress supporting embankments over soft ground. *Geotextiles and Geomembranes*, **9**, 1, 83–98.

Cancelli, A., Monti, R., and Rimoldi, P. (1990). Comparative study of geosynthetics for erosion control. *Proceedings of the 4th International Conference on Geotextiles and Geomembranes*. The Hague, The Netherlands.

Cassidy, P.E., Mores, M., Kerwick, D.J., Koeck, D.J., Verschoor, K.L., and White, D.F. (1992). Chemical resistance of geosynthetic materials. *Geotextiles and Geomembranes*, **11**, 1, 61–98.

Cedergren, H.R. (1987). *Drainage of Highway and Airfield Pavements*. Robert E. Krieger Pub. Co., Inc., Melbourne, Florida, p. 289.

Cedergren, H.R. (1989). *Seepage, Drainage, and Flownets*. Third Edition, John Wiley & Sons, New York, p. 465.

Chaddock, B.C.J. (1988). *Deformation of Road Foundations with Geogrid Reinforcement*. Research Report 140, Department of Transportation, TRRL, Crowthorne, Berkshire, UK.

Chan, F., Barksdale, R.D., and Brown, S.F. (1989). Aggregate base reinforcement of surfaced pavements. *Geotextiles and Geomembranes*, **8**, 3, 165–189.

Chang, D.T.T., Chen, C.A., and Fu, Y.C. (1996). The creep behaviour of geotextiles under confined and unconfined conditions. *Proceedings of the International Symposium on Earth Reinforcement*. Fukuoka, Japan, pp. 19–24.

Chen, Y.H. and Cotton, G.K. (1988). *Design of Roadside Channels with Flexible Linings*. Federal Highway Administration Report, HEC-15/FHWA-1P087-7, McLean, VA, USA.

Christie, I.F. (1982). Economic and technical aspects of embankments reinforced with fabric. *Proceedings of the 2nd International Conference on Geotextiles*. Las Vegas, NV, USA, pp. 659–664.

Christopher, B.R. (1998). The First Step in Geotextile Filter Deign. *Geotechnical Fabrics Report*, IFAI, Roseville, MN, USA, March, pp. 36–38.

Christopher, B.R. and Holtz, R.D. (1985). *Geotextile Engineering Manual*. Report No. FHWA-TS-86/203, Federal Highway Administration, Washington, DC, p. 1044.

Christopher, B.R. and Leschinsky, D. (1991). Design of geosynthetically reinforced slopes. *Proceedings of the Geotechnical Engineering Congress*, Boulder, Colorado, pp. 988–1005.

Christopher, B.R., Gill, S.A., Giroud, J.P., Juran, I., Mitchell, J.K., Schlosser, F., and Dunnicliff, J.D. (1990). *Reinforced Soil Structures, Volume I: Design and Construction*. Federal Highway Administration, Washington, DC, Report FHWA-RO-89-043.

Collin, J.G. (1996). Controlling Surficial Stability Problems on Reinforced Steepened Slopes. *Geotechnical Fabrics Report*, IFAI, Roseville, MN, USA, April, pp. 26–29.

Corbet, S.P. (1992). The design and specification of geotextiles and geocomposites for filtration and drainage. *Proceedings of the Conference Geofad '92: Geotextiles in Filtration and Drainage*, Cambridge, Thomas Telford, UK, pp. 29–40.

Daniel, D.E. (1993). Landfill and Impoundments. In *Geotechnical Practice for Waste Disposal*, Daniel, D.E., Ed., Chapman & Hall, pp. 97–112.

Daniel, D.E., Koerner, R.M., Bonaparte, R., Landreth, R.E., Carson, D.A., and Scranton, H.B. (1998). Slope stability of geosynthetic clay liner test plots. *Journal of Geotechnical and Geoenvironmental Engineering, ASCE*, **124**, 7, 628–637.

Das, B.M., Omar, M.T., and Singh, G. (1996). Strip foundation on geogrid-reinforced clay. *Proceedings of the 1st European Geosynthetics Conference*. Eurogeo 1, The Netherlands, pp. 419–426.

Davies. P.L. (1993). Geomembrane waterproofing of wet hydroelectric tunnel, Drakensberg hydroelectric tunnel, South Africa. In *Geosynthetics Case Histories*, Raymond, G.P. and Giroud, J.P., Eds, on behalf of ISSMFE Technical Committee TC9, Geotextiles and Geosynthetics, pp. 114–115.

de Buhan, P., Mangiavcchi, R., Nova, R., Pellegrini, G., and Salencon, J. (1989). Yield design of reinforced earth walls by homogenization method. *Geotechnique*, **39**, 2, 189–201.

De Garidel, R. and Javor, E. (1986). Mechanical reinforcement of low-volume roads by geotextiles. *Proceedings of the 3rd International Conference on Geotextiles*. Vienna, Austria, pp. 147–152.

den Hoedt, G. (1986). Creep and relaxation of geotextiles fabrics. *Geotextiles and Geomembranes*, **4**, 2, 83–92.

Dikran, S.S. and Rimoldi, P. (1996). Hard facing for steep reinforced slopes: a case history from the UK. *Proceedings of the 1st European Geosynthetics Conference*. Eurogeo 1, The Netherlands, pp. 131–136.

Dixon, J.H. (1993). Geogrid reinforced soil repair of a slope failure in clay, North Circular Road, London, UK. In *Geosynthetics Case Histories*, Raymond, G.P. and Giroud, J.P., Eds, on behalf of ISSMFE Technical Committee TC9, *Geotextiles and Geosynthetics*, pp. 168–169.

Dunnicliff, J. (1988). *Geotechnical Instrumentation for Monitoring Field Performance*. John Wiley & Sons, New York.
Duquennoi, C. (2002). Containment Ponds, Reservoirs and Canals. In *Geosynthetics and Their Applications*, Shukla, S.K., Ed., Thomas Telford, London, pp. 299–325.
Durukan, Z. and Tezcan, S.S. (1992). Cost analysis of reinforced soil walls. *Geotextiles and Geomembranes*, **11**, 1, 29–43.
El-Fermaoui, A. and Nowatzki, E. (1982). Effect of confining pressure on performance of geotextiles in soils. *Proceedings of the 2nd International Conference on Geotextiles*. Las Vegas, NV, USA, pp. 799–804.
ENV ISO 13438. *Standards on Determination of Temperature Stability (Oxidation)*. International Organization for Standardization, Geneva, Switzerland.
Espinoza, R.D. (1994). Soil-geotextile interaction: evaluation of membrane support. *Geotextiles and Geomembranes*, **13**, 5, 281–293.
Espinoza, R.D. and Bray, J.D. (1995). An integrated approach to evaluating single-layer reinforced soils. *Geosynthetics International*, **2**, 4, 723–739.
Fannin, R.J. and Sigurdsson, O. (1996). Field observations on stabilization of unpaved roads with geosynthetics. *Journal of Geotechnical Engineering, ASCE*, **122**, 7, 544–553.
Faure, Y. and Mlynarek, J. (1998). Geotextile Filter Hydraulic Requirements. *Geotechnical Fabrics Report*, IFAI, Roseville, MN, USA, May, pp. 30–33.
Fluet, J.E. (1988). Geosynthetics for soil improvement: a general report and keynote address. *Proceedings of the Symposium on Geosynthetics for Soil Improvement*, Tennessee, USA, pp. 1–21.
Fourie, A. (1998). The Determination of Appropriate Filtration Design Parameters. *Geotechnical Fabrics Report*, IFAI, Roseville, MN, USA, April, pp. 44–47.
Fowler, J. and Koerner, R.M. (1987). Stabilization of very soft soils using geosynthetics. *Proceedings of the Geosynthetics '87*, New Orleans, Louisiana, IFAI, St. Paul, MN, pp. 289–300.
Fox, P.J., Rowland, M.G., and Scheithe, J.R. (1998). Internal shear strength of three geosynthetic clay liners. *Journal of Geotechnical and Geoenvironmental Engineering, ASCE*, **124**, 10, 933–944.
Frobel, R.K. (1996). Geosynthetic Clay Liners, Part Four: Interface and Internal Shear Strength Determination. *Geotechnical Fabrics Report*, IFAI, **14**, 8, pp. 20–23.
Fuller, C. (1992). Concrete Blocks Gain Acceptance as Erosion Control Systems. *Geotechnical Fabrics Report*, IFAI, Roseville, MN, USA, pp. 24–37.
Gautier, J., Colas, S.A., Lino, M., and Carlier, D. (2002). A record height in dam waterproofing with bituminous geomembrane: La Galaube dam on Alzeau river. *Proceedings of the 7th International Conference on Geosynthetics*, France, Calgary, Alberta, pp. 975–978.
Ghiassian, H., Hryciw, R.D., and Gray, D.H. (1996). Laboratory testing apparatus for slopes stabilized by anchored geosynthetics. *Goetechnical Testing Journal*, **19**, 1, 65–73.
Ghoshal, A. and Som, N. (1993). Geotextiles and geomembranes in India – state of usage and economic evaluation. *Geotextiles and Geomembranes*, **12**, 3, 193–213.
Giroud, J.P. (1980). Introduction to geotextiles and their applications. *Proceedings of the 1st Canadian Symposium on Geotextiles*, Calgary, Alberta, pp. 3–31.
Giroud, J.P. (1982). Filter criteria for geotextiles. *Proceedings of the 2nd International Conference on Geotextiles*. Las Vegas, NV, USA, pp. 103–108.
Giroud, J.P. (1984). *Geotextiles and Geomembranes Definitions, Properties and Design*, IFAI, St. Paul, MN, USA.
Giroud, J.P. (1992). Geosynthetics in Dams: Two Decades of Experience. *Geotechnical Fabrics Report*, IFAI, Roseville, MN, USA, July–August; September, pp. 6–9, 22–28.
Giroud, J.P. (1994). Quantification of geosynthetic behaviour. *Proceedings of the 5th International Conference on Geotextiles, Geomembranes and Related Products*. Singapore, pp. 1249–1273.
Giroud, J.P. (1996). Granular filters and geotextile filters. *Proceedings of GeoFilters '96*, Montreal, Canada, pp. 565–680.
Giroud, J.P. (2000). Lessons learned from successes and failures associated with geosynthetics. *Proceedings of the 2nd European Geosynthetics Conference*, Bologna, Italy, pp. 77–118.

Giroud, J.P. and Bonaparte, R. (1989). Leakage through liners constructed with geomembranes – Part I. Geomembrane liners. *Geotextiles and Geomembranes*, **8**, 1, 27–67.

Giroud, J.P. and Carroll, R.G. (1983). Geotextile Products. *Geotechnical Fabrics Report*, IFAI, St. Paul, USA, **1**, 1, pp. 12–15.

Giroud, J.P. and Noiray, L. (1981). Geotextile-reinforced unpaved road design. *Journal of Geotechnical Engineering, ASCE*, **107**, 9, 1233–1254.

Giroud, J.P., Ah-Line, A., and Bonaparte, R. (1984). Design of unpaved roads and trafficked areas with geogrids. *Proceedings of the Symposium on Polymer Grid Reinforcement*, London, pp. 116–127.

Gnilsen, R. and Rhodes, G.W. (1986). *Innovative Use of Geosynthetics to Construct Watertight D.C. Subway Tunnels.* Geotechnical Fabrics Report, IFAI, Roseville, MN, USA, **4**, 4.

Gourc J.P. and Risseeuw, P. (1993). Geotextile reinforced wrap around faced wall, Prapoutel, France. In *Geosynthetics Case Histories*, Raymond, G.P. and Giroud, J.P., Eds, on behalf of ISSMFE Technical Committee TC9, Geotextiles and Geosynthetics, pp. 272–273.

Gray, C.G. (1982). Abrasion resistance of geotextiles fabrics. *Proceedings of the 2nd International Conference on Geotextiles.* Las Vegas, NV, USA, pp. 817–821.

Greenwood, J.H. and Myles, B. (1986). Creep and stress relaxation of geotextiles. *Proceedings of the 3rd International Conference on Geotextiles.* Vienna, Austria, pp. 821–826.

Greenwood, J.H., Trubiroha, P., Schroder, H.F., Frank, P., and Hufenus, R. (1996). Durability standards for geosynthetics: the tests for weathering and biological resistance. *Proccedings of the 1st European Geosynthetics Conference.* Eurogeo 1, Maastricht, The Netherlands, pp. 637–641.

Guido, V.A., Biesiadecki, G.L., and Sullivan, M.J. (1985). Bearing capacity of a geotextile-reinforced foundation. *Proceedings of the 11th International Conference on Soil Mechanics and Foundation Engineering.* San Francisco, CA, pp. 1777–1780.

Guido, V.A., Dong, K.G., and Sweeny, A. (1986). Comparison of geogrid and geotextile reinforced earth slabs. *Canadian Geotechnical Journal*, **23**, 1, 435–440.

Haliburton, T.A. and Wood, P.D. (1982). Evaluation of the US Army Corps of Engineer gradient ratio test for geotextile performance. *Proceedings of the 2nd International Conference on Geotextiles.* Las Vegas, NV, USA, pp. 97–101.

Haliburton, T.A., Douglas, P.A., and Fowler, J. (1977). Feasibility of Pinto Island as a long-term dredged material disposal site. Miscellaneous Paper, D-77-3, US Army Waterways Experiment Station.

Hall, C. (1981). *Polymer Materials.* Macmillan Publisher, London.

Halse, Y.H., Wiertz, J., and Rigo, J.M. (1991). Chemical identification methods used to characterize polymeric geomembranes. Chapter 15, In *Geomembranes: Identification and Performance Testing*, Rollin, A. and Rigo, J.M., Ed., Chapman & Hall, London, pp. 316–336.

Hausmann, M.R. (1990). *Engineering Principles of Ground Modification.* McGraw-Hill, New York.

HB 154-2002. *Technical Handbook: Geosynthetics – Guidelines on Durability.* Standards Australia International Ltd, Sydney, Australia.

Heibaum, M.H. (1998). Protecting the Geotextile Filter from Installation Harm. *Geotechnical Fabrics Report*, IFAI, Roseville, MN, USA, June–July, pp. 26–29.

Hewlett, H.W.M., Boorman, L.A., and Bramley, M.E. (1987). *Design of Reinforced Grass Waterways*, Report 116, CIRIA, London.

Hirano, I., Itoh, M., Kawahara, S., Shirasawa, M., and Shimizu, H. (1990). Test on trafficability of a low embankment on soft ground reinforced with geotextiles. *Proceedings of the 4th International Conference on Geotextiles, Geomembranes and Related Products.* The Hague, The Netherlands, pp. 227–232.

Hoare, D.J. (1982). Synthetic fabrics as soil filters. *Journal of Geotechnical Engineering Division, ASCE*, **108**, GT10, 1230–1245.

Holtz, R.D., Christopher, B.R., and Berg, R.R. (1997). *Geosynthetic Engineering.* BiTech Publishers Ltd., Canada.

Hudson, R.Y. (1959). Laboratory investigation of rubble mound breakwaters. *Journal of Waterways and Harbours Division, ASCE*, September, 93–121.

Hullings, D.E. and Sansone, L.J. (1997). Design concerns and performance of geomembrane anchor trenches. *Geotextiles and Geomembranes*, **15**, 4–6, 403–417.

Hunt, J.R. (1982). The development of fin drains for structure drainage. *Proceedings of the 2nd International Conference on Geotextiles*, Las Vegas, NV, USA, pp. 25–36.

Ialynko, P. and Gonin, H. (1993). The Chanteraines artificial lake: a public park built on a landfill. *Proceedings of the Rencontres '93*. Joué-les-Tours, France, pp. 355–364.

ICOLD. (1991). *Watertight Geomembranes for Dams*. State of the art. International Commission on Large Dams Bulletin 78, Paris, p. 140.

ICOLD. (1993). *Embankment Dams Upstream Slope Protection*. International Commission on Large Dams Bulletin 91, Paris, p. 122.

IFAI. (1992). A Design Primer: Geotextiles and Related Materials. Section 13 Asphalt Overlay. Industrial Fabrics Association International, St. Paul, MN, USA.

Ingold, T.S. (1982a). An analytical study of geosynthetic reinforced embankments. *Proceedings of the 2nd International Conference on Geotextiles*. Las Vegas, NV, USA, pp. 683–688.

Ingold, T.S. (1982b). *Reinforced Earth*. Thomas Telford Ltd., London, UK.

Ingold, T.S. (1994). *The Geotextiles and Geomembranes Manual*. Elsevier Advanced Technology, Oxford, UK.

Ingold, T.S. (2002). Slopes – erosion control. In *Geosynthetics and Their Applications*, Shukla, S.K., Ed., Thomas Telford, London, pp. 223–235.

Ingold, T.S. and Miller, K.S. (1988). *Geotextiles Handbook*. Thomas Telford Ltd., London, UK.

IRC:37. (2001). *Guidelines for the Design of Flexible Pavements*. Indian Roads Congress, New Delhi, India.

IRC:58. (2002). *Guidelines for the Design of Plain Jointed Rigid Pavements for Highways*. Indian Roads Congress, New Delhi, India.

IRC:SP:59. (2002). *Guidelines for Use of Geotextiles in Road Pavements and Associated Works*. Indian Roads Congress, New Delhi, India.

IS:14986. (2001). *Guidelines for Application of Jute Geotextile for Rain Water Erosion Control in Road and Railway Embankments and Hill Slopes*. Bureau of Indian Standards, New Delhi, India.

ISO:12958. (1999). *Geotextiles and Geotextile-Related Products – Determination of Water-Flow Capacity in Their Plane*. International Organization for Standardization, Geneva, Switzerland.

ISO:13439. *Standards on Tests for Hydrolysis*. International Organization for Standardization, Geneva, Switzerland.

Ivy, N. and Narejo, D. (2003). Canal Lining with HDPE. *Geotechnical Fabrics Report*, IFAI, Roseville, MN, USA, June–July, pp. 24–26.

Jay, T. (2002). Improving the performance of geosynthetic materials. *International Railway Journal*, March, 34–36.

Jewell, R.A. (1986). Material properties for design of geotextile reinforced slopes. *Geotextiles and Geomembranes*, **2**, 2, 83–109.

Jewell, R.A. (1990). Revised design charts for steep reinforced slopes. In *Reinforced Embankments, Theory and Practice*, Shercliff, D.A., Ed., Thomas Telford, London, England, pp. 1–30.

Jewell, R.A. (1996). *Soil Reinforcement with Geotextiles*. Construction Industry Research and Information Association, London, CIRIA, Special Publication 123.

Jewell, R.A. and Woods, R.I. (1984). Simplified design charts for steep reinforced slopes. *Proceedings of the Symposium on Reinforced Soil*, University of Missisipi, USA.

Jewell, R.A., Paine, N., and Woods, R.I. (1984). Design methods for steep reinforced embankments. *Proceedings of the Symposium of Polymer grid reinforcement*, Institute of Civil Engineering, London, pp. 18–30.

Jiang, G.L. and Magnan, J.P. (1997). Stability analysis of embankments: comparison of limit analysis with method of slices. *Geotechnique*, **47**, 4, 857–872.

John, N.W.M. (1987). *Geotextiles*. Blackie, London.

Kabir, M.H. and Ahmed, K. (1994). Dynamic creep behaviour of geosynthetics. *Proceedings of the 5th International Conference on Geotextiles, Geomembranes and Related Products*. Singapore, pp. 1139–1144.

Kazarnovsky, V.D. and Brantman, B.P. (1993). Geotextile reinforcement of a temporary road, Smolenskaya region, USSR. In *Geosynthetics Case Histories*, Raymond, G.P. and Giroud, J.P., Eds, on behalf of ISSMFE Technical Committee TC9, Geotextiles and Geosynthetics, pp. 194–195.

Kenney, T.C. and Lau, D. (1985). Internal stability of granular filter. *Canadian Geotechnical Journal*, **22**, 215–225.

Kjellman, W. (1948). Accelerating consolidation of fine-grained soils by means of cardboard wicks. *Proceedings of the 2nd International Conference on Soil Mechanics and Foundation Engineering*, Rotterdam, pp. 302–305.

Koerner, G. (2001). In Situ Temperature Monitoring of Geosynthetics used in a Landfill. *Geotechnical Fabrics Report*, IFAI, Roseville, MN, USA, May, pp. 12–14.

Koerner, R.M. (1984). In-situ soil stabilization using anchored nets. *Proceedings of the Conference on Low Cost and Energy Saving Construction Methods*. Rio de Janeiro, Brazil, pp. 465–478.

Koerner, R.M. (1994). *Designing with geosynthetics*. Third edition, Prentice Hall, New Jersey, USA.

Koerner, R.M. (1996). The state-of-the-practice regarding in-situ monitoring of geosynthetics. *Proceedings of the 1st European geosynthetics Conference*. Eurogeo 1, Maastricht, The Netherlands, pp. 71–86.

Koerner, R.M. (2000). Emerging and future developments of selected geosynthetic applications. *Journal of Geotechnical and Geoenvironmental Engineering, ASCE*, **126**, 4, 293–306.

Koerner, R.M. (2005). *Designing with Geosynthetics*. Fifth edition, Prentice Hall, New Jersey, USA.

Koerner, R.M. and Daniel, D.E. (1993). Manufacturing and construction quality control and quality assurance of geosynthetics. *Proceedings of the MQC/MQA and CQC/CQA of Geosynthetics Seminar*, Philadelphia, PA, USA, pp. 1–14.

Koerner, R.M. and Daniel, D.E. (1997). *Final Covers for Solid Waste Landfills and Abandoned Dumps*. ASCE Press, Reston, VA, and Thomas Telford, London, UK.

Koerner, R.M. and Hwu, B.-L. (1991). Stability and tension considerations regarding cover soils on geomembrane lined slopes. *Geotextiles and Geomembranes*, **10**, 4, pp. 335–355.

Koerner, R.M. and Robins, J.C. (1986). In-situ stabilization of soil slopes using nailed geosynthetics. *Proceedings of the 3rd International Conference on Geotextiles*. Vienna, Austria, pp. 395–400.

Lafleur, J. and Pare, J.-J. (1991). Use of geotextiles in the James bay hydroelectric project. *Geotextiles and Geomembranes*, **10**, 1, 35–52.

Lafleur, J., Mlynarek, J., and Rollin, A.L. (1993). Filter criteria for well graded cohesionless soils. *Proceedings of the Geofilters '92*, Karlsruhe, Germany.

Lambe, T.W. (1958). *The Permeability of Compacted Fine-Grained Soils*. Special Technical Publication 163, ASTM, Philadelphia, PA, pp. 55–67.

Landreth, R.E. (1990). Chemical resistance evaluation of geosynthetics used in the waste management applications. In *Geosynthetics Testing for Waste Containment Applications*, Special Technical Publication STP 1081, ASTM, Philadelphia, pp. 3–11.

Landreth, R.E. and Carson, D.A. (1991). *Inspection Techniques for the Fabrication of Geomembrane Field Seams*. US EPA Technical Guidance Document, EPA/530/SW-91/051, Cincinnati, OH, p. 174.

Lawson, C.R. (1982). Filter criteria for geotextiles: relevance and use. *Journal of Geotechnical Engineering Division, ASCE*, **108**, GT10, 1300–1317.

Lawson, C.R. (1986). Geotextile filter criteria for tropical residual soils. *Proceedings of the 3rd International Conference on Geotextiles*. Vienna, Austria, pp. 557–562.

Lawson, C. (1998). Retention Criteria and Geotextile Filter Performance. *Geotechnical Fabrics Report*, IFAI, Roseville, MN, USA, August, pp. 26–29.

Lawson, C.R. and Kempton, G.T. (1995). *Geosynthetics and Their Use in Reinforced Soil.* Terram Ltd., Mamhilad, Pontypool, Gwent, UK.

Lee, D.H., Tien, K.G., and Juang, C.H. (1996). Full-scale field experimentation of a new technique for protecting mudstone slopes, Taiwan. *Engineering Geology,* 42, 1, 51–63.

Legge, K.R. (1986). Testing of geotextiles. *Proceedings of SAICE Filters Symposium.* Johannesburg, South Africa.

Leshchinsky, D. (1997). Software to Facilitate Design of Geosynthetic-Reinforced Steep Slopes. *Geotechnical Fabrics Report,* IFAI, Roseville, MN, USA, January–February, pp. 40–46.

Leshchinsky, D. and Volk, J.C. (1985). Stability charts for geotextile reinforced walls. *Transportation Research Record,* No. 1131, pp. 5–16.

Leshchinsky, D. and Volk, J.C. (1986). Predictive equation for the stability of geotextile reinforced earth structure. *Proceedings of the 3rd International Conference on Geotextiles.* Vienna, Austria, pp. 383–388.

Ling, H.I. and Liu, Z. (2001). Performance of geosynthetic-reinforced asphalt pavements. *Journal of Geotechnical and Geoenvironmental Engineering,* 127, 2, 177–184.

Lopes, M.L. and Ladeira, M. (1996). Role of the specimen geometry, soil height, and sleeve length on the pullout behaviour of geogrids. *Geosynthetics International,* 3, 6, 701–719.

Love, J.P., Burd, H.J., Milligan, G.W.E., and Houlsby (1987). Analytical and model studies of reinforcement of a layer of granular fill on soft clay subgrade. *Canadian Geotechnical Journal,* 24, 4, 611–622.

Luettich, S.M., Giroud, J.P., and Bachus, R.C. (1992). Geotextile filter design guide. *Geotextiles and Geomembranes,* 11, 4–6, 19–34.

McGown, A. and Andrawes, K.Z. (1977). The influence of nonwoven fabric inclusions on the stress-strain behaviour of a soil mass. *Proceedings of the International Conference on the Use of Fabrics in Geotechnics,* Paris, pp. 161–166.

McGown, A., Andrawes, K.Z. and Al-Hasani, M.M. (1978). Effect of inclusion properties on the behaviour of sand. *Geotechnique,* 28, 3, 327–346.

McGown, A., Andrawes, K.Z., and Kabir, M.H. (1982). Load-extension testing of geotextiles confined in soil. *Proceedings of the 2nd International Conference on Geotextiles.* Las Vegas, NV, USA, pp. 793–796.

McKenna, J.M. (1989). Properties of core materials, the downstream filter and design. *Clay Barriers for Embankment Dams,* Thomas Telford, London.

Madhav, M.R. and Poorooshasb, H.B. (1988). A new model for geosynthetic-reinforced soil. *Computers and Geotechnics,* 6, 4, 277–290.

Madhav, M.R. and Poorooshasb, H.B. (1989). Modified Paternak model for reinforced soil. *Mathematical and Computational Modelling, an International Journal,* 12, 1505–1509.

Marienfeld, M.L. and Baker, T. (1998). Paving fabric interlayer system as a pavement moisture barrier. *77th Annual Meeting, Transportation Research Board,* Washington, DC, National Academy of Sciences.

Marienfeld, M.L. and Smiley, D. (1994). Paving Fabrics: The Why and the How-To. *Geotechnical Fabrics Report,* IFAI, Roseville, MN, USA, June–July, pp. 24–29.

Menoff, S.D. and Eith, A.W. (1990). Quality assurance as a means to minimize long-term liability. *Geotextiles and Geomembranes,* 9, 4–6, 461–465.

Michalowski, R.L. and Zhao, A. (1993). Failure criteria for homogenized reinforced soils and application in limit analysis of slopes. *Proceedings of Geosynthetics '93.* Vancouver, Canada, pp. 443–453.

Michalowski, R.L. and Zhao, A. (1994). The effect of reinforcement length and distribution on safety of slopes. *Proceedings of the 5th International Conference on Geotextiles, Geomembranes and Related Products.* Singapore, pp. 495–498.

Michalowski, R.L. and Zhao, A. (1995). Continuum versus structural approach to stability of reinforced soil structures. *Journal of Geotechnical Engineering Division, ASCE,* 121, 2, 152–162.

Mikki, H., Hayashi, Y., Yamada, K., Takasago, and Shido, H. (1990). Plane strain tensile strength and creep of spun-bonded nonwovens. *Proceedings of the 4th International Conference on Geotextiles, Geomembranes and Related Products*. The Hague, pp. 667–672.

Milligan, G.W.E., Fanin, R.J., and Farrar, D.M. (1986). Model and full-scale tests of granular layers reinforced with a geogrid. *Proceedings of the 3rd International Conference on Geotextiles*. Vienna, Austria, pp. 61–66.

Mitchell, J.K., Hopper, D.R., and Campanella, R.G. (1965). Permeability of compacted clay. *Journal of Soil Mechanics and Foundation Engineering, ASCE*, **91**, 4, 41–65.

Mlynarek, J. and Fannin, J. (1998). Introduction to Designing With Geotextiles for Filtration. *Geotechnical Fabrics Report*, IFAI, Roseville, MN, USA, January–February, pp. 22–24.

Murray, R. (1982). Fabric reinforcement of embankments and cuttings. *Proceedings of the 2nd International Conference on Geotextiles*. Las Vegas, NV, USA, pp. 707–713.

Murray, R.T., McGown, A., Andrawes, K.Z., and Swan, D. (1986). Testing joints in geotextiles and geogrids. *Proceedings of the 3rd International Conference on Geotextiles*. Vienna, Austria, pp. 731–736.

Myles, B. and Carswell, I.G. (1986). Tensile testing of geotextiles. *Proceedings of the 3rd International Conference on Geotextiles*. Vienna, Austria, pp. 713–718.

Narejo, D., Hardin, K., and Ramsey, B. (2001). Geotextile Specifications: Those Vexing Qualifiers. *Geotechnical Fabrics Report*, IFAI, Roseville, MN, USA, pp. 24–27.

NCMA (1997). *Design Manual for Segmental Retaining Walls*. National Concrete Masonry Association, Herndon, VA.

Nishida, K. and Nishigata, T. (1994). The evaluation of separation function for geotextiles. *Proceedings of the 5th International Conference on Geotextiles, Geomembranes and Related Products*. Singapore, pp. 139–142.

Nosko, V. and Touze-Foltz, N. (2000). Geomembrane liner failure: modeling of its influence on containment transfer. *Proceedings of the 2nd European Geosynthetics Conference*, Bologna, Italy, pp. 557–560.

Ochiai, H., Tsukamoto, Y., Hayashi, S., Otani, J., and Ju, J.W. (1994). Supporting capability of geogrid-mattress foundation. *Proceedings of the 5th International Conference on geotextiles, Geomembranes and Related Products*. Singapore, pp. 321–326.

Palmeira, E.M. (2002). Embankments. In *Geosynthetics and Their Applications*, Shukla, S.K., Ed., Thomas Telford, London, pp. 95–121.

Palmeira, E.M. and Fannin, R.J. (2002). Soil-geotextile compatibility in filtration. *Proceedings of the 7th International Conference on Geosynthetics*, France, pp. 853–974.

Paul, J. (1984). Economics and construction of blast embankments using Tensar geogrids. *Proceedings of the Conference on Polymer Grid Reinforcement*, London, UK. pp. 191–197.

Paulson, J.N. (1987). Geosynthetic material and physical properties relevant to soil reinforcement applications. *Geotextiles and Geomembranes*, **6**, 1–3, 211–223.

Paulson, J.N. (1990). Summary and evaluation of construction related damage to geotextiles in reinforcing applications. *Proceedings of the 4th International Conference on Geotextiles, Geomembranes and Related Products*, pp. 615–619.

Perkins, S.W. (2000). Constitutive modeling of geosynthetics. *Geotextiles and Geomembranes*, **18**, 5, 273–292.

Perkins, S.W., Berg, R.R., and Christopher, B.R. (2002). Paved roads. In *Geosynthetics and Their Applications*, Shukla, S.K., Ed., Thomas Telford, London, pp. 185–201.

Pilarzyk, K.W. (1984a). Discussion on revetment design. *Proceedings of the International Conference on Flexible Armoured Revetments Incorporating Geotextiles*, pp. 209–215.

Pilarzyk, K.W. (1984b). Prototype tests of slope protection systems. Discussion on revetment design. *Proceedings of the International Conference on Flexible Armoured Revetments Incorporating Geotextiles*, pp. 239–254.

Pilarczyk, K.W. (2000). *Geosynthetics and Geosystems in Hydraulic and Coastal Engineering*. A.A. Balkema, Rotterdam, The Netherlands.

Pinto, M.I.M. (2004). Reply to the discussion of Applications of geosynthetics for soil reinforcement by M.I.M. Pinto, *Ground Improvement*, 7, 2, 2003, pp. 61–72 by Shukla, S.K. *Ground Improvement*, UK, **8**, 4, pp. 181–182.

Porbaha, A. and Kobayashi, M. (1997). Finite element analysis of centrifuge model tests. *Proceedings of the 6th International Symposium on Numerical Models in Geomechanics (NUMOG VI)*. Montreal, Quebec, Canada, pp. 257–262.

Porbaha, A. and Lesniewska, D. (1999). Stability analysis of reinforced soil structures using rigid plastic approach. *Proceedings of the 11th Pan American Conference on Soil Mechanics and Geotechnical Engineering*. Iguassu Falls, Brazil.

Porbaha, A., Zhao, A., Kobayashi, M., and Kishida, T. (2000). Upper bound estimate of scaled reinforced soil retaing walls. *Geotextiles and Geomembranes*, **18**, 6, 403–413.

Posch, H. and Werner, G. (1993). Geosynthetics used in dry tunnel lining systems, Arlberg Tunnel, Austria. In *Geosynthetics Case Histories*, Raymond, G.P. and Giroud, J.P., Eds., on behalf of ISSMFE Technical Committee TC9, *Geotextiles and Geosynthetics*, pp. 112–113.

Potié, G. (1999). Bitumen geomembranes in canals. *Proceedings of the Rencontres '99*. Bordeaux, France, pp. 201–208.

Qian, X., Koerner, R.M., and Gray, D.H. (2002). *Geotechnical Aspects of Landfill Design and Construction*. Prentice Hall, New Jersey.

Rainey, T. and Barksdale, B. (1993). Construction induced reduction in tensile strength of polymer geogrids. *Proceedings of Geosynthetics '93*, IFAI, MN, USA, pp. 729–742.

Rankilor, P.R. (1981). *Membranes in Ground Engineering*. John Wiley & Sons, Chichester, England.

Rankilor, P.R. and Heiremans, F. (1996). Properties of sewn and adhesive bonded joints between geosynthetic sheets. *Proccedings of the 1st European Geosynthetics Conference*. Eurogeo 1, Maastricht, The Netherlands, pp. 261–269.

Raymond, G.P. (1982). Geotextiles for railroad bed rehabilitation. *Proceedings of the 2nd International Conference on Geotextiles*, Las Vegas, NV, USA, pp. 479–484.

Raymond, G.P. (1983a). Geotextile for railroad branch line upgrading. *Proceedings on Geosynthetics Case Histories, ISSMFE*, TC9. pp. 122–123.

Raymond, G.P. (1983b). Geotextiles for railway switch and grade crossing rehabilitation maritime provinces, Canada. *Proceedings on Geosynthetics Case Histories, ISSMFE*, TC9, pp. 124–125.

Raymond, G.P. (1986a). Performance assessment of a railway turnout geotextile. *Canadian Geotechnical Journal*, **23**, 4, 472–480.

Raymond, G.P. (1986b). Installation factors affecting performance of railroad geotextiles. *Transportation Research Record*, 1071, TRB, National Research Council of USA, pp. 64–71.

Raymond, G.P. (1999). Railway rehabilitation geotextiles. *Geotextiles and Geomembranes*, **17**, 4, 213–230.

Raymond, G.P. and Bathurst, R.J. (1990). Tests results on exhumed railway track geotextiles. *Proceedings of 4th International Conference on Geotextiles, Geomembranes and Related Products*. The Hague, The Netherlands, pp. 197–202.

Repetto, P.C. (1995). Geo-environment. In *The Civil Engineering Handbook*, Chen, W.F., Ed., CRC Press, Boca Raton, FL, pp. 883–902.

Richardson, G.N. (1997a). Swamp Roads and Ramblings. *Geotechnical Fabrics Report*, IFAI, Roseville, MN, USA, May, pp. 17–28.

Richardson, G.N. (1997b). GCL Internal Shear Strength Requirements. *Geotechnical Fabrics Report*, IFAI, Roseville, MN, USA, **15**, 2, pp. 20–25.

Richardson, G.N. (2002). Surface Impoundment Design Goals. *Geotechnical Fabrics Report*, IFAI, Roseville, MN, USA, pp. 15–18.

Richardson, G.N. and Christopher, B. (1997). Geotextiles in Drainage Systems. *Geotechnical Fabrics Report*, IFAI, Roseville, MN, USA, April, pp. 17–28.

Rigo, J.M. and Cuzzuffi, D.A. (1991). Test standards and their classification. *In Geomembranes: Identification and Performance Testing*, Rollin, A. and Rigo, J.M., Ed., Chapman and Hall, London, pp. 22–58.

Rimoldi, P. and Jaecklin, F. (1996). Green faced reinforced soil walls and steep slopes: the state-of-the-art in Europe. *Proceedings of the 1st European Geosynthetics Conference*. Eurogeo 1, The Netherlands, pp. 361–380.

Risseeuw, P. and Voskamp, W. (1993). Geotextile reinforced embankment over tidal mud, Deep Bay, New Territories, Hong Kong. In *Geosynthetics Case Histories*, Raymond, G.P. and Giroud, J.P., Eds, on behalf of ISSMFE Technical Committee TC9, *Geotextiles and Geosynthetics*, pp. 230–231.

Rowe, R.K. and Soderman, K.L. (1985). An approximate method for estimating the stability of geotextile reinforced embankments. *Canadian Geotechnical Journal*, 22, 3, 392–398.

Rustom, R.N. and Weggel, J.R. (1993). A Laboratory Investigation of the Role of Geosynthetics in Interrill Soil Erosion and Sediment Control. *Geotechnical Fabrics Report*, IFAI, Roseville, MN, USA, pp. 16–33.

Sandri, D., Martin, J.S., Vann, C.W., Ferrer, M., and Zeppenfeldt, I. (1993). Installation damage testing of four polyester geogrids in three soil types. *Proceedings of Geosynthetics '93*, IFAI, MN, USA, pp. 743–755.

Sawicki, A. (2000). *Mechanics of Reinforced Soil*. A.A. Balkema, Rotterdam.

Sawicki, A. and Lesniewska, D. (1989). Limit analysis of cohesive slopes reinforced with geotextiles. *Computers and Geotechnics*, 7, 1–2, 53–66.

Schlosser, F. and Vidal, H. (1969). Reinforced earth. *Bulletin de Liaison des Laboratoires des Ponts et Chaussees*, No. 41, France.

Schmertmann, G.R., Chourery-Curtis, V.E., Johnson, R.D., and Bonaparte, R. (1987). Design charts for geogrid reinforced soil slopes. *Proceedings of Geosynthetics '87*. New Orleans, pp. 108–120.

Sellmeijer, J.B. (1990). Design of geotextile reinforced unpaved roads and parking areas. *Proceedings of the 4th International Conference on Geotextiles, Geomembranes and Related Products*. The Hague, The Netherlands, pp. 177–182.

Sellmeijer, J.B., Kenter, C.J., and Van den Berg, C. (1982). Calculation method for fabric reinforced road. *Proceedings of the 2nd International Conference on Geotextiles*. Las Vegas, NV, USA, pp. 393–398.

Selvadurai, A.P.S. (1979). *Elastic Analysis of Soil-Foundation Interaction*. Elsevier, Amsterdam.

Shamsher, F.H. (1992). *Ground Improvement with Oriented Geotextiles and Randomly Distributed Geogrid Micro-Mesh*. PhD Thesis submitted to Indian Institute of Technology Delhi, New Delhi, India.

Sherard, J.L., Decker, R.S., and Ryker, N.L. (1972). Piping in earth dams of dispersive clay. *Proceedings of the ASCE Specialty Conference on Performance of Earth and Earth-Supported Structures. ASCE*, New York, Pt 1, 589–626.

Shukla, S.K. (1997). A study on causes of landslides in Arunachal Pradesh. *Proceedings of Indian Geotechnical Conference*, Vadodara, India, December. Vol. 1, pp. 613–616.

Shukla, S.K. (2002a). Fundamentals of geosynthetics. In *Geosynthetics and Their Applications*, Shukla, S.K., Ed., Thomas Telford, London, pp. 1–54.

Shukla, S.K. (2002b). Shallow foundations. In *Geosynthetics and Their Applications*, Shukla, S.K., Ed., Thomas Telford, London, pp. 123–163.

Shukla, S.K. (Ed.) (2002c). *Geosynthetics and Their Applications*. Thomas Telford, London, p. 430.

Shukla, S.K. (2002d). Slopes – stabilization. In *Geosynthetics and Their Applications*, Shukla, S.K., Ed., Thomas Telford, London, pp. 237–257.

Shukla, S.K. (2002e). Geosynthetic applications – general aspects and selected case studies. In *Geosynthetics and Their Applications*, Shukla, S.K., Ed., Thomas Telford, London, pp. 393–419.

Shukla, S.K. (2003a). Geosynthetics in civil engineering constructions. *Employment News*, New Delhi, 1–7 March, pp. 1–3.

Shukla, S.K. (2003b). How to select geosynthetics for field applications. *Geosynthetic Pulse*, 1, 5, p. 2.

Shukla, S.K. (2004). Discussion of Applications of geosynthetics for soil reinforcement by M.I.M. Pinto, *Ground Improvement*, **7**, 2, 2003, 61–72, *Ground Improvement*, **8**, 4, 179–182.

Shukla, S.K. (2005). Geosynthetic at subgrade level in paved roadways – a summary of functions, design practice and installation, *Civil Engineering and Construction Review*, New Delhi, India (Communicated).

Shukla, S.K. and Baishya, S. (1998). *Documentation and Identification of Geotechnical Parameters Pertaining to Active Landslides on National Highway 52A between Banderdewa and Itanagar.* Report of the project sponsored by the Faculty of Natural Disaster Management, North Eastern Regional Institute of Science and Technology, Itanagar, Arunachal Pradesh, India.

Shukla, S.K. and Chandra, S. (1994a). A generalized mechanical model for geosynthetic-reinforced foundation soil. *Geotextiles and Geomembranes*, **13**, 12, 813–825.

Shukla, S.K. and Chandra, S. (1994b). The effect of prestressing on the settlement characteristics of geosynthetic-reinforced soil. *Geotextiles and Geomembranes*, **13**, 8, 531–543.

Shukla, S.K. and Chandra, S. (1994c). A study of settlement response of a geosynthetic-reinforced compressible granular fill – soft soil system. *Geotextiles and Geomembranes*, **13**, 9, 627–639.

Shukla, S.K. and Chandra, S. (1995). Modelling of geosynthetic-reinforced engineered granular fill on soft soil. *Geosynthetics International*, **2**, 3, 603–618.

Shukla, S.K. and Chandra, S. (1996a). Settlement of embankment on reinforced granular fill – soft soil system. *Proceedings of the International Symposium on Earth Reinforcement*, Fukuoka, Japan, November, pp. 671–674.

Shukla, S.K. and Chandra, S. (1996b). A study on a new mechanical model for foundations and its elastic settlement response. *International Journal for Numerical and Analytical Methods in Geomechanics*, **20**, 8, pp. 595–604.

Shukla, S.K. and Chandra, S. (1997). Effect of densification on the settlement behaviour of reinforced granular fill – soft soil system. *Proceedings of the Indian Geotechnical Conference*, Vadodara, India, December, Vol. 1, pp. 337–340.

Shukla, S.K. and Chandra, S. (1998). Time-dependent analysis of axi-symmetrically loaded reinforced granular fill on soft subgrade. *Indian Geotechnical Journal*, **28**, 2, 195–213.

Shukla, S.K. and Yin, J.-H. (2003). Time-dependent settlement analysis of a geosynthetic-reinforced soil. *Geosynthetics International*, **10**, 2, 70–76.

Shukla, S.K. and Yin, J.-H. (2004). Functions and installation of paving geosynthetics. *Proceedings of the GeoAsia 2004*, Seoul, Korea.

Shukla, S.K., Chauhan, H.K., and Sharma, A.K. (2004). Engineering aspects of geotextile-reinforced roadway of the National Highway, NH-2, Varanasi zone. *Civil Engineering and Construction Review*, New Delhi, India, December, pp. 56–61.

Simac, M.R. (1992). Reinforced Slopes: A Proven Geotechnical Innovation. *Geotechnical Fabrics Report*, IFAI, Roseville, MN, USA, pp. 13–25.

Singh, D.N. and Shukla, S.K. (2002). Earth dams. In *Geosynthetics and Their Applications*, Shukla, S.K., Ed., Thomas Telford, London, pp. 281–298.

Sivaramakrishnan, R. (1994). Jute geotextiles as revetment filter for river bank protection. *Proceedings of the 5th International Conference on Geotextiles, Geomembranes and Related Products*. Singapore, pp. 899–902.

Snow, M.S., Kavazanjian, E., Jr, and Sanglerat, T.R. (1994). Geosynthetic composite liner system for steep canyon landfill side slopes. *Proceedings of the 5th International Conference on Geotextiles, Geomembranes and Related Products*. Singapore, 1994, 969–972.

Sprague, C.J. and Carver, C.A. (2000). Asphalt Overlay Reinforcement. *Geotechnical Fabrics Report*, IFAI, Roseville, MN, USA, March, pp. 30–33.

Staff, C.E. (1984). The foundation and growth of the geomembrane industry in the United States. *Proceedings of the International Conference on Geomembranes*, Denver, CO, pp. 5–8.

Steward, J., Williamson, R., and Nohney, J. (1977). *Guidelines for Use of Fabrics in Construction and Maintenance of Low-Volume Roads*. Report No. FHWA-TS-78-205.

Swihart, J. and Haynes, J. (2002). *Canal-Lining Demonstration Project Year 10 Final Report*. United States Department of Interior, Bureau of Reclamation.

Tan, S.A. (2002). Railway tracks. In *Geosynthetics and Their Applications*, Shukla, S.K., Ed., Thomas Telford, London, pp. 203–222.

Tatsuoka, F., Tateyama, M., Uchimura, T., and Koseki, J. (1997). Geosynthetic-reinforced soil retaining wall as important permanent structures. *Geosynthetics International*, **4**, 2, 81–136.

Tchobanoglous, T., Theisen, H., and Vigil, S. (1993). *Integrated Solid Waste Management, Engineering Principles and Management Issues*. McGraw-Hill, Inc., New York.

Tensar International (2003). Asphalt reinforcement – A587, Fleetwood Road, Cleveleys, Blackpool. Tensar case study, Ref 006.

Terzaghi, K. and Peck, R.B. (1948). *Soil Mechanics in Engineering Practice*. John Wiley & Sons, New York.

Theisen, M.S. (1992). Geosynthetics in Erosion and Sediment Control. *Geotechnical Fabrics Report*, IFAI, Roseville, MN, USA, pp. 26–35.

Theisen, M.S. and Richardson, G.N. (1998). Geosynthetic Erosion Control for Landfill Final Covers. *Geotechnical Fabrics Report*, IFAI, Roseville, MN, USA, pp. 22–27.

Thiel, R.S. and Stewart, M.G. (1993). Field construction quality assurance of geotextile installations. *Proceedings of the MQC/MQA and CQC/CQA of Geosynthetics Seminar*, Philadelphia, PA, USA, pp. 126–137.

Thomas, R.W. and Verschoor, K.L. (1988). Thermal Analysis of Geosynthetics. *Geotechnical Fabrics Report*, IFAI, Roseville, MN, USA, **6**, 3, pp. 24–30.

Tutuarima, W.H. and Van Wijk, W. (1984). Profix mattresses – an alternative erosion control system. *Proceedings of the International Conference on Flexible Armoured Revetments Incorporating Geotextiles*, pp. 335–348.

US Department of the Interior, Bureau of Reclamation. (1992). Design Standards # 13: *Embankment Dams*. Denver.

USEPA. (1988). *Final Cover on Hazardous Waste Landfills and Surface Impoundments*. US EPA Guide to Technical Resources for the Design of Land Disposal Facilities, EPA/530-SW-88-047.

USEPA. (1993). Report of Workshop on *Geosynthetic Clay Liners*. EPA/600/R-93/171, August 1993. Office of Research and Development, US Environmental Protection Agency, Washington, DC.

Van Dine, D., Raymond, G., and Williams, S.E. (1982). An evaluation of abrasion tests for geotextiles. *Proceedings of the 2nd International Conference on Geotextiles*. Las Vegas, NV, USA, pp. 811–816.

Van Santvoort, G. (Ed.) (1994). *Geotextiles and Geomembranes in Civil Engineering*. A.A. Balkema, Rotterdam, The Netherlands.

Van Santvoort, G. (Ed.) (1995). *Geosynthetics in Civil Engineering*. A.A. Balkema, Rotterdam, The Netherlands.

Vidal, H. (1969). The principle of reinforced earth. *Highway Research Record*, No. 282, pp. 1–16.

Voskamp, W., Wichern, H.A.M., and Wijk, W. van (1990). Installation problems with geotextiles, an overview of producer's experience with designers and contractors. *Proceedings of the 4th International Conference on Geotextiles, Geomembranes and Related Products*, The Hague, The Netherlands, pp. 627–630.

Wagner, H. and Hinkel, W. (1987). The New Austrian Tunneling Method. *Proceedings of the Tunnel/Underground Seminar, ASCE*, New York.

Walls, J.C. and Newby, J.E. (1993). Geosynthetics for railroad track rehabilitation, *Proceedings on Geosynthetics Case Histories, ISSMFE*, TC9. Alabama, USA, pp. 126–127.

Wang, D.W. (1994). Filter criteria of woven geotextiles for protective works. *Proceedings of the 5th International Conference on Geotextiles, Geomembranes and Related Products*. Singapore, pp. 763–766.

Wayne, M. and Han, J. (1998). On-Site Soil Usage with Geogrid-Reinforced Segmental Retaining Wall (SRWs). *Geotechnical Fabrics Report*, IFAI, Roseville, MN, USA, pp. 20–25.

Webster, S.L. and Alford, S.J. (1978). *Investigation of Construction Concepts for Pavement Across Soft Ground*. Technical Report S-78-6, US Army Corps of Engineers, Waterways Experiment Station, Vicksburg, MS, USA.

Werner, G., Puhringer, G., and Frobel, R.K. (1990). Multiaxial stress rupture and puncture testing of geotextiles. *Proceedings of the 4th International Conference on Geotextiles, Geomembranes, and Related Products*. The Hague, The Netherlands, Vol. 2, pp. 765–770.

Wewerka, M. (1982). Practical experience in the use of geotextiles. *Proceedings of the Symposium on Recent Developments in Ground Engineering Techniques*. Bangkok, Thailand, pp. 167–175.

Wilson, C.B. (1992). Repair to Reeves lake dam, Cobb County, Georgia. *Proceedings of the 1992 Annual Conference of the Association of State Dam Safety Officials*, Baltimore, Maryland, pp. 77–81.

Wooten, R.L., Powledge, G.R., and Whiteside, S.L. (1990). CCM overtopping protection on three Park way dams. *Proceedings of Hydraulic Engineering National Conference, ASCE*. San Diego, CA, pp. 1152–1157.

Wright, S.G. and Duncan, J.M. (1991). *Limit Equilibrium Stability Analyses for Reinforced Slopes*. Transportation Research Report, No. 1330, 40–46.

Wrigley, N.E. (1987). Durability and long-term performance of 'Tensar' polymer grids for soil reinforcement. *Materials Science and Technology*, **3**, 4–6, 161–170.

Wu, K.J. and Austin, D.N. (1992). Three dimensional polyethylene geocells for erosion control and channel linings. *Geotextiles and Geomembranes*, 4–6, 611–620.

Yin, J.H. (1997a). Modelling geosynthetic-reinforced granular fills over soft soil. *Geosynthetics International*, **4**, 2, 165–185.

Yin, J.H. (1997b). A nonlinear model of geosynthetic-reinforced granular fill over soft soil. *Geosynthetics International*, **4**, 5, 523–537.

Zanzinger, H. and Gartung, E. (2002a). Landfills. In *Geosynthetics and Their Applications*, Shukla, S.K., Ed., Thomas Telford, London, pp. 259–279.

Zanzinger, H. and Gartung, E. (2002b). Performance of a geogrid reinforced railway embankment on piles. *Proceedings of the 7th International Conference on Geosynthetics*, France, pp. 381–386.

Zhao, A. (1996). Limit analysis of geosynthetic-reinforced slopes. *Geosynthetics International*, **3**, 6, 721–740.

Index

AASHTO M288-00 specifications 158–163
Abrasion 83
Abrasion resistance 80, 214
Absorption 38, 50
Advection 242
Aging 43, 78, 91, 153, 250
AGS 301; see also Methods of slope stabilization
Allowable functional properties 40
Allowable loads 81; factors of safety 81; limiting creep strains 81
Allowable properties 92
Anchorage 278
Anchored geosynthetic system (AGS) method 301
Anchored spider netting 301, 302, 357
Anchor trench 240, 250, 251, 278, 308
Anti-clogging criterion 215
AOS see also Apparent Opening Size
Aperture 20
Apparent Opening Size 72, 212, 264, 265
Application areas 105
Application guidelines 263; general guidelines 263; specific guidelines 280
Aspect ratio 52
Asphalt additives 122
Asphalt concrete (AC) 123

Backfill 6, 37, 105
Band drain (Fin drain/strip drain) 22, 42, 131
Barrier 5
Basal geosynthetic layer 112, 182, 188
BCR see also Bearing capacity ratio
Bearing (passive resistance) 32
Bearing capacity ratio 190
Bentonite 24; blankets 24; mats 24
Biaxial geogrid 4
Biaxial tensile test 54, 55, 56
Biological attack 43
Biological degradation 91, 95

Bitumen impregnated geotextile 19
Bitumen retention 96, 125
Bituminous/asphalt overlays 121
Blinding 95, 134
Blocking 95, 134
Bodkin joint 271
Bonded geogrid 4
Bonding 270
Breaking tensile strength 58
Bubble point method 71
Buffer zone 124
Bursting strength 61

Calendering 21, 22
California Bearing Ratio 35, 63
Canal 152, 249
Carbon black 14, 15, 90
Case studies 339; canal 361, 364; dam 364, 366; embankment on soft foundation soil 341, 344; foundation 345; landfills 360; paved road 347, 348, 350; railway track 352; reservoir 361; retaining wall/steep sided slope 339, 340, 341; slope erosion control 354, 355, 356; slope stabilization 356, 357, 358, 359; tunnel 367; unpaved road 346
CBR see also California Bearing Ratio
CBR plunger test 63
CCL see also Compacted clay liner
Certification 279
Chemical degradation 90, 215
Clay blankets 24
Clogging 70, 79, 84
CMD see also Cross machine direction
Code of practice 378
Codole Rockfill dam 155
Compacted clay liner 240
Composite liner 151, 152, 243
Compressibility 50
Concrete bag 296
Conditioning 47
Connection 175, 176, 180, 192

Construction quality assurance 265, 316, 335
Construction quality control 316
Construction-related failures 263; high installation stresses 263; lack of proper overlap 263; loss of strength due to UV exposure 263
Construction survivability 80, 88
Containment 38
Containment facilities 148, 239, 305; canals 152, 153; dams 153, 251; landfills 148, 239; ponds 152, 249; reservoirs 152, 249
Cost analyses 329
Cost analysis experiences 330, 332
Cost-benefit ratio 43, 172, 329
Cost-performance efficiency 329
Coupon 266
CPE 13, 14
CQA *see also* Construction quality assurance
CQA monitor 316, 317
CQC *see also* Construction quality control
Creep 81; creep rupture 81; static fatigue 81; stress rupture 81
Creep behaviour 80
Creep rate 82
Creep rupture 81
Creep under load 60
Cross machine direction 17
Cross-plane permeability 75, 77
CSPE 2, 13
Cushioning 38

Damage assessment and correction 277
Dams 153, 251
Darcy's coefficient 72, 73
Degradation properties 80; abrasion resistance 80; construction survivability 80; creep behaviour 80; durability 80; longevity 80; long-term flow capability 80
Delaying the appearance of reflective cracks 123
Design action effect 175
Design approach 172, 174, 175; limit state 174; working stress 174
Design-by-cost-and-availability 172
Design-by-experience 171
Design-by-function 172
Design-by-specification 172
Design drainage capacity 174
Design flow 174, 293
Design functional properties 40
Design life 40, 70, 87
Design properties 277; chemical resistance 277; tensile modulus 277; tensile strength 277
Design resistance effect 175
Design strength 174

Determining pore size distribution 70; bubble point 70; dry sieving 70; hydrodynamic sieving 70; image analysis 70; mercury intrusion 70; wet sieving 70
Diffusion 79, 80
Direct shear test 65
Double track seams 272
Drain 129, 215, 291
Drainage 5, 22, 29, 42
Drainage composites 7
Dry sieving 70
Dry sieving test method 71
Dual hot wedge seams 272
Durability 43, 80, 87
Dynamic creep 82
Dynamic perforation strength 63
Dynamic puncture strength 61

Earth dam 153, 251
Economic evaluation 329
Electrical leak detection system 321
Elongation 214
Embankment 112, 182, 282; drainage 113; reinforcement 113; separation/filtration 113
Embankment settlement 186
Embankments, potential failure modes 182; embankment settlement 186; lateral spreading 184; overall bearing failure 187; overall slope stability failure 183; pullout failure 187
Endurance and degradation properties 80; abrasion 83; creep 80; durability 87; long-term flow capability 83; stress relaxation 83
Environmental stress cracking 87, 92
EOS *see also* Equivalent opening size
EPS 2, 14
Equivalent opening size 71, 72
Erosion control 135, 226, 294
ESC *see also* Environmental stress cracking
External stability analysis 175
Extruded geogrid 4
Extrusion 21

Fabric 2, 17
Facing design 180
Facing element 106, 108
Factors of safety 40, 81
Fatigue strength 65; mean force 65; number of cycles of force applied 65; range of force 65
Fibre 1, 17; filament 17; slit films 17; staple fibre 17; strand 17
Fibreglass 42

Fibre polymer 213
Fibre size 213
Fibre strength 213
Fibrillated yarn 17
Ficks's first law of diffusion 80
Field performance monitoring 315, 320
Filament 17
Fill material 139; asphalt 139; gravel 139; mortar 139; sand 139
Filter 129, 215, 291; geotextile filter 24, 85; granular filter 129, 131
Filter criteria 72, 156, 215, 216; anti-clogging criterion 215; permeability or permittivity criterion 215; retention or soil tightness or piping resistance criterion 215; survivability and durability criterion 215
Filter-drainage-protection 24
Filtering efficiency 131
Filtration 29, 36
Filtration function, mechanism of 36, 37
Filtration opening size 72
Final cover (top cap) system 151
Fin drain *see also* Band drain (Fin drain/strip drain)
Flexible membrane liner 79
Flexural rigidity 50
Fluid barrier 37
FML *see also* Flexible membrane liner
FOS *see also* Filtration opening size
Friction properties 60
Front blocks method 304
Functional properties 40; allowable functional properties 40; design functional properties 40

Gas collection and control system 151
GCL *see also* Geosynthetic clay liner
GCLs 24, 244; bentonite blankets 24; bentonite mats 24; clay blankets 24
General guidelines 263
Geobag 139
Geocell 5
Geocomposite drain 83, 131
Geocomposites 5, 93; geosynthetic clay liners 93, 244; linked structures 93
Geocontainer 139
Geofabric-wrapped drain 297, 298
Geofabric-wrapped drain method 297
Geofoam 5
Geogrid 2, 4, 93; biaxial geogrid 4; bonded geogrid 4; extruded geogrid 4; uniaxial geogrid 4; woven geogrid 4
Geogrid reinforced modular concrete block wall systems 304
Geojute 16, 138
Geomat 6

Geomembrane 5; natural 93; reinforced 93; reinforced geomembrane 21; textured geomembrane 22; unreinforced geomembrane 21
Geomembrane–clay composites 42
Geomembranes manufacturing 21; calendering 21; extrusion 21; spread coating 21
Geomesh 6
Geometrically open filter 228
Geometry 52
Geonaturals 1
Geonet 5
Geopipe 6
Geospacer 6
Geostrip 6
Geosynthetic clay liner 7, 22, 244
Geosynthetic engineering 25
Geosynthetic failure mechanisms, functions of 173; drainage 173; filtration 173; fluid barrier 173; protection 173; reinforcement 173; separation/filtration 173
Geosynthetic placement 289
Geosynthetic-reinforced granular fill–soft soil system 34
Geosynthetic-reinforced retaining wall 105; backfill 105; facing element 105; reinforcement layers 105
Geosynthetic-reinforced soil slopes, analysis of 234; finite element methods 236; limit analysis method 236; limit equilibrium method 234; slip line method 236
Geosynthetic-related structures, design methods 171; design-by-cost-and-availability 172; design-by-experience 171; design-by-function 172; design-by-specification 172
Geosynthetics 1, 2, 5; application areas 1, 2; application guidelines 263; basic characteristics 6, 7; damage assessment and correction 277; definition 1; description 96; functions 29; growth 7; history 1, 2; identification and inspection 265; installation 268; protection before installation 267; protection during construction and service life 275; sampling and test methods 265; selection 29, 39; types 1–6
Geosynthetics, basic functions of 29; drainage 29, 36; filtration 29, 36; fluid barrier 29, 37; protection 29, 37; reinforcement 29; separation 29, 34
Geosynthetics, effect of 43, 80; aging 43, 91; biological attack 43, 91; chemical 43, 90; creep 43, 80; hydrolysis 43, 90; mechanical damage 43, 83; ultraviolet light exposure 43, 89

Geosynthetics, index properties of 47; specific gravity 47; stiffness 47; thickness 47; unit mass 47
Geosynthetics International 2
Geosynthetics, performance characteristics 38; absorption 38; containment 38; cushioning 38; insulation 38; screening 38; surface stabilization 39; vegetative reinforcement 39
Geosynthetics, selection of 39
Geosynthetics, typical range of specific properties 93
Geotextile 1; bitumen-impregnated 19, 119; knitted geotextile 2; nonwoven geotextile 2; stitched geotextile 2; woven geotextile 1
Geotextile filter 134; blinding 134; blocking 134; clogging 134
Geotextile-related products 5, 8; geocell 5; geofoam 5; geomat 6; geomesh 6; geopipe 6; geospacer 6; geostrip 6
Geotextiles and Geomembranes 2
Geotube 139
Gluing 270
GR *see also* Gradient ratio
Grab tensile test 54
Gradient ratio 83, 84
Gradient ratio test 83
Granular filter 129, 131
Gripping method 52
GRMCBWS *see also* Geogrid reinforced modular concrete block wall systems
Groundwater flow 129
GTP *see also* Geotextile-related products
GWD *see also* Methods of slope stabilization

HCR *see also* Hydraulic conductivity ratio
HDPE 2, 5
High installation stresses 263
Hydraulic conductivity 72, 215
Hydraulic conductivity ratio 84, 85
Hydraulic gradient 74
Hydraulic properties 69; permeability 73; POA (Percent open area) 70; volume flow rate 72
Hydraulic properties of geosynthetics 69; permittivity 69, 74; porosity 69; transmissivity 69, 75
Hydraulic testing 69
Hydrodynamic filtration 71
Hydrodynamic sieving 70
Hydrodynamic test method 71
Hydrolysis 43, 90

Identification and inspection 265
Image analysis 70, 71
Impact strength (dynamic puncture strength) 61
Improved tensile strength 22

Index test (identification test) 40, 60
Index tests, uses of 60; creep under load 60; friction properties 60; wide-width tensile strength 60
Initial preload 52
In-plane coefficient of permeability 213
In-plane permeability 76, 77
Installation damage 50, 88, 108
Insulation 38
Interface friction 65
Interfacial shear stress membrane support 32
Interlocking characteristics 65
Interlocking effect 32
Internal stability analysis 175

Joint efficiency 273
Joint strength 273
Joints 269; bonding 270; gluing 270; overlapping 270; sewing 270; stapling 270; thermal 270
Jute geotextile 234
Jute netting 297

Knitted geotextile 2

Lack of proper overlap 263
Landfill cover system 152
Landfill Leachate collection and removal system 151
Landfill lining system 151
Landfills 148, 239
Lapped seam 295
Lateral spreading 184
LDPE 2, 13
Leachate 41, 85
Lengthening the useful life of the overlay 123
Life cycle cost of a landfill 334; closure 334; construction 334; long-term care 334; operation 334
Limiting creep strains 81
Limit state 174
Limit state design approach 174
Liner 5, 7
LLDPE 13, 14
Load–strain curves 58, 59
Longevity 80, 88, 89
Long-term flow 83; capability 83; characteristics 83
Long-term flow capability 80, 83
Loss of strength due to UV exposure 263

Macrobiological degradation 91
Maintenance 279
Manning's equation 233

Manufacturing quality assurance
  (MQA) 266, 316
Manufacturing quality control 316, 317
Manufacturing specifications for geotextiles
  for railway rehabilitation works 213;
  abrasion resistance 214; colour 214;
  elongation 214; fibre bonding by resin
  treatment or similar 214; fibre polymer 213;
  fibre size 213; fibre strength 213; filtration
  opening size 213; in-plane coefficient of
  permeability 213; mass 213; packaging 214;
  seams 214; type 213; width and length
  without seaming 214; wrapping 214; yarn
  length 213
MARV *see also* Minimum average roll value
MCB *see also* Modular concrete blocks
MD 17, 57
MDPE 13, 14
Mean force 65
Mechanical properties 50; compressibility 50;
  survivability properties 61; tensile properties
  51; toughness 58
Membrane 30
Membrane effect 32
Membrane-encapsulated soil layer 119
Mercury intrusion 70
Mercury intrusion method 71
MESL *see also* Membrane-encapsulated
  soil layer
Methods of slope stabilization 297; anchored
  geosynthetic system (AGS) method 301;
  geofabric-wrapped drain (GWD) method
  297; reinforced soil structures (RSS)
  method 302
Microbiological degradation 60, 87
Micro-reinforcement 105
Mil 22, 48
Minimum average roll value 57, 158, 267
Modified asphalt 122
Modified CBR design method 206
Modified or optimized asphalt 123
Modular concrete blocks 108
Molecular orientation 13
Molecular weight 11
Monofilament yarn 17
Monomers 10
MQA *see also* Manufacturing quality assurance
MQC *see also* Manufacturing quality control
MSW *see also* Municipal solid waste
Mulching mat 140, 141
Multi-axial tensile test 54, 56
Multifilament yarn 17
Municipal solid waste 148

Nonwoven geotextile 2
Normal stress membrane support 32

Offset modulus (working modulus) 58
Opening size 71; AOS 72; EOS 72; FOS 72
Overall bearing failure 187
Overall slope stability failure 183
Overlap 72, 269
Overlapping 270
Overlay 121
Overlay placement 289

PA 2, 13
Packaging 214
Paralink 116
Paved road 117
Paving fabric 96
Paving fabric interlayer 125
Paving roads 288; geosynthetic placement 289;
  overlay placement 289; surface preparation
  288; tack coat application 288
PE 10, 14
Peel strength 274
Percent open area 70
Performance monitoring devices 315, 320
Performance test 40
Permeability 73
Permeability or hydraulic conductivity 72, 215;
  cross-plane permeability 75, 77; in-plane
  permeability 76, 77
Permeability or permittivity criterion 215
Permissible tractive force 233; boundary shear
  stress 233
Permissible velocity approach 233
Permittivity 69, 74
Permittivity criterion 215
PET 2, 13
Physical properties 47; specific gravity 47;
  stiffness (flexural rigidity) 50; thickness 50;
  unit mass (mass per unit area) 48
Piping 84
Piping resistance criterion 215
Plain strain tensile test 54
Plastomeric 93; unreinforced 93
Ply-soil 30
POA *see also* Percent open area
Polyester 43
Polymer 1; effect of temperature 15;
  properties 11–14
Polymerization 10
Pond 152, 249
Pore size distribution 70, 72
Porometry 69
Porosity 41, 69
Portland cement concrete (PCC) paved
  roads 123
PP 2, 13
Prayer seam 270
Prefabricated vertical drain (PVD) 42

Prestressing 279
Protection 29, 37, 267, 275
Protection before installation 267
Protection during construction and service life 275
PS 13
Pullout 30
Pullout (anchorage) test 65
Pullout capacity formula 67
Pullout failure 187
Puncture protection 22
PVC 2, 13

Quality 315
Quality assurance (QA) 24, 60, 249; CQA 265; MQA 266
Quality control (QC) 315; CQC 316; MQC 317

Radial transmissivity method 77
Railway tracks 127, 211, 290
Range of force 65
RECPs see also Rolled erosion control products
Reduction factor 88, 94; biological degradation 91, 95; chemical degradation 90, 215; clogging 70, 79; installation damage 50, 88, 108; intrusion 34
Reeves lake dam 154
Reflective cracking 121
Reinforced drainage separator 7
Reinforced earth 30
Reinforced geomembrane 21
Reinforced retaining walls 302
Reinforced slopes, advantages of 145
Reinforced soil 29
Reinforced soil slopes 302
Reinforced soil structures method 302
Reinforced soil system 35
Reinforcement function 31; tensile member 31; tensioned member 31
Reinforcement interlayer 126
Relative economy 329
Reservoir 152, 249
Retaining wall 105, 175, 280
Retention or soil tightness or piping resistance criterion 215
Revetment 135, 136
Revetment armours 135, 136; asphalt slabs 135, 136; block mats 135, 136; blocks 135, 136; concrete slabs 135, 136; gabions 135, 136; grouted stones 135, 136; mattresses 135, 136; riprap 135, 136
Riprap 130, 136
Roads 117, 119, 196, 207, 286, 288; paved road 119, 207, 288; unpaved road 117, 196, 286
Roll 17, 24
Rolled erosion control products 140, 141
RRR concept 149
RSS see also Methods of slope stabilization
Rupture strain 52

Sampling and test methods 265
Saving of overlay thickness 123
Screening 38
Seam efficiency 273
Seams 269
Seam strength 273, 274
Secant modulus 58
Separation 29, 34
Separation function design method 204
Sewing 2, 270
SFDM see also Separation function design method
Shallow foundation 115, 189, 286
Shear 30, 32
Shear stress 32
Shear stress reduction effect 32, 33
Silt fence 129, 158
SIM see also Stepped isothermal methods
Single cantilever test 50
Slab effect (confinement effect) 32, 33
Sliding 30
Slip sheet 268
Slit films 17
Slit film yarn 17
Slope 135, 226, 294, 297; erosion control 135, 226, 294; slope stabilization 143, 234, 297
Slope reinforcement costs 334
Slope stabilization 143, 234, 297
Soil-burial test 91
Soil cake 134
Soil erosion 135
Soil-geosynthetic interface characteristics 65, 92; interface friction 65; interlocking characteristics 65
Soil-geosynthetic system 40
Soil subgrade description 118; firm 118; medium 118; soft 118
Soil tightness 215
Specification 42, 56, 97
Specific gravity 47
Specific guidelines 263, 280
Specimen 47, 49
Spread coating 21
Stability, basic modes of 237; global stability 237; internal stability 237; surficial stability 237

Stabilization function 118
Standards 377, 378
Stapled seam 271
Staple fibre 17
Stapling 270
Static fatigue 81
Static puncture strength test 61, 63
Stepped isothermal methods 80
Stiffness 47
Stiffness (flexural rigidity) 50
Stitched geotextile 2
Strain rate 52
Stresses, types of 32; normal stress 32; shear stress 32
Stress relaxation 83
Stress-relieving interlayer 124
Stress rupture 81
Strip drain *see also* Band drain (Fin drain/strip drain)
Subgrade soil 35, 118
Surface preparation 288
Surface stabilization 39
Survivability 50, 157
Survivability and durability criterion 215
Survivability properties 61; bursting strength 61; fatigue strength 61; impact strength 61; static puncture strength 61; tearing strength 61
Survivability requirements 157

Tack coat 123, 209
Tack coat application 288
Tangent modulus 58
Tearing strength 61
Tearing strength test 61
Temperature 52
Tensar manufacturing process 20
Tensile member 30, 31
Tensile modulus 51, 58; offset modulus (working modulus) 58; secant modulus 58; tangent modulus 58
Tensile properties 51
Tensile reinforcement 42
Tensile strength 51; breaking tensile strength 58; ultimate tensile strength 58
Tensile strength tests 54; biaxial test 54; grab test 54; multi-axial test 54; plain strain test 54
Tensioned member 29, 31
Test properties 92
Textured geomembrane 22
Textured surfaces 22
Thermoplastics 10, 13, 14
Thermoset 16
Thickness 47

Thin-film geotextile composites 119
Toughness 58
Tractive force 233
Transmissivity 69, 75
Trapezoid tearing strength test 61, 62
TRM *see also* Turf reinforcement mat
Tubular geotextiles 19
Tunnel 157, 254, 308
Turf reinforcement mat 137
Typical geocomposites 7; drainage composites 7; geosynthetic clay liner 7; reinforced drainage separator 7
Typical geogrids 4; biaxial 4; bonded 4; extruded 4; uniaxial 4; woven 4
Typical geotextiles 3; knitted 3; nonwoven 3; woven 3

Ultimate tensile strength 58
Ultraviolet light exposure 43
Uniaxial geogrid 4
Unit mass (mass per unit area) 48
Universal soil loss equation 232, 233
Unpaved age 119
Unpaved road 117, 286
Unreinforced geomembrane 21
USLE *see also* Universal soil loss equation
UV degradation 89, 90
UV light stabilizer 109

Vegetative reinforcement 39
Vertical strip drain 42
VLDPE 13, 24
Volume flow rate 72, 73

Walls retention 175; analysis for the facing system 175; external stability analysis 175; internal stability analysis 175
Warp 17, 57
Warp strength 57
Waterproofing the pavement 123
Waterproofing tunnels 308
Weave 17, 18
Weaving loom 1, 2, 18
Website 385
Weft 18
Weft strength 57
Welding 21
Wet sieving 70
Wet sieving method 71
Wick drain 42
Widening of existing embankments 145
Wide-width strip tensile test 52, 58
Wide-width tensile strength 60
Working stress 174

Working stress design approach 174
Woven geogrid 4
Woven geotextile 1
Wrapping 214

XPS 14

Yarn 17; fibrillated yarn 17; monofilament yarn 17; multifilament yarn 17; slit film yarn 17
Yarn length 213

Zero deformation point 51

eBooks

eBooks – at www.eBookstore.tandf.co.uk

A library at your fingertips!

eBooks are electronic versions of printed books. You can store them on your PC/laptop or browse them online.

They have advantages for anyone needing rapid access to a wide variety of published, copyright information.

eBooks can help your research by enabling you to bookmark chapters, annotate text and use instant searches to find specific words or phrases. Several eBook files would fit on even a small laptop or PDA.

NEW: Save money by eSubscribing: cheap, online access to any eBook for as long as you need it.

Annual subscription packages

We now offer special low-cost bulk subscriptions to packages of eBooks in certain subject areas. These are available to libraries or to individuals.

For more information please contact webmaster.ebooks@tandf.co.uk

We're continually developing the eBook concept, so keep up to date by visiting the website.

www.eBookstore.tandf.co.uk